# Control of Bird Migration

# Control of Bird Migration

P. Berthold

**CHAPMAN & HALL**

London · Glasgow · Weinheim · New York · Tokyo · Melbourne · Madras

**Published by Chapman & Hall, 2–6 Boundary Row, London SE1 8HN**

Chapman & Hall 2–6 Boundary Row, London SE1 8HN, UK

Blackie Academic & Professional, Wester Cleddens Road,
Bishopbriggs, Glasgow G64 2NZ, UK

Chapman & Hall GmbH, Pappelallee 3, 69469 Weinheim, Germany

Chapman & Hall USA, 115 Fifth Avenue, New York, NY 10003, USA

Chapman & Hall Japan, ITP-Japan, Kyowa Building, 3F, 2-2-1
Hirakawacho, Chiyoda-ku, Tokyo 102, Japan

Chapman & Hall Australia, 102 Dodds Street, South Melbourne,
Victoria 3205, Australia

Chapman & Hall India, R. Seshadri, 32 Second Main Road, CIT East,
Madras 600 035, India

First edition 1996

© 1996 Peter Berthold

Typeset in 10/12 Palatino by Best-set Typesetter Ltd., Hong Kong
Printed in Great Britain by St Edmundsbury Press, Bury St Edmunds,
Suffolk

ISBN 0 412 36380 1

A catalogue record for this book is available from the British Library

Library of Congress Catalog Card Number: 95-74634

∞ Printed on permanent acid-free text paper, manufactured in
accordance with ANSI/NISO Z39.48-1992 and ANSI/NISO Z39.48-1984
(Permanence of Paper).

# Contents

# Preface

The topic of the 'control' of bird migration deals with the endogenous mechanisms, environmental stimuli and physiological adaptations which regulate the onset of migration, as well as factors which generate its course and termination in goal areas. While extensive observational studies date back even to Aristotle, the investigation of control mechanisms only began in the early twentieth century with pioneer experiments by Rowan. Since then a worldwide boom of studies on control mechanisms has produced an almost immeasurable wealth of publications and ideas. This is not surprising due to the ubiquity of bird migration and its attractiveness. The aim in this book is thus to summarize these recent developments and try to put them into perspective. It is intended to provide both an interesting and useful compilation for naturalists and teachers, as well as a starting basis for future studies on avian migration and related fields of interest for students and researchers.

The first chapter briefly summarizes general aspects and demonstrates the huge spectrum of migratory phenomena in birds, the history of the study of control mechanisms and the main methods currently applied. It has also been used to point out some critical questions for studies, to give a few important definitions and to compile the most important morphological and physiological preconditions for avian migration. The second chapter, on control mechanisms and ecophysiology, focuses first on three major characteristics: nocturnality, fat deposition and migratory restlessness, and thereafter on biological rhythms, endogenous time-programs, photoperiodism, genetic, endocrine and neuronal control mechanisms, metabolic adaptations, energetics, behavioural aspects and how migrants cross various barriers. Subsequently, more environmental control mechanisms and aspects are treated, in particular the roles of weather, food and population structure. Further information demonstrates how migrants select goal areas and habitats, and focuses on interactions with other annual events. The third chapter briefly summarizes the present knowledge of orientation mechanisms and navigation hypotheses including prepro- grammed directions, compasses, the sensory basis, the redundancy of mechanisms and interactions with environmental factors. In the fourth chapter current microevolutionary processes are treated, and some prospects of human impact on migration systems and their possible future development are briefly addressed. Finally, the information from the previous chapters is outlined in a brief synopsis.

It is hoped that this summary can portray both the enormous pro- gress which has been made in the study of avian migration and also the immense fascination which emanates from this knowledge. It is my hope that this fascination will not only stimulate future investigations but also increase awareness and efforts for the conservation of many endangered migratory species.

This book should be seen as part of a trilogy by the author, the other two being Berthold (1993) *Bird Migration: A General Survey*, Oxford University Press and Berthold (ed.) (1991) *Orientation in Birds*, Birkhäuser, Basel. These three titles form a unity of comprehensive information on our knowledge of avian migration.

Those who are more interested in details of migration patterns of individual species should consult Alerstam (1990), Mead (1983) and Zink (1973–1985).

Although several thousand publications have been issued on the questions treated in this volume, only a fraction of these can be cited in order to keep the bibliography to a reasonable length. For this reason, the most topical secondary review papers are cited wherever possible. Otherwise prominent examples are given.

The taxonomic system used follows Wolbers 1975–82, *Die Vogelarten der Erde*.

# Acknowledgements

Above all, my heartfelt thanks are addressed to Dr John Dittami, Vienna, for his thorough and patient grammatical revision of the first draft of the manuscript. In particular, I also wish to thank the following colleagues who had the kindness to critically read the draft chapters of their special fields of interest: Drs. T. Alerstam, Sweden; F. Bairlein, Germany; H. Biebach, Germany; B. Bruderer, Switzerland; H. Dingle, USA; E. Gwinner, Germany; E. Ketterson, USA; B. Leisler, Germany; T. Piersma, Holland; W. Richardson, Canada; K. Schmidt-Koenig, Germany; S. Terrill, USA; R. and W. Wiltschko, Germany; and J. Wingfield, USA. I am also grateful to Chapman & Hall, especially to Dr Bob Carling and Martin Tribe for their fruitful cooperation. Studies carried out by the author were supported by several grants from the Deutsche Forschungsgeineinshaft.

# 1

# Introduction and general aspects

Before one gets into physiological and ecological aspects of bird migration it is perhaps appropriate for both the newcomer, and even the experienced reader, to first examine what bird migration according to our present knowledge really is and what migrants are able to perform. This may substantially facilitate understanding the often complicated physiological approaches of experiments used to elucidate control mechanisms. Thereafter a short survey and evaluation of current research methods is presented along with a few historical perspectives. This is followed by a discussion of questions crucial for future studies in the field. This is appropriate since investigations often cannot achieve their goal due to difficulties in formulating the questions and determining how to look for the answers. This chapter ends with some definitions, which appear necessary in order to avoid further confusion in the literature, and a compilation of morphological and physiological preconditions for avian migration.

## 1.1 PHENOMENA OF BIRD MIGRATION

Birds are the most mobile group of vertebrates. It is thus not surprising that migration in these 'fugitive' species (Dingle, 1980) plays a more prominent role than in any other group of animals. Roughly half of the approximately 9000 known bird species currently recognized and individuals in the magnitude of 50 thousand million birds perform some type of migratory movements (Berthold, 1993). The ability to fly, the relatively long life-span and the all-year homoiothermy permits birds to migrate regularly to distant areas which are only seasonally inhabitable and to settle in regions with unstable living conditions. An alternative strategy to get around environmental variability, hibernation, widespread in reptiles and mammals, is rare in birds and is known in only one recent species, i.e. the poor-will (*Phalaenoptilus nuttallii*), a nightjar from western North America. Individuals of this species can spend about three months hibernating in rock caves, using body fat deposits, while lowering their body temperature below 10°C (Calder and King, 1974).

Actually, we do not know when migration in birds first appeared. According to Alerstam (1990), 'bird migration has no doubt existed for as long as birds have been present on earth, for more than 100 million years'. In fact, it is likely that flightless *Hesperornis* of the Cretaceous period, which lived about 80 million years ago, were migratory. Several circumstances support the assumption that these toothed birds, like many recent sea birds, migrated (in this case by swimming) to higher latitudes to breed in North America (Tyrberg, 1986). Further, bird migration has most likely evolved independently many times in different genera (Farner, 1955). This is important with respect to studies of phenomena and control mechanisms because it can in part explain different, or sometimes even contradictory, findings among different species, as will be shown later. Orientation mechanisms in birds, as opposed to this, may represent more phylogenetically ancient mechanisms (Terrill, 1990). One would then expect that migratory species in general have not evolved specific orientation mechanisms (perhaps except for a few, like the star compass; section 3.3). More likely, navigational systems may have developed along similar lines in amphibians, reptiles and birds, and common mechanisms may underlie true navigation (section 3.5) in a variety of vertebrate groups (Rodda and Philips, 1992) and may also be similar in long-distance, short-range and local orientation (chapter 3).

Due to the frequency of migratory behaviour and its worldwide distribution in bird species it is not surprising that it is presented in some form on all parts of our planet. Almost all areas between the Arctic and Antarctic are used as breeding grounds. From this huge distribution area for breeding, migrants also regularly reach even the most remote corners of the earth for non-breeding periods. Additionally, there are no geographical barriers such as deserts, mountains, oceans or ice-fields that cannot regularly be crossed by migrants. Thus, routes of long-, middle- and short-distance migrants cover our planet like a network (Figures 1.1 and 1.2), and only a few areas remain untouched. Migration also occurs in all climatic zones, including the tropics, where a variety of intratropical migratory movements have been found (Curry-Lindahl, 1981; Alerstam, 1990). Levey and Stiles (1992) hypothesized that seasonal intratropical movements of Neotropical forest birds predisposed these birds to migration out of the tropics.

With respect to the most likely polyphyletic origin of avian migration, it is not surprising that an almost immeasurable amount of different migratory patterns have evolved. In rather restricted definitions of

**Figure 1.1** Examples of migration routes of long-distance migrants during southbound migration. 1, Alaskan population of Pacific golden plover and other waders to island groups in the Pacific; 2, Swainson's hawk; 3, migration across the Gulf of Mexico of many North American species; 4, Arctic tern; 5, ruff; 6, trans-Saharan migration of many Eurasian species; 7, wheatear (Alaskan population); 8, swallow (three different populations to differing winter quarters); 9, Amur falcon – only the transoceanic migration path is shown; 10, circular return migration of the short-tailed shearwater (*Puffinus tenuirostris*). (Source: Berthold, 1993.)

avian migration the typical and best known annual migration is usually considered. In most cases this is simply the trip from the breeding grounds to the winter quarters and back. Although it is true that these periodic seasonal-return migrations represent a major part of avian migratory movements it is still only a fraction of the types of migration seen across the animal kingdom. According to more comprehensive definitions, for example, Baker (1978), 'Migration: the act of moving from one spatial unit to another', many other kinds of movements can be taken into account. Baker stated that his definition is close to, though even more general than, dictionary definitions (e.g. the act of moving from one place of abode to another). His 'spatial unit' does not include any restriction and thus makes it difficult to discriminate between everyday activities in a given home range and a true migratory movement, at least over short distances. None the less, herein I will generally follow Baker's definition with the constraint that migration is

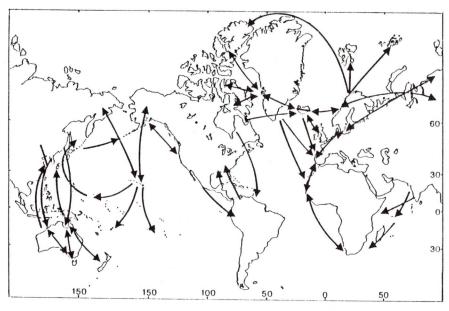

**Figure 1.2** Principal transoceanic migration routes. Northward pointing arrows indicate spring migrations, southward pointing arrows indicate autumn migrations. Actual migrations are usually broad front, hundreds or thousands of kilometres wide. (Source: Williams and Williams, 1990b.)

any movement from one temporarily inhabited home area to another. In so doing, a large variety of both regular and irregular migration forms can be considered. Kennedy (1985) and Dingle (1995) have recently tried to give more specific definitions, and Dingle listed four criteria defining animal migration in general: (1) persistent locomotor activity of greater duration than occurs during foraging; (2) straightened-out movement that differs from foraging searches; (3) initial suppression of responses to stimuli that arrest other movements but with their later subsequent enhancement, and (4) activity patterns particular to departure and arrival. However, this definition is also only partly valid if short-distance vertical migration is considered, a common phenomenon of 'true' migration (below). Dingle (1995) has also attempted to include behaviour and natural selection in a definition of migration as 'specialized behavior especially evolved for the displacement of the individual in space'. However, this definition does not differentiate migratory from non-migratory behaviour and the author finally concluded: 'migration and other forms of behavior, they are but the ends of a continuum. It is not always possible to make a hard and fast distinction between migration and foraging movements'.

To start with, a common type of movement is dismigration, i.e. the post-juvenile dispersal which normally leads to a dispersion of the offspring of a given region in neighbouring areas. Post-juvenile and natal dispersal (the latter as movement from natal to first breeding site, for terminology see Warkentin and James, 1990) extends in small birds commonly over $20-100\,km^2$, but in some species like the barn owl (*Tyto alba*), over distances up to about 2000 km (Bairlein, 1985a). Further, it normally leads to many directions (e.g. 'cannonball effect' in herons; see Berthold, 1993). Dispersal is often not separable from 'true' migration and has so far been relatively little studied and treated by modelling work (North, 1988). According to Dingle (1995), it should not be taken as an alternative term to migration because the separation is mainly due to historical but not biological reasons. He even advocated to drop the term dispersal as a synonym for one-way movement. For more details with respect to biological functions, dispersing distances and speed, etc., see Berthold (1993).

Further, there is a large variety of so-called follow-up migration forms, or pursuing movements, represented by ant-, fire-, flower-, fruit- or seed-followers. Many nectar-feeding species like humming-birds track the altitudinally or latitudinally changes in flowering periods. Many fruit- and seed-eaters behave accordingly in connection with crop seasons. Other species of very different systematic groups, for example, raptors, storks, rollers, bee-eaters and many small birds, regularly follow tropical and subtropical brush and forest fires to prey on fleeing or damaged animals (Berthold, 1993). A peculiar follow-up mechanism has been evolved by the so-called antbirds. Neotropical formicariids especially, but also other species including North American winter visitors, keep more or less in regular touch with neotropical driver or army ants as ant-followers. They usually benefit from darting down to catch food just flushed by the ants (Landsborough-Thomson, 1964; Willis, 1973, 1986; Coates-Estrada and Estrada, 1989). Another common form of regular wandering is moult migration which is wide-spread in larger aquatic species like ducks and geese. These species have a rapid concentrated moult with temporary flightlessness. Moult migration, often northbound, enables individuals to go to areas where they benefit from reduced predation, avoidance of competition, longer days for feeding and/or more favourable food sources (nutrition-quality hypothesis; Jehl, 1990). Other forms of such regular 'intermediate' or 'intermittent' migratory movements, leading to intermediate goals, may bring parts of a population to feeding grounds better than the breeding area before departure to the winter quarters. This, for instance, is the case in European starlings (*Sturnus vulgaris*) in Switzer-land which partly migrate north into the Rhine Valley before they leave to winter in the Mediterranean area (Fliege, 1984). Such intermediate migratory movements may also function as moult migration in European starlings (Schüz *et al.*, 1971) and some small passerines (Ellegren and Staav, 1990). Finally, altitudinal (vertical) migration should be mentioned as a common form of periodic seasonal movement which is formed in many mountainous species which go down to winter in lowlands. Often they cover only very short distances. For instance, some grouse species descend just about 300 m (e.g. Martin, 1987).

Other migration forms are not so regular. Perhaps the most spectacular are the so-called 'irruptions'. These are invasions or immigrations of often huge numbers of individuals, for instance in waxwings (*Bombycilla garrulus*), the former 'plague birds' or 'pest birds', nutcrackers (*Nucifraga caryocatactes*) or tits. In species like tits specific reproductive strategies lead to overcrowding at times (e.g. Berthold, 1993). The emigrations, eruptions or evasions of these species, often popularly referred to as 'vagabonds', belong to a large group of movements called 'escape migrations' or 'escape movements'. In some cases this appears to be due to population density (coincidence with the culmination points of density gradations of breeding populations) and in others to food shortage (sections 2.18 and 2.19). Although they may represent 'death-wandering' of substantial proportions of a population they normally lead to later return movements (Berthold, 1993). Other forms of these 'contingency' movements are, for instance, performed by 'birds of the frost' or 'birds of the cold' which then occur facultatively in certain wintering areas when the weather is too cold in a breeding area. Such irregular weather movements can also be triggered by snow (immediate snow flights), ice (in water birds) or by thunderstorms and cyclones, as in the swift (*Apus apus*) ('cyclonic weather movements', section 2.17). On the other hand, inclement weather may cause reverse migration. During homeward migration individuals may migrate and retreat over larger distances and for quite a period, according to cold and warm spells, until they finally head for their breeding grounds (Berthold, 1993). More or less aperiodic evasive migration is also performed in relation to food shortage, drought, fires and floods, and are ultimately all 'deficiency movements'. Other aperiodic migrations are regular, but unpredictable movements of crossbills in search of fruiting coniferous trees needed for breeding by these extreme food specialists. These 'gypsy birds' may wander thousands of kilometres within a season looking for suitable trees. In case of more favourable conditions they may stay for years in a certain area. This most sedentary behaviour could increase as a result of *Waldsterben* (trees dying due to human activity e.g. pollution) and an advanced fertility of spruce trees (Berthold, 1993). Irregular movements are also a common habit in many species in dry areas, where they are better related to unpredictable wet periods essential for breeding. The phenomenon of abmigration should also be mentioned. In ducks, sedentary individuals, or birds from populations with specific migratory habits, may interbreed with migratory mates with different migratory habits. As a result new habits and movements can evolve when mates (normally males) follow others (females) to unfamiliar areas (male-biased dispersal). Finally, there are different kinds of irregular migration involved in the expansion of distributional areas, the new or resettlement of populations in other breeding areas (for review, see Berthold, 1993).

After these few descriptions a few more basic phenomena have to be treated before one is able to ask proper questions with respect to the control mechanisms of bird migration. The huge masses of avian migrants produce a large variety of different spatial and temporal patterns. Among the spatial patterns, there has to be discrimination between long-distance migration (continent-wide, intercontinental or transoceanic circular return migration, Figure 1.1), middle- or medium- and short-distance (both intracontinental) migration. Many species perform their migratory journeys largely independent of geomorphological and landscape features, i.e. broad-front migration. Others regularly use specific coasts, rivers, mountain ridges or possibly chains of oases as guidelines or leading lines and move on in so-called flyways or migration corridors, i.e. narrow-front or small-front migration. Intermediate behaviour may result in guided broad fronts. In the case of narrow fronts, migrants can reach extreme concentrations in geographical bottleneck areas like spits, valleys or coasts which are well known as mass migration ways. After such a concentrated migration, dispersive migration normally occurs which distributes individuals again over wider areas to winter or breed. An illustrative example is the concentrated funnel-shaped migration of many European species towards southern Spain around Gibraltar and the dispersive fan-shaped migration towards the African winter quarters thereafter. Although we do not know much about the actual routes flown by individual migrants (section 1.3, Figures 1.4 and 1.5), it appears that relatively few birds cover the distance between their breeding grounds and winter quarters in a straight line. On the contrary, there is great variety of forms comprising routes like arrows, bends, hooks or even loops. This demonstrates that the migratory routes and directions are often changed both on a macroscale between to-and-fro migration as well as on a microscale during an individual journey (e.g. Figure 1.5). Some peculiar routes may well represent historically-based detour migration (most likely evolved as resettlement routes after the last ice age; Berthold, 1993).

Another matter of extreme interspecific variation is the length of time used for migration, i.e. the migratory period. For an escape migratory flight from mountains to lowlands less than an hour may be sufficient.

Many sea birds, on the other hand, like albatrosses or shearwaters which practically only land on islands or coasts for breeding, as well as land birds on intercontinental migration, can be *en route* for up to ten and a half months per year. Marsh warblers (*Acrocephalus palustris*), for instance, leave their Central European breeding grounds in mid July for 'autumn quarters' in NE-Africa (Dowsett-Lemaire and Dowsett, 1987). They reach their south African winter quarters around the end of the year and arrive back in the breeding grounds in May (Berthold, 1993, figure 2.65; for other examples see section 2.22). Further, it has to be realized that bird migration comprises a high degree of intraspecific variation. This differential migration (section 1.5) includes population as well as sex and age differences. In many respects it should be carefully considered in studies of control mechanisms of migration. Sometimes it creates peculiar migration patterns like the so-called leap-frog migration that occurs when migrating populations overfly resident ones or overtake others migrating more slowly. This type of migration has been discussed extensively by Salomonsen (1955); he assumed that the underlying selective advantage is to avoid 'devastating competition'. Holmgren and Lundberg (1993) found, on theoretical grounds, that leap-frog migration is the most likely pattern to develop when migration costs are high. Under other circumstances chain migration or no stable migration pattern is more likely to develop. Many species show a remarkable variation in the migratory performance and the spatial and temporal patterns of migration in different populations. A striking example can be seen in the blackcap (*Sylvia atricapilla*; see Berthold, 1993, figure 12).

Perhaps of special interest for migratory movements is actually to consider the extremes. They are not only exciting but also give an impression of the maximum capabilities of migrants. With respect to the distances covered, the record holder is the Arctic tern (*Sterna paradisaea*) which regularly migrates from Arctic breeding grounds to winter quarters in Antarctic waters (Figure 1.1). Individuals which are thought to circle around the Antarctic continent (Schüz *et al.*, 1971; Gudmundsson, 1992) may cover up to 50 000 km per year (individual one-way journeys are documented for up to 18 000–22 000 km; Martin, 1987). With a maximum life expectancy of at least 25 years (like the famous Scandinavian individual 'Silvia', ringed 1969 in Finland and rediscovered in Sweden until 1991), individuals could theoretically reach a lifetime mileage of over a million kilometres – about three times the distance to moon! Annual totals of migrations in the magnitude of 20 000–30 000 km are widespread among long-distance migrants of different groups (Berthold, 1993; Figure 1.1). Maximum non-stop flights range between 5000 and 7500 km in waders which perform long transoceanic and transglacial flights from Alaska to Pacific islands or from Siberia to New Zealand (and which, with the exception of a few species, cannot rest on water for long periods like many other birds; Evans and Davidson, 1990). The flight times for these long non-stop flights are in the region of 80–100 hours. Passerines are capable of making non-stop flights of over 2000 km and hummingbirds up to 800–1000 km, e.g. Williams and Williams (1990a, b). The ruby-throated hummingbird (*Archilochus colubris*), about 4.8 g in weight and 8.5 cm in length, regularly crosses the Gulf of Mexico in a non-stop flight of about 18 hours which requires about 3.2 million wing-beats (Nachtigall, 1993). Maximum distances covered in a short period by geese, waders or thrushes range from 600 to 1000 km in 24 hours and 3000–5000 km in 60–65 hours (Berthold, 1993), and account for up to about 250 km in 24 hours in small songbirds (Berthold *et al.*, 1990b). However, the normal course of migration is much slower (section 2.15).

The maximum altitude at which a bird has so far been observed is 11300 m. The record holder is a Rüppell's griffon vulture (*Gyps rueppellii*) which ran into an aircraft over the Ivory Coast in Africa. Migrants regularly reach altitudes of 4000–6000 m during transoceanic flights and 8000–9000 m across the Himalayas. For more details see Berthold (1993) and section 2.12.

Comparative studies of migratory and non-migratory species indicate that orientation mechanisms are likely to be part of the basic equipment of birds in general and essentially represent phylogenetically ancient mechanisms. These orientation mechanisms may be used for both long-distance migration and the daily routine in the home range. The widespread breeding site fidelity, birth site or even nest site fidelity in adults and to an extent juvenile individuals of all types of migrants (adult and natal philopatry), demonstrate pronounced navigation capabilities. There is a further extraordinary amount of information on resting-place attachment and especially winter recurrence to demonstrate this. Further, there have been normally high return rates with thousands of displacement experiments, most often carried through with domestic pigeons (*Columba livia*), which reaffirm the site fidelity of birds. For a comprehensive recent review see Wiltschko (1992).

Case reports of individual long-distance migrants, like swallows and redstarts (which travel for years between specific points on the globe thousands of kilometres apart), indicate an extremely high navigational precision with point-like accuracy (pinpoint navigation). These, and the results of displacement experiments lead on to the conclusion that transplanted birds are, in principle, capable of returning to a known place from any other place of the earth's surface as long as the return journey lies within the birds' physical limits and a motivation to return exists. For more details and reviews see Berthold (1991b, 1993), Schmidt-Koenig (1975), Wiltschko (1992), Zink (1973–1985), Curry-Lindahl (1981).

Besides spatial precision, migration is also characterized by extreme temporal precision when required. The so-called 'calendar birds' provide the most prominent examples. Above all, long-distance migrants breeding in higher latitudes are known for their capacity to arrive at their breeding grounds year after year so exactly that they can be used to set the calendar. A striking example demonstrating punctuality as well as consistency can be seen in the spotted redshank (*Tringa erythropus*) in Finland. First arrivals in the area of Helsinki over a period of 24 years were between May 1 and 8. The average was May 4, with only ±2.06 days SD (Hildén, 1979).

It would go beyond the scope of this discourse to discuss in detail all possible reasons for the evolution of bird migration and its biological significance. It appears, however, that ultimately all migratory movements are related to providing individuals with sufficient food resources and living space for survival and successful reproduction (Lack, 1968). This is true in both areas with seasonal changes in the nutritional base, as in higher latitudes, and areas with highly unpredictable food situations, as in drought areas. Several authors have formed concepts of how migration might simply have evolved from hunger (Merkel, 1966; for hunger restlessness see section 2.1) or from juvenile dispersal ('exploratory migration model', Baker, 1978; for review, see Gauthreaux, 1982). Terrill (1990) has developed a theoretical approach to the evolution of complex endogenously programmed long-distance migration from dispersal and nomadism through facultative finally to obligate migration (see also section 1.5). The former 'glacial theories', assuming that migrants originated in ancestral 'homes', and the hypothesis that migration routes followed from continental drift (Drift Theory; Dingle, 1980, for review) pale in the light of recent findings of how rapidly migration patterns can adapt to actual ecological situations through fast microevolutionary processes (section 4). Holmgren and Lundberg (1990) stated that latitudinal-dependent survival and reproduction in combination with the magnitude of migration costs seem to be important determinants of migration patterns.

With respect to the risk of migration, I think the prevailing opinion is that migration is one of the most hazardous stages of avian life cycles. This view may have a personal bias in many people due to occasionally bitter experiences during travelling. Still, the fact that migration is such a common strategy among birds already indicates that it must be a rather successful proposition (Mead, 1983). In contrary to the view that migration is a risk already, Kipp (1943) has pointed out that many long-distance migrants [like the marsh warbler or pied flycatcher (*Ficedula hypoleuca*)] are able to maintain stable populations on only one brood per season, whereas non-migrants, like tits or finches, in the same areas often have to raise several broods, above all with respect to high winter mortality. Mönkkönen (1992) has shown that nearctic and palaearctic tropical migrants are characterized by a smaller number of broods and in part also by smaller clutches than more sedentary birds. In eastern nearctic tropical migrants a higher survival rate could also be shown. On the other hand, O'Connor (1990) found in an analysis of the breeding biology of species breeding in Britain no evidence that multiple brooding would be more common among residents than among migrants. Hence, the intriguing question as to whether migratory or resident behaviour is generally more risky and would thus require higher reproductive output can presently not be satisfactorily answered. It should be treated in a special comprehensive study.

However, we have to bear in mind that in no way can migrants escape every threatening condition by migrating. Thus, both regular normal losses as well as occasional mass accidents occur. Direct information on normal losses during, and thus the risks of, migration are scarce. For migratory shorebirds Evans (1991) stated that no estimates are yet possible of the extent of losses of juveniles during their first migration but they appear to be considerably higher than in adults. According to Evans and Davidson (1990) losses may arise from directional errors during flight, as well as from inadequate physiological preparation before departure or choice of inappropriate weather conditions in which to set out. Several studies on shorebirds indicate that the survival in species migrating short distances is higher than that of long-distance migrants. Barnacle geese (*Branta leucopsis*), for instance, suffer in some years a juvenile mortality rate of up to 35% between Spitsbergen and Scotland during their autumn migratory period (Owen and Black, 1991). Ralph (1978) has estimated the theoretical maximum loss of young autumn passerine migrants along the eastern coast of the United States due to disoriented offshore movements to be in the magnitude of 1–10%. Lindström (1989) estimated that about 10% of the finches that he observed resting in S. Sweden were killed by raptors during autumn migration.

There are, of course, reports of occasional mass accidents of migrants during migration as well as during wintering. Examples are given in section 3.7 and by Suter and van Eerden (1992). Overwater migration can especially be hazardous; many birds are regularly killed during flights over the Gulf of Mexico. In May 1976, in one of the largest warbler disasters recorded, as many as 200 000 birds were washed up on the shores of Lake Huron, having been caught aloft in a spring storm (Morse, 1989). But even such disasters do not appear to be of great overall importance.

In conclusion, the available facts demonstrate that migration is a rather safe enterprise but also bear in mind the necessity of safe and reliable control mechanisms to insure its success (sections 2 and 3). This view is also supported by the finding that globally threatened species in the western Palaearctic are no more or less migratory than the complete fauna. It suggests that being migratory does not necessarily impose a special risk of threat (Bibby, 1992). However, regionally threatened species are often significantly more likely to be migratory than the complete fauna (Bibby, 1992; Berthold *et al.*, 1993). They may be affected by a series of recent human impacts which can act in the breeding grounds, migratory resting areas and winter quarters. Another risk related to bird migration is the bird strike problem in aviation. Collisions of flying objects and migrants do kill only small numbers of birds but often heavily damage aircrafts and their pilots and passengers. In small countries like Israel, with dense aviation through a migratory flyway, the bird collision problem has stimulated a series of detailed bird migration studies from which our knowledge of avian migration has greatly benefitted (e.g. Leshem, 1989).

## 1.2 HISTORY OF THE STUDY OF CONTROL MECHANISMS

Farner (1955) divided the history of migration studies into two sections. The beginning has been arbitrarily set with Aristotle (Stresemann, 1951). The two sections are the period of observations until 1925, which was especially intensive after 1825, and the period of experimental investigations, which began with pioneer experiments by Rowan (1925). The second section has been characterized by an uninterrupted sequence of increasing experimentation. Some inaccurate initial ideas of the control of bird migration, like the transmutation theory (assuming that species present in the summer could change into others in winter), or that of widespread hibernation, which go back to Aristotle, persisted for a long time. The latter was still propagated by Linné. Emperor Frederick II was obviously the first to develop reliable concepts of how migrants come and go 'following food supply and heat' in his famous work 'De arte venandi cum avibus' in the 13th century. He also realized that in formations of flying migrants regular changes in leadership occurred indicating that the leader is subject to special demands (Stresemann, 1951). Later, it was von Pernau who in 1702 questioned whether the departure of early migrants at the height of the summer was triggered by deteriorating environmental factors. He postulated a 'hidden urge for migration' (in the bird itself) responsible for the onset of autumn migration and thus recognized the urge to migrate as an innate behaviour. Similarly, Legg (1780a, b) postulated for migrants 'in all probability the innate knowledge which prompts them to make these yearly excursions'. The first simple experiments of far-reaching significance were actually conducted by Johann Andreas Naumann (1795–1817). He kept migrants in a bird room and observed their nocturnal migratory activity through a small hole in the door. For the golden oriole (*Oriolus oriolus*) and the pied flycatcher, in which migratory restlessness was maintained up to the winter, he concluded that these species 'migrate far away, presumably as far as to Africa'. He thus was the first to relate the amount of observed migratory activity to the distance covered. In continuation of this idea, von Lucanus (1923) and Stresemann (1934) hypothesized that an endogenous migratory urge or migration drive existed for as many days as necessary in inexperienced first-year migrants to fly a population-specific distance from the breeding grounds to the winter quarters.

The period of concentrated and directed experimentation on control mechanisms was then initiated by William Rowan in 1925. He combined the traditions of natural history and experimental biology in his pioneering research into the external and internal factors involved in bird migration (Ainley, 1988). Rowan subjected dark-eyed juncos (*Junco hyemalis*) and American crows (*Corvus brachyrhynchos*) to an artificially increasing day length in winter producing an early gonadal recrudescence and migratory behaviour. He first developed a complex theory of the control of avian migration by sexual hormones, but later (Rowan, 1946) he extended it suggesting that the entire physiology of the animal may be involved in the stimulation of migration. Here, we were left to find the crucial control mechanisms and agents in protracted studies.

## 1.3 CURRENT APPLIED METHODS

The easiest way to obtain bird migration data is to make sight observations on passing migrants as first systematically applied by Thienemann (1927), the founder of the Vogelwarte Rossitten. When systematically performed, they not only yield daily and seasonal patterns of migration but also elucidate orientation behaviour and flight strategies in different species in relation to varying weather conditions and other environmental circumstances. Well-planned migration studies are therefore a major task of many permanent or temporal bird observatories (for review see, for example, Bub, 1983). Their research is of prime importance with respect to species in which individual banding is difficult. For the visual studies of the overhead passage of nocturnal migrants, the moon-watching technique (essentially limited to full-moon periods without cloud cover), as well as a ceilometer technique, using powerful light beams, have been applied to some extent (e.g. Gauthreaux, 1969; Balciauskas and Zalakevicius, 1991). The most important method

applied to the whole field of bird migration studies is individual banding (Figure 1.3). Bird banding, although present in ancient times
(reports of Plinius; Bub and Oelke, 1989), was first institutionalized in
1903 by the Vogelwarte Rossitten. Over the past 90 years about 120
million individuals have been ringed in Europe and the world total
may be in the magnitude of 200 million. From the results of recoveries,
spatial and temporal patterns of migration have been characterized and
compiled in atlases of bird migration, e.g. Zink (1973–1985). Unfortunately, the recovery rates of banded birds are normally very low
(e.g. Berthold, 1993). However, the information relating to control
mechanisms of banded birds is greatly enhanced when regular retraps
can be investigated, above all on permanent trapping stations which
exist in a number of countries (Bub, 1983). In the history of the development of appropriate capturing methods and trapping techniques the
'mist' nets – hardly visible nylon nets – have brought a revolutionary
upswing in the trapping figures since the 1950s. They are now regularly
used in long-term trapping programs aimed at the detailed study of
problems of bird migration [like the 'Mettnau–Reit–Illmitz Program'
(MRI-Program) started in 1974, Berthold *et al.*, 1991; or in similar
programs run by the British Trust for Ornithology, in the Operation
Baltic and by other organizations, Berthold, 1993]. Besides the normal
metal rings, colour rings, auxiliary colour leg markers, leg flags, wing
tags, back tags, neck collars, nasal markers, colour-dyeing of the
plumage, miniature data logging devices (altimeters, depth gauges,
speedometers, activity recorders) and radiotransmitter packages have
been used for special studies (Figure 1.3; for review, see, for example,
Calvo and Furness, 1992; Berthold, 1993). In some recent studies,
trapping figures were significantly increased by the play-back of song,
and also in nocturnal migrants during the night when powerfully
amplified tape-lures for artificially induced landfall were used (e.g.
Herremans, 1989). A recent highly promising method is satellite-
tracking of individual birds equipped with transmitters (Figure 1.3).
Most likely this new approach will soon yield virtually uninterrupted
descriptions of complete individual migratory journeys as well as
physiological data recorded during actual migration (Nowak and
Berthold, 1991; Berthold *et al.*, 1992b; Figures 1.4 and 1.5). A recently
tested flight-path recorder can store angles of homing pigeons' flights
and allows for the reconstruction of flight paths (Dall'Antonia *et al.*,
1990). Similarly, Wilson *et al.* (1991) proposed dead reckoning as
an alternative to telemetry. They tested Jackass penguins (*Spheniscus
demersus*) in Africa by deploying a simple compass that recorded bird
heading over specified time intervals which then was related to penguin
swimming speeds. So far, more direct recording of migratory flights
have been limited to a few cases where migrants were followed by

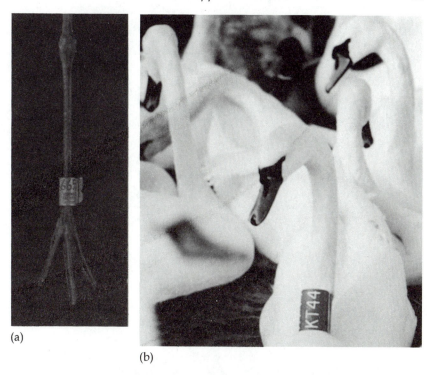

(a)

(b)

(c)

**Figure 1.3** (a) Leg of a white stork with a metal ring for distant identification; (b) mute swan with neck collars (Source: Rutschke, 1992); (c) white storks with transmitters for satellite-tracking.

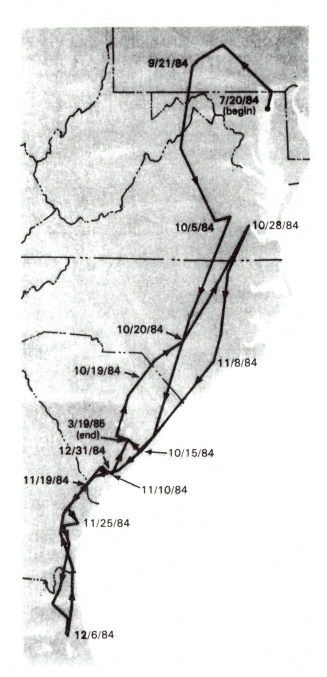

**Figure 1.4** Route of a bald eagle in the USA from July 21, 1984 to March 19, 1985. The eagle was banded in South Carolina as a fledgling, and was caught four years later at Chesapeake Bay, when the transmitter was attached. It was in South Carolina when the transmitter stopped. (Source: Strikwerda *et al.*, 1986.)

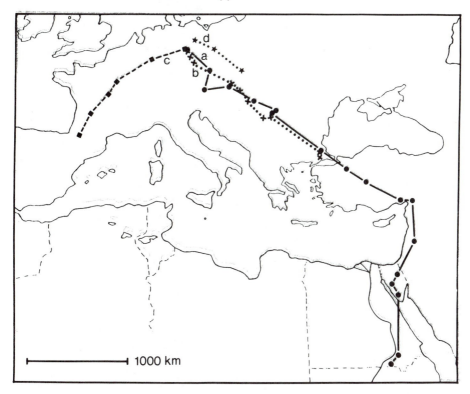

**Figure 1.5** Migratory routes of four white storks (a–d) recorded by satellite-tracking. (Source: Berthold *et al.*, 1992b.)

cars, planes or ultralight gliders (e.g. Berthold, 1993). However, studies from ultra-light aircraft have recently been greatly extended in Israel (Leshem, 1989).

A milestone in the study of bird migration, especially nocturnal migration, was the increasing application of the radar technique after the Second World War. Echos obtained by surveillance, tracking and height-finding radar elucidated the incredible multitude of patterns in diurnal and nocturnal migration and have also made nocturnal migration a 'visible' migration (Figure 1.6). Radar studies further produced quantitative information about altitudes used for migration, differences between tracks (the actual migratory movement over the ground) and headings (the courses taken by migrants; section 1.5), between ground speed (migratory speed relative to the ground) and air speed (speed relative to the surrounding air). Radar ornithology further demonstrated main migratory directions and the extent and compensation of wind drift, among other factors (e.g. Eastwood, 1967;

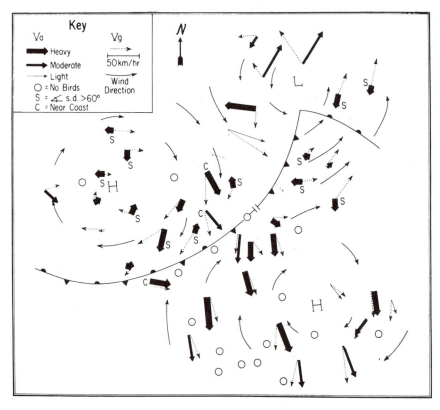

**Figure 1.6** Average direction and density of bird migration over the North Atlantic Ocean as detected by radar during an 8 hour observation period. Light, moderate or heavy migration is indicated by the width of the arrows. Data plotted on a weather diagram. (Source: Williams and Williams, 1978.)

Richardson, 1990; Berthold, 1993). Unfortunately, radar echos usually allow only identification to the level of broad taxonomic groups but rarely to the level of species. As a rule, four categories, i.e. waterfowl, shorebirds, sea birds and passerines, can be identified (Williams and Williams, 1990a, b). Intensity variations of radar echos dependent on wing-beat frequencies characterize only a few species sufficiently for identification (e.g. Bruderer and Jacquat, 1972). This difficulty to determine species is a major handicap for most radar studies. However, attempts are being made to improve species determination by using, for instance, amplitudes of radar echos and Doppler effects (Birjukow and Netschwal, 1988). Only recently (Larkin, 1991), slow flying radar targets were attributed to insects, which were previously believed to represent bird migration. It should be added that formerly assumed negative effects of radar studies on migratory birds (electrophone effects; Schüz *et al.*, 1971) have not been confirmed. Recently, another method of studying nocturnal migration has been developed: to record the flight calls of migrants automatically on tape in connection with a voice-activated switch (Dierschke, 1989). Lastly, numerous studies have been, and still are, based on bird casualties at lighthouses, TV towers, oil flares, etc. The accidentally obtained samples from actual migration can be used to gather details of migratory movements and the extent of fat deposits which are important for energetic calculations (e.g. Terrill, 1990).

After it had become clear that the behaviour of caged migratory birds during the migratory periods could reflect migratory habits (sections 1.2 and 2.1), bird keeping and detailed studies of caged individuals became an important approach to the analysis of control mechanisms in avian migration. Registration cages were used to obtain quantitative patterns of migratory activity ('migratory restlessness' or 'zugunruhe'; section 2.1). These cages are equipped with microswitches which respond to weight changes (hopping) on moveable perches or to air movements produced by wing 'whirring', with photo sensors or ultrasonic systems and various kinds of registration devices (e.g. Biebach *et al.*, 1985). Migratory activity can also be more directly assessed by video-recording with infrared illumination. A variety of circular cages have been developed to serve as 'orientation cages' in which directional choices or preferences can be determined. Initially, training cages with feeders were used, currently octagonal types with tangentially or, more often, radially positioned perches are widespread, along with so-called Emlen funnels. The latter initially consisted of a blotting paper funnel with an ink-pad base. Footprints of the experimental birds showed their directional hopping. Currently, funnels have been lined with typewriter correction paper and scratches are analysed on a light screen. The funnel cages are covered with glass or other transparency material (Figure 1.7; for more details see, for example, Berthold, 1993). Scratches can now be counted and analysed in a fully automatic computer-connected system (Berthold *et al.*, unpublished).

The investigation of caged birds has a number of advantages: individuals of known origin can be investigated; long-term studies on the same individuals are possible; experimental conditions can be varied and crucial factors held constant. Various experimental procedures can also be applied, like injections, implantations and castration, as well as stimulus conditioning. With the help of special equipment heart rate measurements can also be used. Modern respirometers, i.e. gas analysers for the determination of oxygen consumption and carbon-dioxide production, have considerably improved metabolic investigations, as have recent developments in wind tunnel technology.

(a)

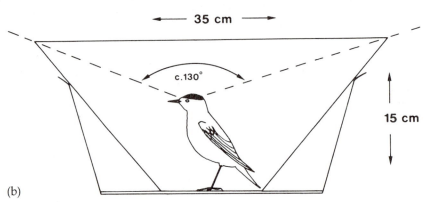

(b)

**Figure 1.7** (a) Funnel-orientation cage lined with typewriter correction paper. (b) Schematic cross-section showing dimensions and field of vision of test bird. (c) Correction paper with scratches of a typical 1.5 hour test. The paper is placed on a light screen and a template dividing it into 24 sectors of 15° each. (d) Numbers of scratches counted perpendicularly across each sector from the paper shown in (c). Sector 1 is north. Sum of scratches, $m$, 463; mean angle, $a$, 226°; vector length, $r$, 0.61. (Source: Helbig, 1991b.)

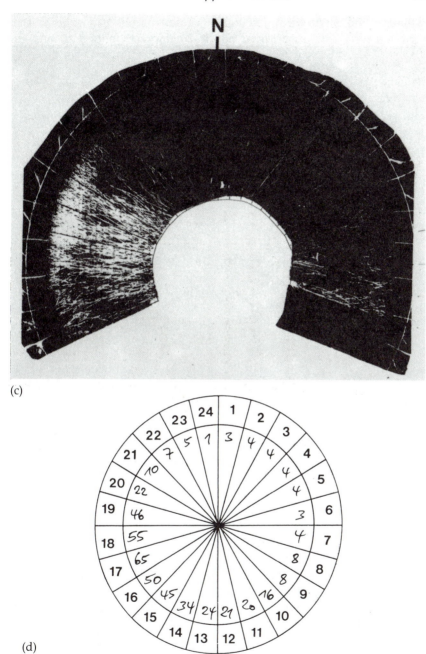

(c)

(d)

**Figure 1.7** *continued*

Specially trained, or even bred, birds (domestic pigeons, 'Grippler') can be investigated when flying uninterruptedly in a wind tunnel for several hours while wearing respiratory masks and having parameters like respiratory frequency, heart rate and body temperature monitored by telemetric means (e.g. Nachtigall, 1990; Norberg, 1990). Wind tunnel experiments with migrating birds are also possible in principle, and preliminary tests with garden warblers (*Sylvia borin*) have been made (Nachtigall, 1990). The structure of the wake behind a flying bird can be analysed by letting it fly through a cloud of helium-filled soap bubbles. By stereophotogrammetry the three-dimensional pattern can be assessed and the total vortex circulation calculated. The lift:drag ratio can be estimated with an ornithodolite (a computerized optical rangefinder-theodolite that measures bird flight speed) and a video recording of the flight (for review, see Alerstam, 1991b). Climbing performance can be experimentally tested in trained raptors carrying loads of various masses (Pennycuick *et al.*, 1989). Statements on the energetics and water economy of flying birds and reasonable estimates for migrants have considerably improved by combining data from free-flying birds followed by radio-tracking, by catching exhaled air in bags, using wind-tunnel methods or double-labelled water (Nachtigall, 1990). The use of the doubly labelled water technique (e.g. Norberg, 1990; Berthold, 1993), or *in vitro* studies (Ramenofsky, 1990), have substantially improved metabolic studies including estimates of the mean daily energy expenditure (e.g. Masman *et al.*, 1989). Combined methods of faecal analysis, stomach irrigation, alimentary flushes (simply with water), as well as the use of marker isotopes, now allow detailed studies of food intake and food choice in migrants (Brensing, 1977; Jenni *et al.*, 1990; Gauthier and Thomas, 1990). Caged experimental birds can also be tested for orientation behaviour in planetaria. In a few migratory species, like the blackcap, large-scale breeding in captivity has become possible and allowed a new research field, combining experimental genetics and migration (section 2.4), to open up. Subtle techniques have greatly facilitated neuroendocrine research. For instance, from small blood or faecal samples, in microlitres ($10^{-6}$ l) up to 10–15 hormones can quantitatively be determined, partially using immunological techniques. At the same time, NMR (nuclear magnetic resonance) techniques allow analysis of neuroendocrinological and metabolic activity in various brain regions in living birds (e.g. Dittami, 1981; Wingfield and Farner, 1993).

Research on orientation has brought about a large variety of mixed-field and laboratory approaches. Both displacement experiments of free-living species and releasing experiments with domestic pigeons, especially carrier pigeons, after transport from their lofts, have been carried out thousands of times and over displacement distances of thousands of kilometres. In these experiments time lags, delayed releases, clock shifts, detour translocations, conditioning and other treatments have been applied. Here orientation ability and performance (homing speed and success), initial orientation, erroneous initial orientation and their consequences, vanishing points and directions, as well as directedness during homeward journeys, have been used to quantify orientation and to establish sensory performance (Helbig, 1991a–d; Wiltschko, 1992; chapter 3).

The quantitative determination of fat (or lipid) stores in migrants is of crucial importance in, for example, the calculation of flight ranges, metabolic water production or flight speeds (sections 2.8, 2.10 and 2.14). Whereas fat extractions of carcasses yield fairly reliable results (Blem, 1980, 1990), indirect methods are still intensively debated. Assessment of lipid reserves by distinguishing arbitrary 'fat classes' ('fat scoring'), especially when based on fat accumulation in the furcular region and on intraperitoneal fat reserves [introduced by Helms and Drury (1960)], has been refined considerably by Kaiser (1992) to give a scale of nine main classes with up to four intermediate classes each, and also by Möhring (1992) in a similar fat-scoring technique. Roby (1991) has compared two non-invasive techniques to measure total body lipid in live birds, TOBEC (total body electrical conductivity) and IRI (near infra-red interactance). TOBEC (that can be applied with a commercial fat scanner, e.g. EM-SCAN; Möhring, 1992), provided an accurate estimate of total body lipid in live bobwhites (*Colinus virginianus*; Roby, 1991), but a false estimation in starlings (*Sturnus valgaris*) in the magnitude of 40% (Möhring, 1992), and various values in migrating sandpipers (Skagen *et al.*, 1993). Certainly, with respect to exact estimations of fat stores, the end of the matter is not yet reached. In most studies fat deposits are related to fat-free (dry) body mass. Determination of fatfree mass and non-fat components is presently reconsidered (e.g. Lindström and Piersma, 1993), since protein storage during migration appears to be more important than formerly assumed (section 2.7.2).

## 1.4 CRUCIAL QUESTIONS

Actually, crucial questions that guarantee purposeful investigations in the field of control mechanisms of avian migration are easily formulated. But surprisingly, the approach to answer them is not infrequently confusingly formulated. For instance, in the same way that causal and final or teleonomical aspects of questions are confounded, proximate, ultimate and permissive factors of migration are confused (for detailed definitions see, for example, Baker, 1938; Thomson, 1950; Farner, 1955). Simple correlations are often postulated and tested as basic cause-and-effect relationships (for instance, section 2.6). In this way, problems of actual migration and evolutionary processes in their development are also confounded. What we are normally interested in when studying control mechanisms of avian migration can be summarized in the following seven questions:

1. What makes individuals, populations or species either migratory or resident? This is a very basic question which can be relatively easy tested in partially migratory populations. In both forms – facultative and obligate partial migration (section 1.5) – young birds have to initially decide whether to migrate and older individuals whether to change their initial habit in subsequent periods. Here, one can approach the question of endogenous and exogenous control and their relative importance. These questions are the basis for the 'genetic' and 'behavioural–constitutional' hypotheses (section 2.4; Berthold, 1988a). The question why migration in individual bird species has evolved may be related to these considerations but has to be treated on a different level of analysis beyond those discussed in section 1.1 and chapter 4.

2. When does a migrant start to prepare for migration and what are the cues for starting? These questions are especially meaningful in species which depart extremely early after fledging and which need at least some preparation before take-off. It is surely of similar importance in forms which start extreme flights across large ecological barriers which require special preparations.

3. How is migration initiated? The main problems are differentiating between socially-induced migration and individual decisions, which are especially relevant in solitary migrants. A major aspect of this question is the relative role of endogenous and exogenous factors. In addition, it may be important in determining to what extent experience or maturation are involved and how the control of departure is altered.

4. How does a migrant know where to fly? Again, this is easy in social migrants with experienced conspecifics. Still, they may also have individual orientation mechanisms. However, for solitary migrants, the question is more difficult. Here one should ask how they manage to reach species- or population-specific resting areas and to what extent experience alters orientation in subsequent migratory periods.

5. How is the course of migration controlled? Migrants could theoretically cover their distances in a number of ways: small hops with many stopover periods, in big jumps, or in the extreme case in a single non-stop flight. All types actually occur and are discussed under aspects of selection for optimization with respect to time, energy and predation, or undisturbed courses of endogenous pro-

grams (e.g. Piersma, 1987; Alerstam and Lindström, 1990). The basic question is to what extent they reflect species-specific strategies or, rather, how they depend on environmental considerations.

6. When, where and why should a migrant stop for resting? A number of factors are particularly important in the timing and habitat choice for resting, i.e. energetic considerations, structure and nutritional aspects of habitats and social aspects such as competition. 'Optimal migration' theory has recently shed new light on such considerations (e.g. Alerstam, 1991a).

7. How does a migrant know when to end its journey? In social migrants a socially-induced resting area could end the first migratory journey, and return to the known breeding area the second. In inexperienced lonely migrants, on the other hand, innate recognition of resting-area characteristics or factors in the migrant itself, like time or energy programs, could make it stop in a given area.

The data presented in the following sections will be treated within the framework of these questions.

## 1.5 SOME DEFINITIONS

As already outlined in section 1.1, the 'normal' migration type is represented by annual (biannual or seasonal) migration most commonly commuting between breeding grounds and wintering areas. Terrill and Able (1988) propose 'annual migrant' for this type which has been termed 'obligate', 'true' or 'regular migrant'. The opposite behaviour is represented by the year-round resident. Between these extremes are migratory movements in one direction, like dispersal, specific movements, like moult migration, or more irregular movements, such as irruptions, nomadism and partial migration (section 1.1). When describing annual migration one should keep in mind that the goal of the journey from the breeding grounds does not necessarily represent 'winter quarters', e.g. in intratropical movements. In many cases the neutral term 'resting area' is much more appropriate. The same holds true for 'autumn' and 'spring' migration, which often are far better ascribed as 'there-and-back' migration, 'to-and-fro', 'out-and-return' or 'away-and-homeward' migration.

Before a migrant departs, an individual normally gets into a certain 'migratory disposition', i.e. a distinct physiological state which includes metabolic adaptations which enhance energy supply and preparations for the onset of migratory activity. In more detail, this 'zugdisposition', or state of readiness, to migrate comprises deposition of fat as a fuel (section 2.1), integration of enzyme systems for storage and rapid utilization of energy stores (section 2.7.1), development of migratory behaviour (sections 2.1 and 2.13), and, to some extent, hypertrophy of flight muscles and increased hematocrit (packed or centrifuged cell fraction of the blood in relation to plasma; sections 2.7.2 and 2.11). 'Migratory restlessness', occasionally 'nocturnal unrest' (zugunruhe) is exclusively related to the migratory activity patterns in captive migrants, irrespective of nocturnal or diurnal activity. The terms do not apply to the behaviour of free-living migrants before departure, as occasionally erroneously stated. A lot of confusion has been caused by a lack of clear definitions for the various types of 'partial migration'. This was pointed out by Terrill and Able (1988). They proposed the term 'obligate partial migration' to refer to the behaviour of individuals of a partial migrant population that migrate each year regardless of annual environmental or population changes. They also suggested the term 'facultative partial migration' to be used for individuals that may or may not migrate in a given year, where movements appear to depend largely, if not entirely, on environmental conditions. Lastly, they recommended that the term 'partial migrant' only be used to reflect the behaviour of individuals within a population and not as a description of a species. The terms 'obligate phase' and 'facultative phase' were their terms to describe the behaviour of individual annual migrants where the initial portion of migration from the breeding area is more endogenously controlled and the final part more environmentally. This fits, to some extent, an earlier concept of Helms (1963) termed as 'motivational migratory restlessness' ('zugunruhe, *sensu stricto*') and 'adaptational migratory restlessness' ('zugunruhe, *sensu lato*'). The term 'differential migration' refers to those numerous cases in which migration in some classes of individuals (ages, sexes, races) differs with respect to timing and distance, or both (Terrill and Able, 1988).

In early literature, there was a rather strict separation between 'weather birds' or 'less typical migrants' (in which migration appeared to be primarily exogenously controlled) and 'instinct birds' or 'typical migrants' (endogenously controlled, for more details, see Berthold, 1975). Terms as such are still used, but since they obviously characterize only extremes on a large scale of differently controlled migrants (chapters 2–5), they are of little importance.

Orientation studies have brought along a series of rather strict definitions. A 'migratory drive' or 'migration divide' often separates populations with different migratory directions, or is at least characterized by distinct population-specific concentrations of directions in a broader fan of routes. Many migrants follow 'main migration routes', which often represent their 'preferred migratory directions' and, from the view of their control, their 'primary', 'normal' or 'principal' directions. Often, these directions are altered to 'secondary directions', e.g. in the case of 'reverse migration' (or 'reverse orientation', i.e. active migration in the seasonally inappropriate opposite direction; section 1.1), or due to wind drift. Mixed orientation behaviour may then lead to directional preferences in a 'bimodal axial' fashion instead of the normal 'unimodal sectorial' one. In the case of drift, the 'track' or 'track direction' (flight direction over the ground) is always different from its 'heading' (direction of the body axis). 'Ground speed' refers to the velocity of a bird relative to the ground and 'air speed' to the velocity of a bird relative to the air mass in which it flies. 'Instantaneous migration speed' has been defined by Gudmundsson *et al.* (1991) as a function of the rate of fat deposition and the current fat load in models of time- and energy-selected migration (section 2.8). Extreme drift may rarely result in 'retrograde migration' (being blown backwards whilst heading forwards). Less effective drift can be compensated by 'partial' or 'total compensation', depending on how much the body axis (heading) is turned towards the lateral wind component. Also, 'over-compensation' occurs, most commonly during later parts of a migratory flight in order to make up for drift experienced earlier. 'Reorientated migration' or 'reorientation', characterizes above all the change of heading in birds towards the land while migrating (or being drifted) offshore. 'Redetermined migration' or 'delayed correction' term rather late compensatory headings, normally occurring in a subsequent flight after an intermediate landing. Migrants can show 'downwind orientation' or 'upwind orientation' (anemotropism, anemotaxis; Schüz *et al.*, 1971) when migrating with or against a given wind direction; downwind migration may exhibit 'pseudodrift' (apparent, but not real, drift; section 3.7). For more details see, for example, Richardson (1990); Berthold (1991b, 1993).

## 1.6 MORPHOLOGICAL AND PHYSIOLOGICAL PREREQUISITES FOR MIGRANTS

The essential morphological and physiological characteristics delineating birds from other vertebrate groups are those associated with flight, and this form of locomotion is also the most important for migratory movements (section 2.14). Flight requires specific adaptations in the flying apparatus, i.e. the wings, and extremely efficient respiratory, cardiovascular, muscular and metabolic systems to enable this highly demanding mode of locomotion to occur. This is easy to understand if one keeps in mind that this kind of movement involves extremes in exercise, as birds are forced to pass through large horizontal and altitudinal differences even when performing everyday tasks. For instance, foraging flights in albatrosses may cover distances up to 15 000 km per month, or up to 14 000 km in half a month under extreme weather conditions (Jouventin and Weimerskirch, 1990; Prince *et al.*, 1992). In this section the most important morphological and physiological adaptations serving as prerequisites for migrants, above all with respect to flight performance, will be treated.

Flight strongly influences structural and functional aspects of the bird's whole body. Optimizing flight performance has led to a remarkable similarity in basic body plan and biology in most species. Basic prerequisites associated with flight include a skeleton with a strong central 'box' similar in nature to the box girders of structural engineering. This 'box' ensures a high degree of central stability, which protects internal organs during aerial manoeuvres, produces a streamline overall body contour and provides a large attachment area for the huge flight muscles, pectoralis and supracoracoideus, the 'flight motor'. Horn as material for the wings, and light toothless bills, pneumatization of bones (being hollow, containing air-sacs) and unilateral development of ovaries all contribute to a reduction in body mass and a facilitation of flight characteristics (Perrins, 1990). These basic adaptations to aerial movements in birds are in general a starting point for migratory movements, especially compared to amphibians, reptiles and mammals. Thus, although birds are actually predisposed to migratory movements, long-distance migration in particular requires many additional specific morphological and physiological features. The most important ones are adaptations in the flight apparatus and other aspects which affect flight performance.

Most of the investigations to date have compared the relationship between wing morphology and migration performance. The reason is clear. A simple comparison, for instance, between the elegant flight of a swift and the more clumsy aerial movements of jays and tits demonstrates impressively that long, thin pointed wings can improve flight and migratory performance. Indeed, among birds of similar mass and wing length, forms with long, thin wings, and a high aspect ratio (ratio of wing span to wing width, or wing area or wing length to wing width; Kerlinger, 1989) have greater mechanical efficiency and thus lower cost of transport and, theoretically, a higher potential for migratory movements. A pointed wing produces a low air resistance and, at the same time, a good supporting power (owing to a small induced drag; Alerstam, 1990). Kerlinger (1989) pointed out that birds with long, pointed wings generate more power at a lower cost because pointed wings are more able to shed tip vortices that create drag because they must be pulled along by the bird. Rayner (1990) summarized details on the aerodynamic properties of small, round wings versus long, pointed wings according to his vortex theory. According to this theory, the so-called continuous wake gait is confined to cruising flight in birds with longer, more pointed wings, which makes these birds such effective flyers. In raptors, slotted wing-tips may compensate to an extent for the lack of long or pointed wings, as slots may permit slower flight by reducing tip vortex phenomena or may promote faster acceleration than is possible without slots (Kerlinger, 1989). Tucker (1993) obtained strong evidence that the tip slots of soaring birds reduce induced drag in the sense that the separated tip feathers act as winglets and increase the span factor of the wings. Alerstam (1991b) has summarized recent interesting findings and discussions on wing action and vortices in the wake of flying birds. The wake behind a fairly rapid flying bird consisted of a pair of continuous undulating wing-tip vortices. These continuous trailing vortices of a concertina wake are radically different from the closed-loop (ring) vortices demonstrated earlier for birds hovering or in slow flight and may be of crucial importance for the excellent flight performance of migrants. Perhaps, migrants can also arrange their feathers to achieve a 'favourable roughness' on the plumage surface (with a favourable Reynolds number), which could promote supercritical flow in the same way as controlled surface roughness favours the flight of golf balls or model aircraft.

It had been formerly assumed that migratory performance is associated with longer, more pointed wings (Schüz *et al.*, 1971). Pointing of the wing can be expressed in the 'wing-pointing index' (or 'primary projection', i.e. the extent to which the primaries of the wing-tip project beyond the secondaries or tertials, expressed as a percentage of the total wing length). Indices vary from only about 15% in wrens to over 70% in swifts. In wheatears (*Oenanthe oenanthe*) the indices range from the most migratory forms in Greenland and North America (35%) to resident conspecifics in Africa (22%) (Alerstam, 1990). A number of authors have tried to develop adequate indices (known as Kipp's wing pointing index etc; see Winkler and Leisler, 1992; Berthold, 1993). However, some authors, like Keast (1980), Niemi (1985) and Oehme (1990b), were not able to demonstrate even simple relationships between wing pointedness and migratory performance. Leisler (1990) and Winkler and Leisler (1992) reassessed these and other relationships on large data sets from several taxonomic groups of Old and New World species. The main results of these studies were as follows. In general, migrants tend to be more uniform, in some groups they are smaller than equivalent residents, and their bills are shorter. The aspect ratio, as an important variable for flight cost, is generally higher in migrants. In passerines with more uniform flight style than in nonpasseriformes, more attenuated wings with longer distal wing segments were found (Figure 1.8). According to Rayner (1988), some nonpasserines like ducks and swans, show clear wing specialization for migration. They have a relatively high aspect ratio, but notably small wings. This trend is most

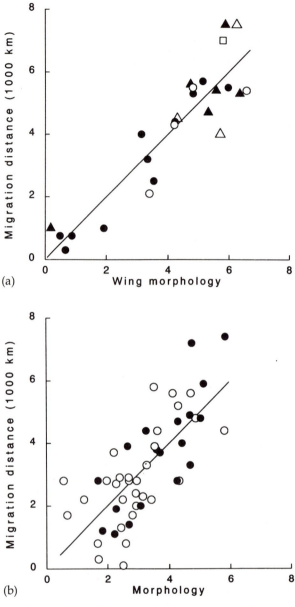

**Figure 1.8** (a) The relationship between wing morphology and migration distance in 25 species of sylviids. Morphology represents the prediction of migration distance through a multiple regression with the number of notched primaries and carpometacarpus length as predictors ($R^2 = 0.831$, $P < 0.0001$). (b) The relationship between wing morphology and migration distance in 50 species of parulids. Morphology represents the prediction of migration distance through a multiple regression with femur, ulna and carpometacarpus length as predictors ($R^2 = 0.538$, $P < 0.0001$. (Source: Winkler and Leisler, 1992.) Symbols denote species of several genera.

marked in species which breed in high Arctic latitudes (e.g. eider ducks) and would be of adaptive value here in that it would allow birds to move rapidly in response to adverse weather conditions. A relative lower body weight as a characteristic for tropical migrants, as mentioned above, was also reported by Mönkkönen (1992) for nearctic and palaearctic species. Winkler and Leisler (1992) also stated that migration sets the upper limits to the hind limb development, the supracoracoideus and other minor flight muscles. Why femur and pelvis length should be related to migration is not understood. No clear relationship between mass and migration has been found in the small bird species analysed.

As discussed by Eaton *et al.* (1963), low-cost transport over large distances may require various morphological adaptations which conflict with possible adaptations for foraging and habitat use, and all aspects of flight performance cannot be favoured simultaneously. With data on *Sylvia* species, Winkler and Leisler (1992) showed that migration and habitat use could affect each other, e.g. short round wings which increase manoeuvrability and thus enhance flight capacity in more dense habitats are incompatible with high migratory performance. The demands of migration may even restrict adaptive radiation in groups of temperate zone birds – a matter that certainly needs further investigation.

It is well known that many passerines and nonpasserines have shorter wings than older conspecifics during their first year of life. Alatalo *et al.* (1984) suggested that young birds may benefit from the possession of short wings and increased manoeuvrability. Assuming a similar energy output, however, this developmental difference should result in a slower course of migration among first-year birds. These age-dependent differences in the course of migration have been found, or assumed, to occur in many species (Berthold, 1993) (for possible compensatory mechanisms of juveniles see section 2.8). In raptors, smaller wings, lighter wing loadings and larger tails of immature birds appear to be adaptations for slower and more manoeuvrable flight which may prevent younger birds from crashing into obstacles (Kerlinger, 1989).

The major flight muscles in birds, the pectoralis and supracoracoideus, responsible for upward and downward stroke of the wing, form a complex, highly specialized 'aero-engine'. Butler and Woakes (1990) summarized its form and function as follows. The pectoralis as the main part can constitute up to 35% of total body mass. In volant birds it consists predominantly of fast-contracting, but highly-oxidative, fibres ('red fibres'), which are required to beat the wings continuously at high frequencies, sometimes for many hours (section 1.1). These fast oxidative glycolytic (FOG) muscle fibres have a relatively small diameter and being highly oxidative provide the basis for extended flights. A small part of the muscle is composed of fast glycolytic (FG) fibres, which are larger and possess small amounts of mitochondria. These 'white' fibres can metabolize glycogen anaerobically and are used predominantly during take-off, landing and for sudden changes in direction. Finally, a few species of birds (most probably those that glide; Butler, 1991) have also some slow oxidative (SO) fibres which are normally involved in postural support. In line with the extensive oxygen demand in the pectoralis of flying birds there is a higher blood capillary:fibre ratio and thus greater oxidative capacity than is found in the leg muscles. This, along with the smaller cross-sectional area of the FOG fibres in the pectoralis, ensures shorter diffusion distances for the oxygen supply. Lundgren (1988) and Lundgren and Kiessling (1988) found, in comparative studies, that the distance of migration is inversely related to fibre size and positively related to capillary density in the pectoralis. Hence, long-distance migrants have higher oxidative capacities and also a higher capacity for fatty acid oxidation (section 2.7.1). For more details on mitochondrial density see Gaunt *et al.* (1990), for details of the neural control of flight muscles see Wilson (1990), for processing of information of feather-associated receptors in the domestic pigeon as adaptations to flight see Necker (1991). Suarez (1992) emphasized that in hummingbirds maximal capacities for $O_2$ and substrate delivery to muscle mitochondria, as well as mitochondrial oxidative capacities, may be at the upper limits of what are structurally and functionally possible given the constraints inherent in vertebrate design.

In many migrants, particularly small birds, body mass can change enormously during a long flight (section 2.8). As a consequence, the balance between the mechanical power output by the flapping wings and the power input by the flight muscles must shift. Pennycuick (1978) suggested that birds may compensate for this by consuming protein from the flight muscles, rather than having to transport excess muscle towards the end of a migratory flight. Evidence for this suggestion is still weak (section 2.7.2). According to Rayner (1990), 'bounding' flight (with pauses between bouts of flapping in which the wings are folded against the body) may be a mechanism to level the required power. This type of flight will be treated in more detail in section 2.14.

Further adaptation to flight can be found in the cardiovascular system. Birds have 1.4–2.0 times larger hearts and lower resting heart rates than mammals of similar body size. They also have a greater cardiac output for a given oxygen consumption than similarly sized mammals. The higher cardiac output may be an important factor in maintaining a higher oxygen uptake during flight than similar-sized mammals do when running (Butler and Woakes, 1990). In addition, there are other cardiovascular and blood adaptations in migrants which are required during high altitude flights: these are discussed in section 2.11.

As another adaptation to flight, the structure of, and breathing mechanism in, birds' lungs result in a much higher efficiency for gas exchange than in the mammalian alveolar lung, or in any other vertebrate lung. Bird lungs are interconnected to a system of thin-walled air sacs by a complex array of tubes. Air is continuously directed through the lungs during inspiration as well as expiration (Goslow *et al.*, 1990). Thus, the avian lung has a cross-current arrangement between the convective gas flow through the parabronchi and the returning mixed venous blood (Fedde, 1990). As has been shown in starlings, effective lung ventilation during flight can be increased above that required by metabolic rate only. This hyperventilation may improve gas exchange considerably under extreme conditions and also serves a thermoregulatory function (Butler and Woakes, 1990, sections 2.10 and 2.11). The attachment of the pectoralis muscles to the sternum suggests that they could compress the thorax and assist ventilation during flight (Banzett *et al.*, 1992). A comprehensive review of ventilation, gas exchange and oxygen delivery in birds has been given by Brackenbury (1991), and more details with respect to requirements during migration are mentioned in sections 2.11 and 2.14.

A number of further adaptations in birds exist concerning metabolism, nutrition and digestion. These include seasonal shifts from carbohydrate to lipid metabolism or from carnivorous/insectivorous to omnivorous/vegetable diets. Lastly, these adaptations may involve seasonal changes in the morphology of the digestive tract (sections 2.7 and 2.9).

# 2

# Control mechanisms and ecophysiology

If one had to name the most obvious criteria of migratory physiology among them would certainly be fat deposition and the pronounced nocturnal activity during migration in normally diurnal species, often expressed as migratory restlessness in caged individuals. These criteria will first be treated in the following section.

## 2.1 NOCTURNALITY, FAT DEPOSITION AND MIGRATORY RESTLESSNESS

### 2.1.1 Nocturnality

It is not surprising that nocturnal bird species, like nightjars, or both diurnal and nocturnal species, like waders (e.g. McNeil, 1991), migrate at night. It is, however, surprising that so many purely diurnal species during the non-migratory seasons become more or less nocturnal during migration, like many insectivorous long-distance migrants of higher latitudes. Evidence for this pronounced nocturnality has been obtained in a number of ways: specific analysis of calls of nocturnal migrants, nocturnal lighthouse casualties, extensive use of the moon-watch methods and the absence of many migratory species during diurnal migratory studies (Dorka, 1966; Biebach *et al.*, 1991; Dolnik, 1990). It has to be kept in mind that systematic surveys of nocturnal–diurnal migratory behaviour showed that the majority of species may migrate by night and day and that it may be impossible to allot them to a diurnal or nocturnal migratory category. However, of 147 species from 16 orders surveyed in the British Isles, 75% of all species had been recorded as a night migrant at some time, including the chaffinch (*Fringilla coelebs*) and the swallow (*Hirundo rustica*) (and the white stork

(*Ciconia ciconia*) in Africa) (Elkins, 1988b; Martin, 1990). Some groups, like, for instance, thrushes and *Sylvia* warblers, seem to be more likely to travel at night than others. The finding that most nocturnal migrants may also travel by day suggest that there are barely no 'exclusively nocturnal' migratory species but doubtless many travel predominately at night. The question why so many migrants migrate nocturnally in the first place has often been discussed (Stresemann, 1934; Dorka, 1966; Kerlinger and Moore, 1989; Terrill, 1990; Berthold, 1993). The following seven hypothetical explanations have been offered.

1. Predator-avoidance hypothesis. In line with the common idea that migration is risky (section 1.1), it was proposed that nocturnal migration could have evolved in order to reduce predation pressure from diurnal raptors. Their influence appears, however, apart from a few species like the Eleonora's falcon (*Falco eleonorae*) or the sooty falcon (*Falco concolor*), in the Eurasian–African migration system, not to be higher than for resident species. And during the night, owls are active and well known to predate on migrants as well, although these predator–prey relationships are not well understood during migration.
2. Forced-flight hypothesis. Many species migrate day and/or night, depending on local or seasonal circumstances (e.g. weather, required migration speed; Dorka, 1966). When migrating over vast distances, for instance, the ground conditions may be so disadvantageous that the bird is forced to migrate into the night. This factor, however, cannot explain the ubiquity of nocturnal migration or its widespread use over short distances and suitable areas.
3. Daytime feeding hypothesis. Although nocturnal feeding during migration might play a role in groups of open-country species, like waders or waterfowl (e.g. Evans and Davidson, 1990; Martin, 1991), it is not important in the majority of migrants which, in general, feed exclusively during the day. Long-distance migration requires considerable stores of energy reserves, especially for ocean and desert crossings where no replenishment is possible. Theoretically, if such energy loading required a whole day of feeding, migration would have to take place at night. Both field observations and experiments, e.g. with reduced feeding periods [for instance, in the white-crowned sparrow (*Zonotrichia leucophrys*); King, 1961a], indicate that foraging over only part of the day is necessary for extensive fattening in some species. In other species, daytime appears to be a critical factor. In sanderlings (*Calidris alba*), for instance, extension of the overall daily foraging period was observed during spring migratory fattening in order to increase the foraging time (Lindström, 1991). Similarly, day length appears to be crucial

for fattening in Whimbrels (*Numenius phaeopus*; Zwarts, 1990; section 2.1.2) and other species (Piersma *et al.*, 1991).

4. Time-gain hypothesis. Even if daytime were not entirely necessary for feeding it is rather doubtful whether the remaining daytime slot alone would be sufficient for long-distance migrants to cover tens of thousands of kilometres in the appropriate time period. Time limitations could be an essential factor here in the evolution of nocturnal migration in otherwise diurnal birds (e.g. Stresemann, 1934).

5. Energy-saving or atmospheric hypothesis. As Kerlinger and Moore (1989) have pointed out, there are four characteristics of the nocturnal atmosphere that make migration at night more favourable for migrants than during the day: flying in cooler, more dense air at night favours the lift:weight and drag ratios so that migratory stages performed by flapping flight are less costly; horizontal winds are slower at night so that in head wind or crosswind less energy is needed and, in addition, birds are not as likely to be blown off course; winds at night are less variable in direction which allows for more undisturbed migration; and vertical wind currents caused by thermal convection are absent at night. This alleviates the need for corrections, compensation and additional energy expenditure (section 3.7).

6. Increased physical safety or dehydration avoidance hypothesis. Night air temperatures are generally lower and relative humidity tends to be higher. Both promote convective cooling and may prevent dehydration due to evaporative water loss. In highly demanding flights there is also a risk of hyperthermy and exhaustion which may be reduced at night (section 2.10).

7. Star compass allowance hypothesis. In spite of the fact that a wide variety of animals are mobile at night, stellar orientation has only been demonstrated in birds and a few other animals. It may then be largely a novel evolutionary development for birds (Able, 1980; Terrill, 1990; section 3.3). We do not know how important the use of the stars is in migratory bird species as a compass or landmarking system. Thus, the question of whether stellar orientation accelerated the evolution of nocturnal migration or vice versa remains open.

At present it appears that the main forces in the evolution of nocturnal migration are related to hypotheses 3–6; the foraging hypothesis, as well as possible time gains, energy output reduction, physical safety increase and an optimization of permissive conditions. Perhaps, the analyses of the migration strategies over large barriers, like deserts, may provide a better understanding of nocturnality. Data from such studies may demonstrate why many birds equipped with sufficient fuel reserves prefer to spend the day roosting in the open desert instead of migrating diurnally (section 2.16).

### 2.1.2 Fat deposition

Examination of migratory bird casualties at lighthouses and trapped individuals at ecological barriers, like deserts or oceans, have exposed researchers to the phenomenon of how fat migratory birds can be. This is unavoidable as fattened long-distance migrants are tactily soft and spongy. Their cutaneous and subcutaneous fat stores often coalesce into an essentially continuous sheath around the torso to such an extent that not even the medial part of the pectoral muscles remains exposed (Figure 2.1). In penguins and a few other aquatic bird species fat stores are so extreme that they can be described as blubber. In most birds, however, subcutaneous fat is more or less patchy (Pond, 1978). Cutaneous and subcutaneous fat (or lipid, which is used as an interchangeable word) is stored in a number of morphologically distinct fat 'organs'. Sixteen such organs have been described for chicken; they are identifiable in the embryo at 16 days (Pond, 1978). Fifteen of them (with the exception of the superficial pectoral fat body) have also been found in a migratory species, the white-crowned sparrow, by McGreal and Farner (1956; Figure 2.1). These fat organs are apparently homologous in several groups of birds (King and Farner, 1965), particularly the claviculo–coracoid, the spinal, synsacral, ischio–pubic and lateral thoracic fat bodies. Some fat bodies, like furcular and intraperitoneal, are of prime importance for estimating lipid reserves as a fat index. For details of judging fat reserves see section 1.3 and recent discussions in Blem (1990) and Kaiser (1993b). Lipid in birds is also stored at a number of intraperitoneal sites and in organs including the digestive tract and the bone marrow (above all in waterfowl). There is less lipid storage in muscle or liver, and no lipid increase in the heart during migration. For more details see Pond (1978), Blem (1990) and Napolitano and Ackman (1990).

Adipose tissue in birds is metabolically specialized for lipid storage like that in other vertebrates (section 2.7.1). Lipid is added to vacuoles of adipocytes, specialized cells for lipid storage, obviously without an increase in the number of cells (Blem, 1980). Birds lack, however, typical brown adipose tissue which is found in some mammals, mainly for the production of large amounts of heat (Blem, 1990). Birds do sometimes develop 'brownish' adipose tissue in winter, after exposure to cold. Here, multilocular fat cells are surrounded by numerous capillaries, high mitochondrial density, but no sympathetic nerve endings are present as is the case for mammals (Hissa, 1988; Blem,

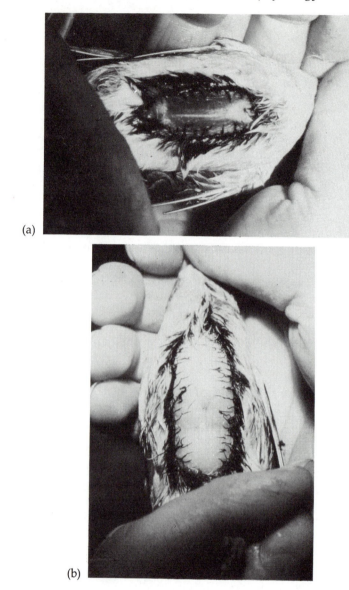

(a)

(b)

**Figure 2.1** Two garden warblers. (a) In premigratory condition without fat deposits. (b) Within the autumn migratory season with large fat deposits. (c) Major subcutaneous fat organs in the white-crowned sparrow. (Source: King and Farner, 1965.)

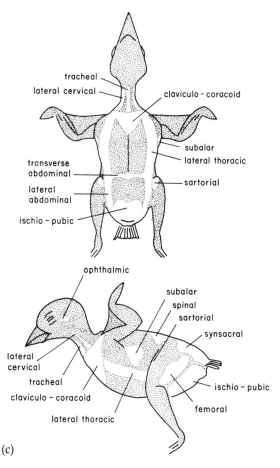

(c)

**Figure 2.1** *continued*

1990). Adipose tissues in birds appear to be functionally 'white' (Saarela *et al.*, 1991).

Since lipid is the most important energy base for all aspects of avian life history, like breeding, moulting or wintering (Blem, 1990), it is not surprising that it also plays a major role as the fuel for migration. Lipid has by far the highest caloric content of potential energy carriers that can be stored in the body. This is usually 9.0–9.5 kcal/g = 37.7–39.7 kJ/g, compared to only 4.0–4.5 kcal/g = 16.7–18.8 kJ/g in storable carbohydrates, or 4.3 kcal/g = 18.0 kJ/g in proteins (Blem, 1990). Adipose tissue also appears to function on a 'tank' principle in migratory birds. Here, storage and mobilization involve only the addition or subtraction of dry lipid (King and Farner, 1965), whereas storage of carbohydrates like glycogen is usually accompanied by extensive water storage (which may account for up to 3–5 g water per gram carbohydrate; Blem, 1990). Another positive aspect of lipids is that fatty acid molecules can be used efficiently by β-oxidation. This is an extremely efficient metabolic reaction because of the low number of intermediate steps required to produce energy (Blem, 1990; section 2.7.1). Thus, it is not surprising that fat has evolved as the fuel for avian migration in all parts of the world, including the tropics (e.g. Jones and Ward, 1977). Carbohydrates are only of less importance in migration since they only provide short-term energy sources. Although they may be the substrate for thermogenesis in small birds, glycogen levels are often suppressed during migration (Blem, 1990; Marsh and Dawson, 1982; Whittow, 1986; section 2.7.2). Protein, finally, is used as an energy source only under extreme conditions. Its metabolism is complex and inefficient and, unlike lipid and carbohydrate, results in toxic by-products (Blem, 1990). The premigratory build-up of protein reserves found in a number of species (i.e. in breast muscles hypertrophy) may be useful in the muscle maintenance and repair during migration (Piersma, 1990; section 2.7.2).

Since fat metabolism represents a central issue in avian migration, a number of aspects should be considered in detail in this sense. Sections in this chapter have therefore been devoted to: basal physiological aspects of fat metabolism and the relationship between flight muscle and fat metabolism (section 2.7.1), hyperphagia, utilization of foodstuffs and nutritive strategies (section 2.9), changes in body composition during the migratory period (section 2.7.2), and, finally, the amount of fat deposits and specific migratory requirements (sections 2.8 and 2.15). Section 2.8 includes calculations of theoretical flight ranges of migrants and energy balances during migratory periods as well as over the course of the year.

However, one additional point should be stressed here. That is the mechanical aspect of fat deposition. Many long-distance migrants accumulate very large fat deposits for extended non-stop flights (section 1.1). In some cases this amounts a lean body mass increase of more than 100% (sections 2.7.2 and 2.8). Doubling body mass during the long flights may raise substantial problems in relation to flight mechanics. It is surprising in this respect that the question of what factors determine the fat location have rarely been addressed (Piersma, 1990). Pond (1978) has discussed this issue to a certain extent. One hypothesis is that body insulation is a determinant of fat distribution. There is little support for this in data on distribution, dynamics of deposition and the reabsorption of lipid deposits. Insulation effects of subcutaneous fat layers in birds appear to be rather low and not essential to thermoregulation. Varying the extent of these layers has little effect in either nestlings or adult birds (Biebach, 1977; Pond, 1978). Further, no adaptations in subcutaneous fattening, according to ambient temperature, have been reported. Here, one can compare birds migrating to relatively hot areas and returning to Arctic breeding grounds. Another hypothesis emphasizes mechanical support and protection of organs. There is some relevant data under discussion for mammals but whether internal fat deposits play such role in migratory birds during fast and far locomotion remains open. The third hypothesis treats mechanical effects on balance and buoyancy in locomotion. Unfortunately, the matter has not been adequately investigated in migratory bird species. There is some evidence in the white-throated sparrow (*Zonotrichia albicollis*; Pond, 1978) and in the European blackbird (*Turdus merula*; Biebach, 1977) that fat pads are larger in the peritoneum of the abdomen during premigratory fattening than during pre-winter fattening when subcutaneous deposits are larger. Extensive fattening at, or close to, the centre of gravity of the bird during migratory periods could have the effect that the flight disturbance due to extra weight would be minimized. The relationships between migratory fat deposition, fat depletion and the mechanics of flight require more detailed study. The last hypothesis stressed by Pond (1978) concerns sexual and social signals and appears to be irrelevant with respect to migration phenomena.

### 2.1.3 Migratory restlessness

In addition to fat deposition, in captive migratory birds the occurrence of 'migratory restlessness' or 'zugunruhe' is a typical characteristic. It was first described in detail by Johann Andreas Naumann (1795–1817) as flying, hopping, fluttering and wing whirring at night for nocturnal migrants. Migratory restlessness was probably well known among bird keepers long before Naumann's description, but was considered more to be a bad habit leading to feather abrasion. Thus, preventive means like tying the wings to the back were discussed (e.g. Johann Friedrich Naumann, 1897–1905). Because migratory restlessness is the expression of an instinct (section 2.4) such attempts were none the less doomed to failure. Johann Andreas Naumann was the first to actually discuss the function of migratory restlessness and he also related its amount to the distance which had to be covered by the bird. He examined migratory restlessness in two long-distance migrants, the golden oriole (*Oriolus oriolus*) and the pied flycatcher, in his bird room by sight observations and listening to the birds' nocturnal movements. Realizing that migratory restlessness persisted into the winter he concluded that these species 'migrate far away, presumably as far as Africa', which later proved to be right. As mentioned in section 1.5, it should be emphasized again that migratory restlessness or zugunruhe has been defined to be the expression of migratory activity in caged migratory individuals and thus, by definition, is not a pre-take-off restlessness of birds in the wild as has been repeatedly proposed (e.g. Faaborg and Chaplin, 1988: 'The term used to describe this premigratory restlessness is the German term zugunruhe. When this state of restlessness is strong enough and the bird is exposed to the appropriate set of short-term cues, it initiates flight'). The relationship between migratory restlessness and initiation of migration is discussed in sections 2.3 and 4.1.

Migratory restlessness has actually been qualitatively and quantitatively recorded in over 100 migratory species (Berthold, 1988b, 1993), including large species like geese (Bruns and Ten Thoren, 1990). Surprisingly, until recently, insufficient was known about the quantitative nature of migratory restlessness to expand on Naumann's original hypothesis, with the exception of a few reports on occasional observations (Kramer, 1949). Recent experiments with nocturnally migrating garden warblers (*Sylvia borin*) and a few other species based on recording of perchhopping activity and accompanied by video records of nocturnal movements (under infra-red illumination; Berthold and Querner, 1988) have resolved the original query. Video recording elucidated that practically all migratory restlessness involves wing whirring, i.e. a characteristic wing beating with high frequency and low amplitude in a sitting position. When individuals in migratory disposition wake up during the night, they normally start wing whirring. Subsequent movements like perch-hopping, body turning and grounding are accompanied by wing whirring of varying intensities. Thus, migratory restlessness appears to be essentially a cage-adapted migratory activity expressed as a reduced flight behaviour, i.e. migration in a sitting position. The normal cages are simply too small to allow for normal flights. Initially, caged migrants do appear to try to take-off as can also be the case when migrants are stimulated by specific calls [e.g. bobolink (*Dolichonyx oryzivorus*); Hamilton, 1962; section 2.13]. Most migrants then, rapidly adapt their flight behaviour to cage conditions by selftraining.

Migratory restlessness is easily discernible in nocturnal migrants which, outside migratory periods, are exclusively diurnally active (Figure 2.2). Here, the observed nocturnal locomotor activity is of a primarily migratory nature (for other kinds of restlessness see below). Migratory restlessness is, however, also easily recognizable and, in addition, quantitatively measurable, in diurnal migrants. Here, the alteration of diel patterns of locomotor activity, above all the increase of activity peaks at species-specific diurnal migration periods, indicates migratory restlessness (Figure 2.2). By subtracting activity patterns

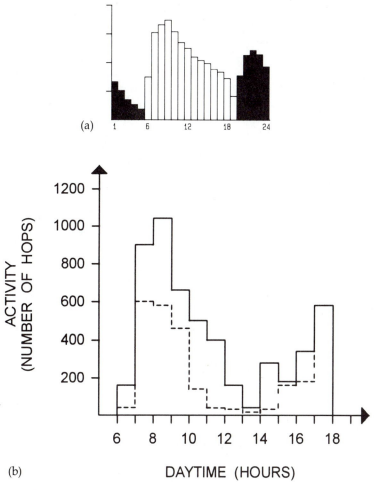

**Figure 2.2** (a) Daily activity patterns of blackcaps with nocturnal migratory restlessness (black columns) (Source: Brensing, 1989). (b) Daily acitvity patterns of a goldfinch (*Carduelis carduelis*): solid line, within the autumn migratory season; broken line, after the migratory season (Source: Glück, 1978).

obtained in non-migratory periods from those recorded during migratory seasons, migratory restlessness can be quantitatively determined (Berthold, 1978a).

Generally, migratory restlessness, when properly delimited from other types of unrest, is a fairly good reflection of the species- and population-specific migratory behaviour and the course of migration. This view is supported by an increasing body of evidence within three main criteria. First of all, migratory restlessness occurs, as expected, almost exclusively in caged individuals of migratory forms and is (almost) entirely absent in non-migratory forms like, for example, house sparrow (*Passer domesticus*), Clark's nutcracker (*Nucifraga columbiana*), crested tit (*Parus cristatus*) and non-migratory black-cap populations from the Cape Verdes and Madeira (for review, see Berthold, 1975, 1988b). The occasional occurrence of nocturnal restless-ness in some caged non-migrants as, for example, in the white-crowned sparrow (*Zonotrichia leucophrys nuttallii*), may possibly be the expression of acclimatization restlessness (below) or perhaps migratory activity due to a degree of obligate partially migratory behaviour. This should be tested before an atavistic remnant of ancestral migratory behaviour can be assumed (Berthold, 1975). Secondly, in more than 25 species and populations it has been demonstrated that migratory restlessness is closely related to the specific migratory seasons, the length of the migratory period, the distance to cover or the ecological characteristics of the migratory journey (Figure 2.3; Berthold, 1993; for more details see section 2.3). Lastly, migratory restlessness occurs in purely night migrants at night, in diurnal migrants during the day and in species migrating at dawn at that time [e.g. reed bunting (*Emberiza schoeniclus*); Berthold, 1978a]. It appears as if migratory restlessness in diurnal,

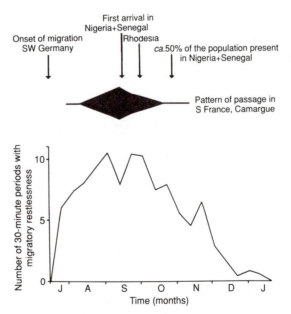

**Figure 2.3** Comparison of the annual pattern of migratoy activity (zugunruhe) of caged garden warblers and data on the migration course of conspecifics in the wild. (Source: Berthold, 1993.)

nocturnal and dawn/dusk migrants are essentially exhibited at the appropriate times used by free-living conspecifics (Berthold, 1978a, b; Glück, 1978; Brensing, 1989), although more quantitative comparisons are needed.

Unfortunately, as Helms (1963) pointed out, not all nocturnal loco-motor activity in normally diurnally active birds is migratory restless-ness. Five (or six) other kinds of restlessness activities have been established in the meantime, and others cannot be excluded. Figure 2.4 represents five. Three of them are evident in a comparative study of hand-raised, trapped young and trapped adult blackcaps. Trapped individuals show, in contrast to hand-raised conspecifics, premigratory nocturnal activity. This is particularly marked in adult birds. It per-sists even throughout the complete post-nuptial moult. This unrest may be related to trapping and caging and the loss of the previously experienced freedom, home range and habitat. It has been termed 'acclimatization restlessness' (Berthold, 1980, 1988b). That it may play a common role in experimental birds which were trapped, has rarely been considered and may often falsify results of tests in orientation cages for preferred migratory directions. 'Winter restlessness' was found in adult trapped blackcaps between the periods of autumnal and vernal migratory restlessness. It may either simply represent a continuation of acclimatization restlessness or an appetitive behaviour related to a hitherto experienced wintering area. It may be an explanation for some of those cases in which an extension of the period of 'migratory rest-lessness' over the species-specific autumnal migratory period has been reported (Gwinner and Czeschlik, 1978; and below). A third type, described by Merkel (1956), is the widespread 'summer restlessness' ('pseudorestlessness' or 'pseudo-zugunruhe'). It has been discussed extensively by Gwinner and Czeschlik (1978), where three hypothetical explanations were offered: (1) it is not related to migration and reflects other nocturnal activities; (2) it is related to migration and results from the inability of caged birds to respond to stimuli of their home area to which they are imprinted; or (3) it results from the inability of caged birds to perform reproductive activities. According to the evidence at that time, the authors favoured the third explanation. This view has recently been supported by observations on blackcaps. For instance, individuals exhibiting summer restlessness in cages ceases immediately when they are transferred to aviaries where they can breed. Individuals engaged in breeding in aviaries also start, as a rule, to develop summer restlessness immediately they are returned to cages. Lastly, blackcaps from the Cape Verdes which, as residents, do not exhibit any migratory restlessness develop summer restlessness during the period of main gonadal development and song (Berthold, 1988b). All these observa-tions indicate that summer restlessness is an appetitive behaviour

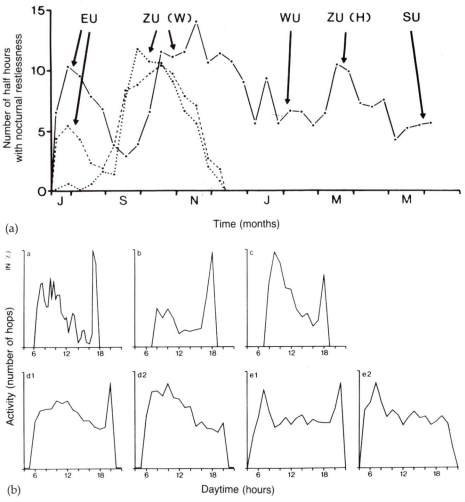

(a)

(b)

**Figure 2.4** (a) Nocturnal restlessness of blackcaps. Solid line, adult trapped; broken line, first-year trapped; dotted line, first-year hand-raised individuals. EU, acclimatization restlessness; ZU, zugunruhe; W, autumn migratory period; H, spring migratory period; WU, winter restlessness; SU, summer restlessness. (b) Diurnal activity patterns of caged birds with roost-time restlessness. a, American robin; b and c, song thrush and reed bunting; d1 and e1, great tit and blue tit, respectively, d2 and e2, as d1 and e1, but roost-time restlessness disappeared when a sleeping box was offered. (c) Diurnal activity patterns of 10 garden warblers. The birds were kept in a 12 hours light–dark ratio during the autumn migratory period. Solid line, activity pattern with *ad libitum* food; broken line, with two interposed periods of 2 hours of food deprivation (black bars) and 'hunger restlessness'. (Source: Berthold, 1988b.)

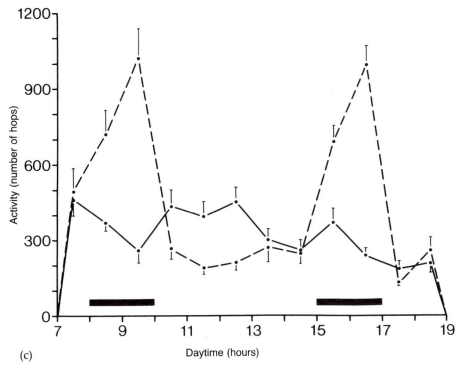

(c)

**Figure 2.4** *continued*

related to conditions permissive for breeding and has little, if anything, to do with migratory activity. This view is supported by findings in some other species. In white-crowned sparrows, chaffinches and indigo buntings (*Passerina cyanea*) 'nocturnal restlessness is less consistently oriented during the late part of spring migratory restlessness than earlier in spring' (Gwinner and Czeschlik, 1978), or summer restlessness is not oriented (indigo bunting, Emlen *et al.*, 1976; *Sylvia*, Terrill and Berthold, unpublished). Hence, true summer restlessness might be distinguishable from true migratory restlessness by non-directedness or, in addition, by lower intensity (Figure 2.4). In white-crowned sparrows, experimentally elevated circulating levels of testosterone did not suppress summer restlessness nor did exposure to estradiol-treated females. It is therefore suggested that unknown environmental or neuroendocrine cues suppress summer restlessness during breeding. Testosterone treatment at the end of the reproductive period, however, postponed the termination of summer restlessness (Schwabl and Farner, 1989a). According to the abovementioned results obtained from blackcaps, the availability of an appropriate nesting site appears to be the crucial factor for the termination of summer restlessness.

For a few species, 'roost-time restlessness' is known as a movement of caged individuals during late afternoon and evening when freeliving conspecifics head for common roosts. It occurs either all year round or during those seasons when common roosts are visited. It appears to be a preprogrammed appetitive behaviour (Berthold, 1988b), first described by Graczyk (1963) for the European blackbird and song thrush (*Turdus philomelos*), and by Eiserer (1979) for the American robin (*Turdus migratorius*). It was also found in other European species [great tit (*Parus major*), blue tit (*P. caeruleus*), reed bunting; Figure 2.4; Berthold *et al.*, 1986; Brensing, 1989].

It is well known from studies using various kinds of animals that food deprivation may immediately result in increased diurnal activity. This 'hunger restlessness' (Berthold, 1988b) is a well known phenomenon in birds since its first description for a migratory species, the white-crowned sparrow, by Merkel (1966). Although it is mainly expressed as diurnal activity (Figure 2.4) it can also influence nocturnal activity, including migratory restlessness. This may be of prime importance as 'a model of how primitive migration flights were caused' (Merkel, 1966). Hence, it will be treated in detail in section 2.19.

Ritchison *et al.* (1992) monitored activity levels in captive screech-owls (*Otus asio*) and found evidence suggesting that natal dispersal is influenced by intrinsic factors. Natal dispersal could be initiated by an endogenous 'dispersal restlessness' or 'dispersal drive', and short-and long-distance disperses could develop different amounts of restlessness. Herremans (1990a) assumed nocturnal dispersal of juvenile reed warblers (*Acrocephalus scirpaceus*). In songbirds, any kind of dispersal restlessness has so far not been demonstrated (Berthold, 1988b).

To summarize, as Helms stated, it is clear that not all nocturnal restlessness in migratory bird species can be considered as migratory. But, if nocturnal migratory restlessness is carefully analysed in comparative studies, true migratory restlessness may be satisfactorily distinguishable from other types of nocturnal unrest. Autumn migratory restlessness in this way is a fairly good reflection of migratory activity *per se*, but spring/summer restlessness is not because it often lasts throughout the breeding season until the post-nuptial moult (Gwinner and Czeschlik, 1978). The latter actually has to be reconsidered. It is not unlikely that in studies where true spring migratory and summer restlessness are separated (by differences in intensity, orientation behaviour or other criteria) a fairly good correlation between the termination of spring restlessness and the migratory period may be found. Also, migratory restlessness in the autumn migratory period is thought to occasionally considerably exceed the normal period of autumn migration (Gwinner and Czeschlik, 1978). Recently, however, there is an increasing body of evidence that migratory movements towards the end of the autumn migratory period often last much longer than had been expected. For instance, wintering outside the moult period in many Palaearctic migrants in Africa appears to be more of a dynamic process than a static event (Curry-Lindahl, 1981; Berthold, 1988c; Terrill, 1990). Southward migration of Eurasian passerines to African winter quarters may last for up to five months, up to the end of February (Pearson and Lack, 1992; Pearson, pers. comm.). For the present it remains open to what extent facultative migratory activity (section 2.19) might be involved in these 'winter' movements, but it is likely that programmed migratory restlessness for such movements may also last considerably longer than formerly concluded. This should be kept in mind when migratory restlessness is discussed in terms of 'endogenous time-programs for migration' (section 2.3). For more details on restlessness behaviour in birds see Berthold (1988b).

## 2.2 CIRCADIAN AND CIRCANNUAL RHYTHMS

As a response to the diurnal and seasonal periodic fluctuations of our planet, plants and animals have developed a variety of 'biological clocks' or 'internal clocks', of which endogenous diurnal, so-called 'circadian' rhythms, and endogenous annual, so-called 'circannual' rhythms (from the Latin *circa*, about; *dies*, day; and *annus*, year) are the most important. Circadian rhythmicity is ubiquitous to the extent that it has to be considered as one of the basic characteristics of life (e.g. Aschoff, 1960), and circannual rhythmicity also appears to be widespread in long-lived creatures (Gwinner, 1986). According to Pittendrigh (1981) and Gwinner (1986), these biological clocks are important by 'assuring an appropriately stable temporal sequence in the program's successive events, they, in effect, measure the lapse of (siderial) time'; and 'by providing for a proper phasing of the program to the cycle of environmental change, they, in effect, recognize local time'.

Surprisingly, few studies have concentrated on circadian aspects in migratory bird species. McMillan *et al.* (1970) maintained white-throated sparrows initially in constant light–dark ratios where they exhibited diurnal activity and, due to the spring migratory period, some nocturnal migratory restlessness. During subsequent periods of constant dim light, activity patterns, including migratory restlessness, showed a typical free-run. With the resumption of the light cycle the activity pattern, including migratory restlessness, was re-entrained (Figure 2.5). This experiment clearly demonstrated that the expression of migratory restlessness can be based on an endogenous circadian

**Figure 2.5** The perch-hopping records of a white-throated sparrow under various light regimes. To emphasize the nocturnal migratory restlessness, the activity records are arranged so one day is measured from noon (12) on the left to noon (12) on the right, with midnight (24) in the centre. SR is sunrise and SS is sunset. Day–night transitions were not gradual but abrupt. (Source: McMillan *et al.*, 1970.)

rhythm in constant conditions where no direct environmental stimuli are present. Such an endogenous control of migratory activity had been proposed earlier by Wagner (1930), Palmgren (1944), Wolfson (1945) and others. Despite the fact that there had been studies only in about 20 bird species, which demonstrated its persistence under free-running conditions (e.g. Aschoff, 1979), there is no doubt that migratory restlessness is largely based on circadian rhythms. This conclusion is justified by the fact that it has been found to occur in a significant number of bird species under constant experimental conditions where either its control by circannual rhythms (Gwinner, 1986, 1990; and below) or its daily pattern in relation to the patterns in free-living conspecifics was demonstrated (Berthold, 1978b). Dyachenko and Dolnik (1990) found an interesting dependence of the circadian period, $\tau$, from the migratory state in chaffinches; $\tau$ was 22.6 hours in fat birds but 24.2 hours in lean birds. The greater $\tau$ in lean birds may be related to a general higher locomotor activity during the diurnal period, as observed in a number of species (section 2.13). Kumar *et al.* (1991) have shown that photoperiodic effects on body fattening and metabolic events in migratory buntings are mediated by circadian rhythms. Further studies of circadian aspects of avian migratory behaviour would appear to be valuable with respect to remaining fundamental physiological, as well as ecological, problems, of which a few more examples will be given.

Circadian rhythms are generally assumed to be self-sustained oscillators due to the realization decades ago that regularities of physical oscillation theory also apply to these rhythms (Aschoff, 1964). Originally, a one-oscillator model system was considered, but soon the general understanding shifted to models of two or more oscillators (multi-oscillator system). Gwinner (in Aschoff, 1967) proposed that the circadian basis of migratory restlessness could not rely on a simple one-oscillator model. He therefore proposed a model with two coupled oscillators, which are correlated to light intensity with opposite signs (Aschoff, 1967). Later results obtained by McMillan *et al.* (1970) appear to be compatible with this model. The authors suggested that the phase relationships of the two oscillators were linked in a light cycle or constant dim light, whereas in constant bright light the 'nocturnal' migratory activity became arhythmic and led to arhythmic 'daytime' activity. After the migratory period, when nocturnal restlessness had ceased, the daytime component of activity in constant bright light resumed a circadian periodicity. It thus appeared that bright light can affect the nocturnal oscillator but does not influence the diurnal one.

In mammals, as well as in birds, the phenomenon of 'splitting' locomotor activity into two components is well known, and in European starlings such splitting of the circadian locomotor activity rhythm was found to be induced by testosterone (Gwinner, 1975; section 2.6). It has been proposed that the development of nocturnal migratory activity in otherwise day-active birds may partly be a result of these hormonal effects on the coupled oscillators. Here, endocrine factors simply shift one of the components to the night (Gwinner, 1975). In this context, it should be mentioned that migrants often strongly reduce the second peak in their daytime activity pattern so that the premigratory bigeminus (bimodal peak) changes into a more or less one-peaked pattern (Brensing, 1989; Figure 2.42).

Brensing (1989) has carried out an extensive comparative study of the diurnal activity patterns of migratory songbirds whilst they are staying in resting areas, based on trapping data and measured locomotor activity patterns in caged birds. He found that species staying in the same habitats may differ in their daily activity patterns, which possibly contributes to reduce competition (Figures 2.43 and 2.44; sections 2.13 and 2.21). These differences in circadian organization and their adaptive significance need to be studied in more detail. The same holds true for characteristics as the so-called 'rest for sleeping' (einschlafpause), which is typical for many migrants before the nocturnal take-off (section 2.13). Genetic aspects of circadian rhythms in migratory bird species have not yet been treated at all (Hall, 1990).

Daily activity patterns are only one temporal aspect of migratory regulation. Seasonal endogenous timing mechanisms in the control of avian migration have also been assumed. For instance, von Pernau (1702) wrote that typical migrants were 'driven at the proper time by a hidden drive' (section 1.2). Later, Rowan (1926) stated that 'those species that breed in the northern hemisphere and winter on the equator or cross it and winter in the southern hemisphere, make necessary the assumption that there is another and internal factor, a physiological rhythm'. In spite of these early suggestions it took a long time to produce convincing evidence for true endogenous seasonal rhythms, so-called circannual rhythms. Attempts to test the endogenous rhythm hypothesis have been listed by Gwinner (1986). Why they were more or less unsuccessful, according to him (Gwinner, 1990), was due to the fact that they suffered from a lack of planing. A model was needed to describe the phenomena, to define the relevant questions and to design the appropriate experiments. Such a model of oscillation theory became available when Aschoff (1960) and Pittendrigh (1960) proposed to treat endogenous rhythms like physical self-sustaining oscillators that free-run with their own natural 'circa'-period in constant conditions and are synchronized with appropriate environmental zeitgebers in the natural situation (Gwinner, 1990).

The first fully convincing proofs for true circannual rhythms in birds

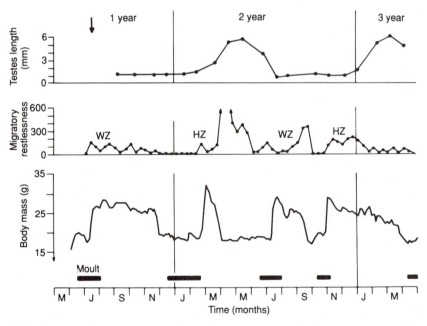

**Figure 2.6** Endogenous annual periodicity (circannual rhythms) of (from top to bottom) testes length, migratory restlessness (zugunruhe), body mass and moult, of a garden warbler. The hand-raised bird, hatched at the end of May, was transferred to constant conditions (a light–dark ratio of 10:14 hours) in June (arrow) and kept there for 10 years; results are depicted from the first three study years. HZ, return migration; WZ, outward migration period. (Source: Berthold, 1993.)

were obtained in Old World warblers (*Phylloscopus*, Gwinner, 1967a; and *Sylvia*, Berthold *et al.*, 1971). Figure 2.6 shows some of these results. Circannual rhythmicity in four annual cycles can be seen. They are demonstrated in moult, body mass, migratory restlessness and testis length in an individual male garden warbler which has been raised and kept under constant experimental conditions (including a 10:14 light–dark cycle). As in most tests of 'circa'-rhythms, the events of the first cycles occurred at the appropriate times of year whereas the following cycles somewhat deviated from them, in this case advanced with respect to the natural calendar. This typical deviation of these

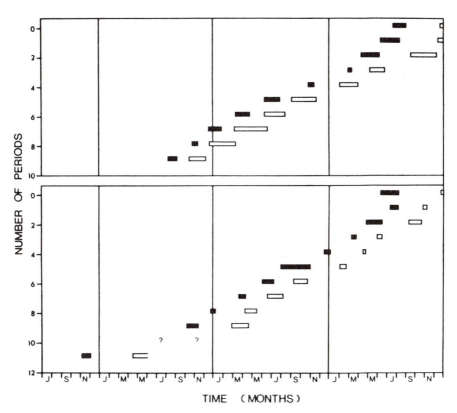

**Figure 2.7** Circannual rhythms of moult in an individual hand-raised German blackcap (above) and a hand-raised German garden warbler (below) kept for 8 and 10 years, respectively, under constant experimental conditions (daily light–dark ratio 10:14 hours). Black bars, 'summer' moults; white bars, 'winter' moults; ?, left-out moulting period. Subsequent moulting periods are plotted from top to bottom, and the calendar year is plotted 3.5 times in order to show the 'free-run' of the moult rhythms. (Source: Berthold, 1988a.)

'circannual' rhythms from the calendar year can result in a so-called free-run which can be taken as a proof of their endogenous nature. The individual whose data are shown in Figure 2.6 had produced such a free-run. It had actually been kept together with nine conspecifics and 10 blackcaps, continuously under constant conditions for 10 years (in the same experimental room). In this long-term study, circannual rhythms continued with their own characteristic periods of about 9–10 months. Thus, individuals here went through nine subjective annual cycles within eight calendar years. This was demonstrated well in the moult cycles where periods, and onsets and ends, were easily determined (Figure 2.7; Berthold, 1978b); it was also present to some extent in the body mass cycle (Figure 2.8). Cycling occurred up to the tenth experimental year. The interval between body mass peaks was also shorter then 12 months. Due to the endogenous nature of circannual rhythms and individual variation in their spontaneous period, lengths from 7 to 12 months, the 20 experimental birds in the experimental chamber were clearly desynchronized after several years. At various times, some were singing in full breeding condition, while others were

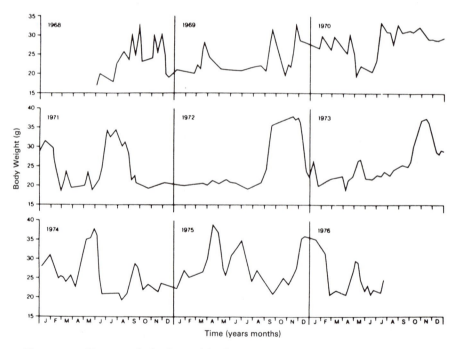

**Figure 2.8** Circannual rhythm of body weight in an individual hand-raised German garden warbler kept for 9 years under constant experimental conditions (daily light–dark ratio 10:14 hours. (Source: Berthold, 1988a.)

still on 'homeward migration', in 'post-nuptial' moult or performing 'autumn migration'. This was the best possible proof that the cycles could be expressed independently from any seasonal cues from the environment. Additionally, the data indicate the absence of any substantial group synchronization in these species. The behaviour of the birds was entirely the outcome of their individual-specific endogenous rhythms. Complete free-run of circannual rhythms has recently also been shown for another species [stonechat (*Saxicola torquata*), moult and gonadal cycles, Gwinner and Dittami, 1990].

Circannual rhythms involved in the control of migratory events have been demonstrated so far for 14 bird species from four different continents: Europe (mainly), Africa, Asia and North America (Gwinner, 1990; Holberton and Able, 1992). For five additional species of migrants there is circumstantial evidence for these rhythms (Gwinner, 1990). With respect to migration, circannual rhythms have been shown to control migratory activity and migratory fattening, perhaps the control of food preferences and utilization and migratory orientation. These rhythms also certainly involve other physiological processes, including basic endocrine systems etc., but these cycles have not been intensively investigated. Circannual rhythms are not restricted to typical migrants but have also been demonstrated in less typical migrants like crossbills (*Loxia curvirostra*), partial migrants and resident forms (e.g. Gwinner, 1986). How widespread circannual rhythms in birds may be and to what extent they may be generally involved in the control of migration remains to be shown. As Nolan and Ketterson (1990a, 1991) have reported, darkeyed juncos caught at their regular winter sites and then held outdoors, fail to become restless or to fatten the following autumn. This suppression of the migratory state is not attributable to the shorter day lengths the birds experienced but may raise the question whether the control of migration in that species may be based on any circannual rhythms at all. However, in this species Holberton and Able (1992) recently obtained data which provides just about the strongest evidence for the existence of endogenous, free-running long-term circannual rhythms in birds. Dark-eyed juncos subjected to constant dim light exhibited up to three cycles of gonadal development, moult, migratory fattening and zugunruhe.

The deviation of circannual rhythms from the calendar year under constant experimental conditions indicates that under natural conditions some factors must adjust these rhythms to the calendar year, i.e. the appropriate biological seasons. These synchronizers or 'zeitgebers' are, above all, photoperiod, in tropical species perhaps rain, and to some extent other factors like social bonds. They are discussed in section 2.5.

One puzzling characteristic of these cycles is that their period lengths are generally less than 12 months. This common shortness may be adaptive. Many migrants return to their breeding grounds earlier in later years than in the first year (e.g. Boddy, 1992). When breeding and autumn migration are delayed in a given year birds should nevertheless be ready for the homeward migration at the appropriate time. When, after a normal return, especially favourable circumstances allow for exceptionally early breeding an advanced cycle would allow them to do so. In all these cases either some acceleration of physiological processes or prompt readiness for subsequent processes is required, and that may best be reached by a circannual physiological system that would tend to be somewhat ahead of the specific situation and thus can easily be advance-shifted by synchronizers when appropriate (Berthold, 1974).

An intriguing question is whether endogenous factors linked to circannual rhythms can trigger migratory processes without additional influences of environmental factors? There is now convincing evidence that the programmed endogenous migratory urge can immediately trigger migrants to initiate at least their first departure from the breeding grounds. A comparative study in 19 species of European leaf and reed warblers, small thrushes and some others, showed a very high correlation coefficient of 0.967 between onset dates of migratory activity in hand-raised caged birds and actual migration in conspecifics of the same populations in the wild. This finding implies that the endogenous mechanisms triggering migratory activity in captive individuals also initiates actual migration, and that environmental factors in these cases are of minor, if any, importance for the initiation of departure (Berthold, 1990a). This view is also supported to some extent by the observation that, in a number of species, individual migration schedules are fairly constant over several years (Maisonneuve and Bédard, 1992). Wood (1992) concluded, from field studies, that in yellow wagtails (*Motacilla flava*) the onset of homeward migration in west Africa is probably endogenously controlled. A similar conclusion has been drawn by Curry-Lindahl (1981) for central Africa. Rohwer and Johnson (1992) obtained evidence that differences in scheduling of autumn migration in two subspecies of orioles are under endogenous (genetic) control. Johnson and Herter (1990) summarized the following observations. Passage migrations of Siberian knots (*Calidris canutus*) through Norway were remarkably similar in years when the weather was very different. Male Lapland longspurs (*Calcarius lapponicus*) arrived at nesting sites in Greenland at about the same time each year regardless of weather conditions. The same was found in many Arctic species in the Russian Arctic, and for the spotted redshank (see section 1.1). All these cases indicate rigid endogenous control of migratory behaviour.

## 2.3 ENDOGENOUS TIME-PROGRAMS

Naumann's (1795–1817) idea that migratory restlessness persisting into the winter meant that birds migrate long distances, e.g. into Africa (section 1.2), was expanded on by von Lucanus (1923) when he actually speculated that 'the winter quarters are not searched purposefully by a migrant but the goal of the migratory journey is reached when the urge to migrate ceases'. Stresemann (1934) later interpreted migratory activity explicitly in terms of a time-program for migration. In his words, 'the bird moves on until its urge to migrate ceases; and the migration drive is active during as many days as on the average are necessary to cover the distance between birthplace and winter quarters'. These ideas, however, were first experimentally tested by Gwinner (1968) and Berthold *et al.* (1972a). In the two comparative studies, migratory rest-lessness in caged *Phylloscopus* and *Sylvia* warblers was quantitatively measured. Comparisons of the time span and the amount of migratory restlessness with respect to the naturally occurring migratory distance indicated that a time-program for migration existed which, when coupled to programmed migratory directions (and a more or less pro-grammed course of migration including migratory speed), could guide inexperienced first-time migrants to certain winter quarters. This hypo-thesis of an innate time-and-direction program (section 3.5) should not be confounded with a formerly given distance–time hypothesis (Morse, 1989), which stated that migration date is a function of distance travelled. A subsequent comprehensive study of the genus *Sylvia*, including 13 species and populations, demonstrated a clear relationship between the amount of migratory restlessness found in caged individuals and the distance to be covered. The longer the normal migratory distance, the more migratory restlessness one finds in caged individuals (Figures 2.9 and 2.10). This is also the case under constant experimental conditions (Berthold *et al.*, 1972a).

None the less, simple correlations between the amount of migratory activity (restlessness) and the distance to cover give only a rough framework for the rate of an endogenous time-program in the control of migration. Gwinner (1972) has tried to expand on this with the following calculation. For the willow warbler (*Phylloscopus trochilus*), a long-distance migrant, and the chiffchaff (*P. collybita*), a short-distance migrant, the distances travelled per time period during the autumn were calculated from recoveries of banded individuals. These distances were then compared with the amounts of migratory restlessness displayed by caged conspecifics during exactly the same time intervals. He then went on to calculate what distance the experimental birds would have migrated over using all their migratory restlessness. The results demonstrated that the birds would have actually ended up in their species-specific winter quarters (Figure 2.10).

In another approach, Berthold and Querner (1988) recorded the total amount of wing whirring activity in garden warblers with video recordings under infra-red illumination (section 2.1). An experimental group of Central European birds showed a total average whirring time of 165 hours. When this total was multiplied by the estimated average flight speed of the species during migration, about 30 km/hour, a theoretical flight distance of 4950 km was obtained. This corresponds to the distance between the breeding grounds and the central African winter quarters (Berthold, 1988a). Thus, migratory restlessness, or zugunruhe, in this case appears as 'migration on perches' or 'migration in a sitting position' (section 2.1).

The idea that migratory activity (restlessness) functions to determine the course of migration is also supported by the fact that the patterns in a number of species appear to be adaptively preprogrammed to meet specific demands of the migratory journey. For instance, Central European species which normally cross the Mediterranean and the Sahara, like the garden warbler, initiate migration in short stages (Kaiser, 1992), accelerate it towards the Mediterranean (Klein *et al.*, 1973) and cross the ecological barriers in a few extended stages (Biebach, 1990). Thereafter they slow down when moving on further in central and southern Africa (Bairlein, 1990). The pattern of migratory restlessness of garden warblers corresponds very well to that course of migration (Figure 2.3). Other European species which normally cross the Sahara also show approximately similar patterns or patterns skewed to the right. In species like subalpine warbler (*Sylvia cantillans*), lesser whitethroat (*S. curruca*), Orphean warbler (*S. hortensis*), redstart (*Phoenicurus phoenicurus*), nightingale (*Luscinia megarhynchos*, Berthold, 1973, 1985b, unpublished) or willow warbler (Gwinner, 1968) peaks of activity occur in the middle of the migratory period or somewhat

———————————————————————————————→

**Figure 2.9** (a) Patterns of migratory activity (zugunruhe) of caged individuals of six *Sylvia* species correlated with the migration distances of conspecifics in the wild (solid bars: km) during first southbound migration. (Source: Berthold, 1993.) (b) Relationship between the amount of migratory restlessness displayed in experimental groups and the migratory distance to be covered in free-living conspecifics, in 13 species and populations of *Sylvia* warblers. 1, Spectacled warbler (*Sylvia conspicillata*); 2, Marmora's warbler; 3, blackcap from Canary Islands; 4, Sardinian warbler; 5, Dartford warbler; 6–8, blackcaps from S France, S Germany and S Finland, respectively; 9, whitethroat; 10, subalpine warbler; 11, lesser whitethroat; 12, garden warbler; 13, barred warbler. (Source: Berthold, 1984b.)

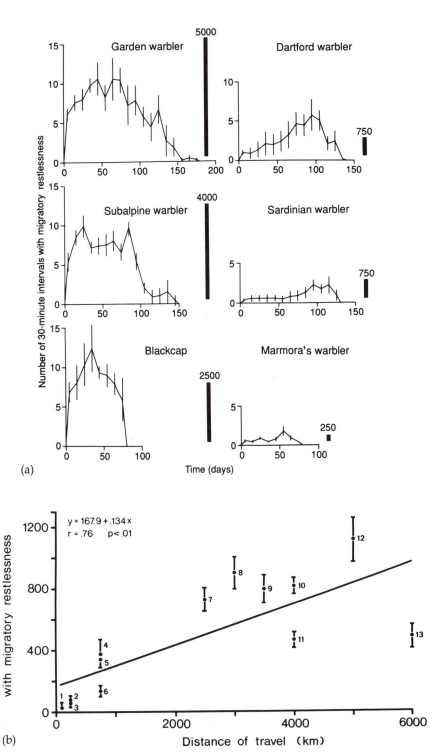

(a)

(b)

before, and correspond very well with the time when the Sahara desert is crossed. In contrast, short-distance and partial migrants which migrate late in the season often have peaks towards the end of the migratory season. Thus, their restlessness patterns are skewed to the left (Figures 2.9 and 2.10). Patterns of this type have been obtained in the Mediterranean warblers, e.g. Sardinian warbler (*Sylvia melanocephala*), Dartford warbler (*S. undata*) and Marmora's warbler (*S. sarda*), the black redstart (*Phoenicurus ochruros*), the European robin (*Erithacus rubecula*, Berthold, 1973, 1985b, unpublished), and the chiffchaff (Gwinner, 1968). In another long-distance migrant of the *Sylvia* species, the barred warbler (*S. nisoria*), which winters in SE Africa but circumvents the Mediterranean by land and avoids central desert areas, the pattern of migratory restlessness is also skewed to the right, but flatter compared to species crossing the Sahara more centrally (Figure 2.11). Finally, even an extremely unusual pattern of migration in the marsh warbler is reflected to a considerable extent in the pattern of migratory restlessness. As already mentioned in section

**Figure 2.10** Breeding areas and winter quarters of chiffchaff and willow warbler. Numbers refer to various races of either species (numbers encircled represent winter quarters). Large solid circles are calculated end-points of migration (with standard errors) for chiffchaffs and willow warblers kept under three different experimental conditions. (Source: Gwinner, 1972.)

1.1, marsh warblers leave Central Europe in about mid-July. They move rapidly on to NE Africa where they either stop or move on slowly to their south African winter quarters, which are reached from December to February (Berthold and Leisler, 1980; Figure 2.65). Caged individuals correspondingly show a two-peaked pattern of restlessness with an extremely long extension (Figure 2.11). Hence, there is evidence in a number of species supporting the view that the temporal patterns of migratory activity (restlessness) are detailed time-programs adapted to specific migratory journeys and particular ecological conditions.

If patterns of migratory activity (restlessness) functioned as specific time-programs for the course of migration, they should actually be under fairly rigid endogenous control. This indicates the necessity to be insensitive to a number of environmental variables. About 10 relevant tests have been made on garden warblers where effects of complete darkness, reducing body mass and adverse rainy weather conditions on initial and central parts of the migratory restlessness were examined. In these cases, temporary depression of activity was found. In addition, patterns of migratory restlessness (and of body mass changes, i.e. fattening) were compared in experimental groups raised in light conditions normally experienced by other populations and in a variety of constant light–dark ratios. Outside of the temporary depression, these disturbances of the endogenously-controlled patterns of migratory restlessness (as well as of body mass changes) had no effects on the termination and thus the duration of the patterns or their shapes after treatment (Figure 2.12; for review see Berthold, 1985b; Gwinner *et al.*, 1992a). There were, however, two exceptions. First, when birds were kept in long-term, constant photoperiodic conditions, prolonged and flattened patterns of both restlessness and body mass changes were found (Figure 2.13). Secondly, when spotted flycatchers (*Muscicapa striata*) and garden warblers were temporarily starved to the extent that they showed a marked decline in body mass, migratory restlessness increased during the decline and was reduced thereafter during refeeding (Biebach, 1985; Gwinner *et al.*, 1985; Figure 2.63). These last short-term alterations after severe weight loss appear to reflect an adaptive mechanism for birds in a critical energetical situation. Starved migrants may be stimulated to move on to areas which allow refeeding where migration may be shortly interrupted (sections 2.8 and 2.19).

Apart from these two exceptions, the endogenously controlled patterns of migratory activity (restlessness) and of body mass changes (fattening) in the garden warbler appear to be under a fairly rigid endogenous control. Even when large parts of such a pattern, or the entire initial phase, were suppressed neither a prolongation of the patterns nor another compensatory mechanism occurred. This test

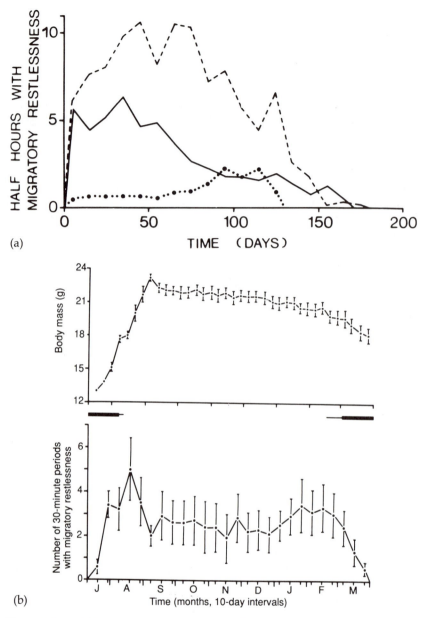

(a)

(b)

**Figure 2.11** (a) The pattern of migratory restlessness of barred warblers (solid line) in comparison to those of garden warblers (broken line) and of Sardinian warblers (dotted line). (Source: Berthold, 1979b.) (b) The migratory activity pattern (below) of caged German marsh warblers. During the fast course of migration to northeast Africa, higher values of zugunruhe are reached on average than during the considerably slower migration towards South Africa. Also depicted are changes in body mass (upper graph) as well as juvenile moult and part of winter moult (solid bars). (Source: Berthold, 1993.)

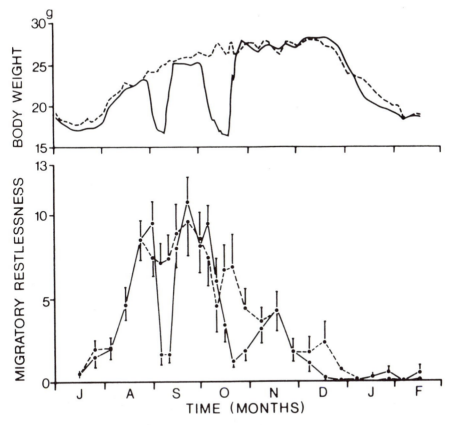

**Figure 2.12** Seasonal patterns of body weight (above) and of migratory activity (below) in two groups of hand-raised German garden warblers. Broken line, control group; solid line, experimental group with two interposed periods of temporal malnutrition; vertical lines, standard error. (Source: Berthold, 1988a.)

has further shown that the two endogenous programs for migratory activity (restlessness) and body mass changes (fattening) are largely independent of each other. Even complete suppression of body mass increase from the premigratory period does not prevent normal onset and at least initial development of migratory restlessness. The latter course of restlessness is only disturbed if body weight is strongly reduced by starvation to a critical premigratory level (Berthold, 1985b). Therefore, it appears likely that, at least, in typical migrants like the garden warbler the endogenously programmed pattern of migratory activity (restlessness) could well function as a time-program for migration and that at least inexperienced first-time migrants with undisturbed fat reserves may essentially be guided by that program to their winter

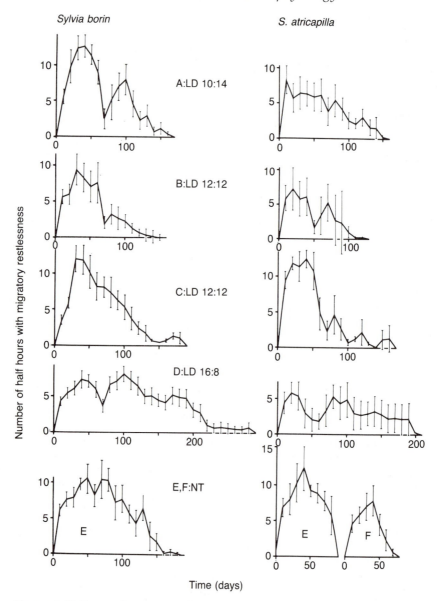

**Figure 2.13** Seasonal patterns of migratory activity in various experimental groups of garden warblers (left) and blackcaps (right). A–D, various constant daily light–dark ratios; E and F, natural light conditions in the blackcap for individuals hatched early (E) and late in the season (F). Vertical lines, standard error. (Source: Berthold *et al.*, 1972a.)

quarters. The endogenously controlled body mass changes and energy storage may contribute to this time-program, but they appear less important since their expression is not a prerequisite for at least large parts of the course of the pattern of migratory activity (restlessness).

Recent studies also indicate that differential migration can be based on different endogenous time-programs. Terrill and Berthold (1989) demonstrated that female hand-raised blackcaps held in constant conditions showed significantly more, and significantly longer, autumnal migratory activity than males from the same population (Figure 2.67), which is in accordance with differential migration in the wild. Similarly, Holberton (1993) found that captive female dark-eyed juncos initiated zugunruhe earlier than males and exhibited a longer period of migratory activity. Holberton concluded that these differences in patterns of zugunruhe support the hypothesis that sex-biased differential migration is heritable in dark-eyed juncos. For discussion of data in this species see also Ketterson and Nolan (1985). It is likely that many cases of differential timing of migration between sex and age classes, as they are especially obvious in raptors (Kjellén, 1992), are similarly preprogrammed. The same may hold true for the few cases in which it could be shown that females commence their homeward migration later than males, or accumulate fat later than males [e.g. eastern great reed warbler (*Acrocephalus orientalis*), Nisbet and Medway, 1972; and orchard oriole (*Icterus spurius*), Rogers and Odum, 1966].

As Gwinner (1990) pointed out, some results suggest similarities in the temporal patterns of autumn migration in the field and measured restlessness, but generally these correlations are not as pronounced as those between overall distance and duration. He further stated that the correlative data obtained are consistent with the hypothesis that a time-program can determine distance in inexperienced migrants, but they do not prove the mechanism. He stressed the possibility that the circannual migratory time-programs do not actually determine migratory distance but rather just provide a 'temporal window' within which other mechanisms may act. Certainly, more rigorous testing of the hypothesis is necessary to compare the performance of caged birds with the actual migratory course of free-living conspecifics. This requires particularly detailed studies of how migrants manage their daily migration stint in relation to their endogenous programs.

To sum up: in a few investigated species, like the garden warbler, willow warbler, chiffchaff and blackcap, the available evidence suggests that the endogenous patterns of migratory activity (restlessness) reflect time-programs for migration. It appears likely that such (inherited; section 2.4) programs could provide specific spatiotemporal programs to guide at least inexperienced migrants from their breeding area to unknown winter quarters, in the sense of a vector-navigation hypothesis (section 3.5 and chapter 4). This is, at present, the only viable hypothesis to explain the phenomenon of how unknown winter quarters can be reached in singly migrating species. There is good agreement between the duration of migratory activity (restlessness) in captive migrants and that of the migratory period in free-living conspecifics. The idea that patterns of migratory restlessness tend to last longer in late autumn and winter than actual migration, does not coincide with the fact that actual migration often appears to end much later than previously expected (and that non-migratory 'winter restlessness' may have been overlooked; section 2.1). There is also agreement to some extent between the amount of restlessness and the total flight duration of the autumn migratory route, but the exact relationship between restlessness and the daily migration stint needs to be clarified. It is also open to what extent time-programs of migratory activity may be supported by endogenous energetic programs (expressed in body mass changes and fattening). Still, the loose coupling between the two in constant conditions makes strong support improbable. Nevertheless, the size of migratory fat reserves also appears to be strictly controlled by a genetically determined annual set-point mechanism, like migratory activity (next section). Their construction may involve short-term oscillations of body mass as demonstrated in the garden warbler (Bairlein, 1990). The time-program hypothesis has also been supported by a few field experiments. In a displacement study with European starlings first-time migrants travelled to an atypical, but predictable, area over about a corresponding distance (section 3.1). In a retention experiment with blue-winged teals (*Spatula discors*) their subsequent migratory distances were shortened (Bellrose, 1958). To what extent endogenous time-programs are used in migration-experienced migrants after the first migratory period is unknown. Experiments by Nolan and Ketterson (1990a,b) with dark-eyed juncos and Schwabl *et al.* (1991) with dunnocks (*Prunella modularis*), in which migrants were exposed to previous wintering sites, modified or even suppressed migratory activity. This indicates that winter-site recognition may possibly influence migratory activity in experienced migrants. However, more studies of this type are needed before other currently overlooked effects can be excluded (see also section 2.20). For a final consideration, see section 3.5: vector-navigation.

As mentioned in section 1.1, dispersal may well be included in the class of migratory movements. There is considerable evidence that dispersal behaviour does not only depend on the condition of birds at the time of independence (Gauthreaux, 1988) but appears to be based on endogenous factors. The distance dispersed, which in general is greater for female birds, is thought by many investigators to be determined by natural selection (Ketterson and Nolan, 1990). In detailed studies with indigo buntings, Payne (1991) found that the variation in distance within the natal area appeared to be independent of local population density, social competition, active kin recognition and avoidance of incest, but was affected by the date of hatching. The results obtained are inconsistent with predictions of social and avoidance-of-inbreeding hypotheses, but are consistent with a model of neutral dispersal within a genetically open population. Thus, there may well also be specific endogenous programs for avian dispersal and it may be a rewarding task to look for more cases of possible 'endogenous dispersal restlessness' as mentioned in section 2.1. Morton (1992), however, proposed, according to studies on white-crowned sparrows, that intrasexual competition in females may contribute importantly to dispersal patterns in migratory passerines.

## 2.4 GENETIC CONTROL

Until recently, the genetic basis of animal migration was a matter of pure speculation (e.g. Dingle, 1991, 1994). The possible role of genetic control mechanisms in bird migration, however, has been considered for a long time (e.g. Nice, 1933, 1934; Lack, 1943/1944). After a period of correlative work where, for example, polymorphisms in egg albumen protein were related to migratory habits in partial migrants (Milne and Robertson, 1965), experimentation in migration genetics in birds began at approximately the same time as in insects and fish, about 15 years ago. The blackcap, an Old World warbler with a wide range of migratory habits between residency and long-distance migration, using various migratory directions and wintering areas (section 1.1), was a most suitable species for these genetic experiments. It is common all over its distribution area, so experimental birds were easily obtained. This species is also relatively easy to maintain and breed in captivity (Berthold *et al.*, 1990b). Aviaries cross-breeding, selection experiments on migration features (1977–1992), data from a few other experimental studies and field investigations have indicated that immediate genetic effects are involved in the control of avian migration. The main results are briefly summarized in the following, although more details can be found in other recent reviews (Berthold, 1990b; 1991a–c; Berthold and Helbig, 1992; Berthold and Querner, 1982d; Dingle, 1991, 1994).

### 2.4.1 Urge to migrate

The inherited character of migration was shown when exclusively migratory blackcaps from Central Europe and resident conspecifics from Africa were cross-bred; 33% of the F1-hybrids exhibited migratory activity and demonstrated that migratory activity and, thus, the urge to migrate can be transmitted rapidly into the offspring of a non-migratory bird population (Berthold *et al.*, 1990c; Figure 2.15). This fact was further examined in a two-way selection experiment with migratory and non-migratory fractions of two partially migratory species. The percentage of migratory individuals increased in the selection line of migrants and decreased correspondingly in non-migrants (below). This result indicated that the migration drive was heritable, even in partially migratory species.

According to the previous sections, the production of migratory activity appears to be closely linked to endogenous circannual rhythms. Since these rhythms are also expressed in inexperienced individuals, raised in constant experimental conditions, it is likely that these rhythms are, on the whole, inheritable, including the urge to migrate and the temporal course of migratory activity. However, in contrast to mammals (Pengelly and Asmundson, 1971), the inheritance of these rhythms has not yet been demonstrated in birds. The fact that cross-breeding experiments transmit migratory activity only to some of the offspring indicates that the urge to migrate is obviously quantitatively inherited, according to a threshold model (e.g. Falconer, 1981). Its control is, therefore, most likely polygenic (Berthold *et al.*, 1990c).

A particularly interesting case indicating genetic control of the migration drive is described by Harris (1970), who cross-fostered young of the normally sedentary (in Britain) herring gull (*Larus argentatus*) with the migratory lesser black-backed gull (*Larus fuscus*). A significant portion of these *L. argentatus* migrated, although not as far as their foster parents, suggesting both gene and environmental influences. In the reverse experiment, cross-fostered *L. fuscus* still migrated, indicating that genetic influences on migration were dominant.

### 2.4.2 Urge to migrate in partial migrants

Two main contradictory and controversial hypotheses have been proposed about the control of partial migration in birds – the 'genetic' hypothesis and the 'behavioural–constitutional' hypothesis. They have to be understood as extreme attempts to explain a control which, in fact, may well include both genetic and environmental factors. The genetic hypothesis was favoured by Lack (1943/1944) following earlier ideas of Nice (1933). This hypothesis essentially states that the decision to migrate in individuals is made at fertilization by the combination of the parental genes. Genetically-determined residents (permanently) stay in or close to the breeding ground, whereas genetically-determined migrants take-off for the population-specific migratory journey at the appropriate time based on their inherited urge to move. The behavioural–constitutional hypothesis was proposed in detail by Kalela (1954) and Gauthreaux (1978), following earlier ideas of Miller (1931). Miller investigated the loggerhead shrike (*Lanius ludovicianus*); his results indicated a relatively weak physiological migration drive which could be modified by 'psychic' differences. In this way certain individuals, which failed to correspond to the changing seasons, remained as residents while others reacted and migrated. Kalela believed 'that true migration originating as a consequence of autumnal territorial behaviour (or increased aggressiveness in general) is not uncommon in birds'. He proposed that conflicts in autumn may allow strong individuals to stay on the breeding grounds as residents whereas losers are forced to leave as dispersed individuals or as true migrants – resulting in what is termed partial migration. In the 'dominance rank' hypothesis of Gauthreaux this simply meant that dominance influenced dispersal, irruptions, short- and long-distance migration directly, which results in partially migratory behaviour. Finally, various authors (e.g. Cornwallis and Townsend, 1968) have evaluated the immediate role of food supply (section 2.19). Still, constitution, aggressiveness and dominance can also be controlled by genetic traits, although environmental factors may be of great importance (as in the dominance rank of great tits; Westman, 1990). The behavioural–constitutional hypothesis should not be taken as a complete non-genetic alternative to the genetic hypothesis. The difference is that the genetic hypothesis implies a direct genetic control of migration whereas the latter hypothesis portrays migratory movement as dependent on individual constitution and that of other individuals of the population, as well as population density and food supply. These mechanisms are thus open to considerable variation depending on the given situations.

Unfortunately, during the extensive theoretical discussion of possible control mechanisms of partial migration, there has rarely been clear separation of obligate and facultative migration strategies, and proximate and ultimate factors (section 1.5). This has led to a lot of confusion with regard to differing control mechanisms (e.g. Berthold, 1993).

Experimental and observational evidence for the control of obligate partial migration has been obtained in about five species. In the black-cap, selective breeding experiments with partial migrants from southern France for up to six generations (Figure 2.14) and additional field studies, have shown that: (1) Both behavioural traits, migratoriness and residency, have a genetic basis (Berthold and Querner, 1982a; Berthold *et al.*, 1990a). (2) Heritability estimates for both traits are high, with heritability values ranging from 0.58 to about 1 (Berthold, 1988d; Berthold *et al.*, 1990a). (3) Exclusive migratoriness or residency can rapidly be selected for within a few (about three to six) generations

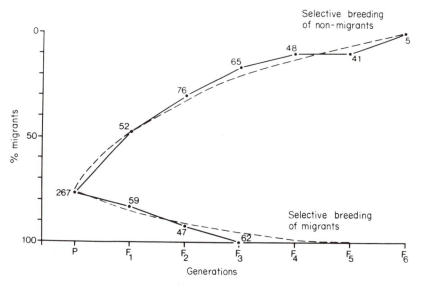

**Figure 2.14** Results of a two-way selective breeding experiment with partially migratory blackcaps (from the Mediterranean). Non-migrants were bred up to the $F_6$-generation and migrants up to the $F_3$-generation. Numbers indicate individuals tested in each generation. The broken lines represent mathematical functions that best fit the selection response. (Source: Berthold *et al.*, 1990a.)

(Berthold *et al.*, 1990a). (4) Most likely, both traits are so-called 'threshold characters', which are determined by multiple loci. This mode of inheritance, based on quantitative genetics, outweighs the idea of single-locus determination and of a genetic dimorphism (polymorphism), since the offspring of given pairs normally represent both behavioural traits (Berthold *et al.*, 1990a). (5) Consistency tests show that captive individuals, being non-migratory, do not change to migratoriness in subsequent years, although a certain proportion go the other way (as it often occurs in many species in the wild; Berthold, 1993). Since this happened in unaltered, constant conditions, it is likely that this change is an effect of maturation, which corresponds to age-dependent changes in the wild (Berthold, 1990b). Comparably, a number of long-lived species start periodic seasonal migrations only when due to species-specific maturation processes, breeding maturity is reached after wintering and 'summering' for several years on the wintering grounds. (6) Migratoriness is expressed in a higher proportion of females (as is the case in many partial migrants; e.g. Berthold, 1993). This bias towards females also occurs in selective breeding experiments demonstrating sex-linked inheritance (Berthold, 1986). (7) Migratoriness and residency appear to be balanced to some degree by assortative mating as a result of habitat segregation. Non-migrants tend to breed more in evergreen vegetation where they stay all year round. Here they may have a higher breeding success but may suffer from more resource competition, especially during the winter, than the migratory fraction which tends to breed in more deciduous vegetation and return from less severe wintering conditions (Berthold, 1986).

There is almost identical evidence for genetic control mechanisms involved in partial migration in the European robin from another two-way selective breeding experiment (Biebach, 1983; heritability value 0.52). Conditional factors involved in this species will be treated in section 2.18. In the blackbird, Schwabl (1983) raised nestlings from pairs in a German population which, due to extensive field studies, were known to be either residents or migrants. When tested in registration cages, migratoriness was more pronounced in the offspring from migrant parents. The author concluded that wintering strategies in this species appear to be genetically determined, at least in the first year. In an earlier study, Graczyk (1963) recorded a stronger tendency towards migratory activity in offspring of forest blackbirds compared to urban blackbirds, which may differ locally in their degree of migratoriness and residency (Schwabl, 1983; Luniak *et al.*, 1990). Walasz (1990) did not find corresponding differences in the migratory restlessness between urban and forest blackbirds.

One of the most detailed field studies of wintering strategies in a partial migrant is still that of Nice (1933, 1937) on the song sparrow (*Melospiza melodia*). She was able to establish genealogies for about 65 individuals through up to three generations. Her results demonstrated that most of the individuals had either a definite migratory or sedentary habit. Yet, as she did not detect pure lines of migrants and non-migrants, she dismissed her initial believe in a genetic control mechanism. However, 24 of her recorded pairings (of migrants, non-migrants and mixed pairings), and the known wintering strategies of the offspring, can actually be used for a statistical analysis. The data treated accordingly show that migratory parents do produce predominately migratory offspring and vice versa (Berthold, 1984a). For the stonechat there is also some indication of genetic control of partial migration. Dhondt (1983) reported that the number of individuals wintering in western Europe is significantly correlated with winter temperature in the previous winter and suggested that the migratory behaviour in this species could also have a genetic basis. Adriaensen *et al.* (1993) analysed the increase of the Dutch population of the great crested grebe (*Podiceps cristatus*). The results from ringing recoveries suggested that the increase in the proportion of locally wintering individuals may reflect changes in the genetic composition of the breeding population. In a number of field studies on other partial migrants the possible role of genetic factors involved in the control remains open [e.g. silvereye (*Zosterops lateralis*), Lane, 1972; Mees, 1974; goldcrest (*Regulus regulus*), Hildén, 1982]. For environmental control of partial migration see sections 2.17–2.19, for the overall control see section 2.18.

In the blackcap, Pulido and Berthold (1991) studied the effects of different migration habits on genetic population structure using an electrophoretic survey of 39 protein loci in three natural and two laboratory populations, representing all possible migration patterns. Genetic variability was highest in the migratory population and decreased with the proportion of migrants in the populations. Of the eight polymorphic loci found, one enzyme polymorphism proved to be strongly associated with non-migratory behaviour. The 'slow' allozymic variant of glycerol-3-phosphate dehydrogenase (G3PD) was exclusively found in sedentary populations. Moreover, this allele showed a strong response to artificial selection for non-migration: its frequency increased significantly in three generations. There is circumstantial evidence for the assumption that G3PD is functionally linked to the display of migratory behaviour. This entails that the allele frequency increase is attributed to selection rather than to genetic hitchhiking. Different selection-migration scenarios by which the distribution of the G3PD allozymes in natural populations could have evolved, and is currently maintained, are presently investigated in view of the prevailing theories on the origin of migratory behaviour.

### 2.4.3 Time-programs for migration

In the previous section, it was demonstrated that many migrants show species- and population-specific patterns of migratory activity (migratory restlessness), even under constant experimental conditions, which are related to the length and circumstances of the migratory journey and can be interpreted in terms of endogenous time-programs

---

**Figure 2.15** (a) Time course of migratory activity in groups of hand-raised blackcaps from four populations (above) and of hybrids and their parental stocks (below). SFi, S Finland; SG, S Germany; SFr, S France; CI, Canary Islands, Africa. Vertical lines: standard error. (Source: Berthold and Querner, 1981.) (b) Patterns of migratory activity during the first autumn migratory period of three groups of hand-raised blackcaps: SG, data from S German birds; CV × SG, data from $F_1$-hybrids of S German birds with those from the Cape Verde Islands; a, under simulated photoperiodic conditions of S Germany; b, under photoperiodic conditions of Cape Verde (mean values and standard errrors of examples). (Source: Berthold *et al.*, 1990c.)

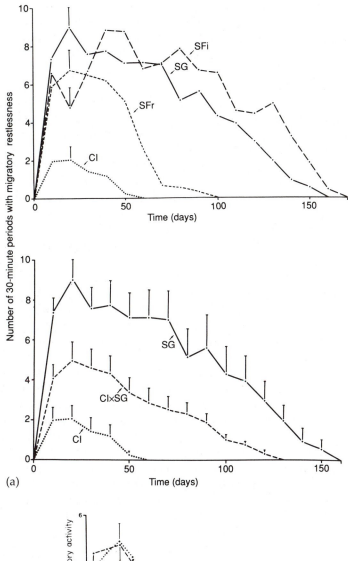

(a)

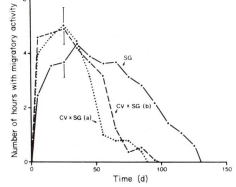

(b)

for migration. This hypothesis suggests that genetic factors are directly involved in the expression of these activity patterns. This possibility has so far been demonstrated in three cross-breeding studies, two carried out with blackcaps and one with redstarts. The parental stocks for studies with blackcaps were Central European birds which migrate regularly to the Mediterranean and display large amounts of migratory restlessness in captivity consistent with their long migratory routes. They were crossed with African conspecifics from the Canary Islands (which only migrate to some extent within the archipelago or possibly to the west African coast and exhibit only small amounts of restlessness) and with those from the Cape Verde Islands where blackcaps are resident and do not display migratory restlessness. In both cross-breeding studies, F1-hybrids exhibited intermediate patterns of migratory restlessness (Figure 2.15). These results show that in the populations studied, migratory activity, expressed as migratory restlessness, is a quantitatively inherited population-specific characteristic. The partial genetic transmission of migratory activity into the offspring of the non-migratory population implies a polygenic control of these activity patterns as mentioned above. Corresponding results appear to emerge from a current cross-breeding study with redstarts and black redstarts. These interspecific hybrids appear, in addition, to demonstrate intermediate dates for the onset of autumn migration in comparison to their parental species.

At present, the genetic variation of the amount of migratory activity (restlessness), and of the preferred migratory direction, is being studied comparatively in a Central European (German) and a Mediterranean blackcap population (from southern France). The German blackcaps regularly migrate over distances ranging from at least about 700 to 4500 km to various winter quarters which extend at least over an area from southern France to the Ivory Coast using a fairly narrow angle of migratory directions of about 90° (Berthold *et al.*, 1990b). The amount of migratory restlessness displayed individually varies accordingly by a factor of about six (between approximately 150 and 900 hours, respectively). Secondly, in the Mediterranean partially migratory population, the migratory fraction most likely travels mainly very short distances and reaches only up to about 1300 km maximum. Migratory restlessness among individuals varies accordingly by a factor of 100. Migratory directions range in this population from west over south to east over about 180° (Zink, 1973–1985; Berthold, 1986). The first results indicate that there is a high degree of genetic variation in the amount of migratory activity (restlessness) in the German population ($h^2 = 0.37 - 0.46$; Berthold and Pulido, 1994) and much less genetic variation in the Mediterranean birds (Berthold, in preparation). (For heritability estimates of migration times in swans see the next section.)

So far, it has not been reported whether patterns of body mass changes and fat deposition also depend on direct genetic control. However, since these patterns can be endogenously controlled, as shown above, direct genetic control appears likely. A current cross-breeding experiment with redstarts as long-distance migrants with marked fat deposition and black redstarts as short-distance migrants with less pronounced fat deposition (Berthold, 1985a) has indicated that inheritance does play a major role: F1-offspring appear to show intermediate patterns of body mass cycles and migratory fattening.

### 2.4.4 Migratory directions

Studies in avian orientation have accumulated indirect evidence for the genetic aspects of migratory directions over the past few decades. In general, the following facts have emerged: (1) even inexperienced migrants tested in orientation cages demonstrate regular species-and population-specific migratory directions; (2) displacing first-time migrants in the field (white storks, Schüz *et al.*, 1971; European starlings, Perdeck, 1958) resulted in migratory directions parallel to their parental populations (section 3.1); (3) irregular migrants like facultative partial migrants and individuals performing irruptive movements also migrate quite regularly in specific directions; (4) inexperienced cuckoos (*Cuculus canorus*) migrate from Eurasia to typical African winter quarters independent of the migratory practices of the fosterparents (Sutherland, 1988); (5) inexperienced *Sylvia* warblers and pied flycatchers also show population-specific seasonal changes in their directional preferences comparable to the course of migration when kept in captivity (Figure 2.16; below). All these findings indicate that many migrants possess preprogrammed migratory directions or directional programs ('directional sense', innate 'sense of direction') which appear to be under direct genetic control.

Convincing evidence for genetic mechanisms involved in orientation has been obtained in three recent experiments with blackcap populations. In the cross-breeding experiment (reported above) between migratory and resident blackcaps (from Germany and the Cape Verdes) six hybrids displaying migratory activity were tested in orientation cages. These birds showed a significant directional preference which corresponded to the principal axis of migration in the parental population (Figure 2.17: Berthold *et al.*, 1990c). This result indicated that migratory directional preferences in this species can be inherited. At present it is not clear whether solely an axial preference was inherited or whether a proper northward and southward discrimination occurred (Berthold *et al.*, 1990c). It is also unclear whether the inheritance of the

**Figure 2.16** Orientation behaviour of garden warblers kept in circular cages in S Germany. Arrows, mean directional preferences. They are plotted in areas where free-living birds of the same populations would be found, on average, at corresponding times during outward migration. Winter quarters in Africa are indicated by the hatched areas. (Source: Gwinner and Wiltschko, 1978.)

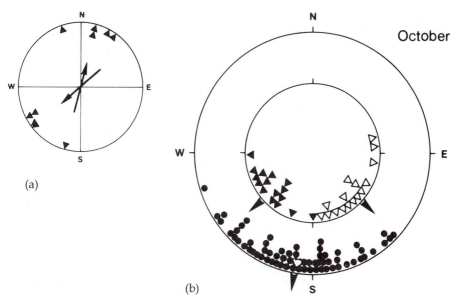

**Figure 2.17** (a) Orientation behaviour of hybrids of migratory (S German) and resident (Cape Verdean) blackcaps in orientation cages. Triangles, mean direction of individual birds during return migration (upper semicircle) and during outward migration (lower semicircle); arrows, average direction of all individuals for both periods. (Source: Berthold *et al.*, 1990.) (b) Directional choice of blackcaps in orientation cages. Solid triangles, birds from S Germany; open triangles, birds from E Austria; dots, hybrids of both populations; long solid triangles, average direction. (Source: Helbig, 1991a.)

preferred direction is coupled to that of the migratory urge and which parts of the orientation systems are inherited by F1-hybrids.

In the second experiment, cross-breeding of migratory blackcaps using different migratory directions was carried out. Blackcaps have two migratory divides in Europe, the main one more or less separates individuals migrating to either eastern or western parts of the Mediterranean or Africa. When individuals from both populations were hand-raised and tested, they showed preferences in their migratory directions like free-living conspecifics. Subsequently cross-bred, they produced phenotypically intermediate offspring (Figure 2.17). There was no indication of sex-linked inheritance nor increase in scatter in the F1-generation. A tendency was found for siblings to orient more similarly than non-siblings but full siblings sometimes differed significantly in both mean directions and the accuracy of orientation (Helbig, 1991a). Thus, migratory directions appear to be quantitatively genetic characters controlled by a number of genes as other migratory features treated above. The investigation of a small sample of F2-birds indicated that the number of genes involved may be rather small (Helbig, 1991b).

The data depicted in Figure 2.16 indicates that, in the garden warbler, seasonal changes in the migratory direction are also genetically encoded; blackcaps and pied flycatchers show similar shifts under experimental conditions (Helbig, 1991b). The results of the above-mentioned experiment confirm the assumption of a genetic control of such shifts. The blackcaps migrating to the east (Figure 2.17) shifted their preferred direction from October to November from 134 to 185° (in order to reach African winter quarters in the wild). The blackcaps migrating to the west, wintering in the Mediterranean area, do not show a comparable shift, neither in the wild or in the experiment. The hybrids of both groups, however, performed an intermediate shift of only 10° compared with 51° in the SE migrating parental group. Formerly, such shifts were believed to be versified through certain 'programmed bearing points' along programmed routes or by a special 'species memory' (Creutz, 1987).

In the last experiment, blackcaps were trapped during the winter in England and used for breeding and cross-breeding experiments in southern Germany. Blackcaps wintering in England originate from Central Europe, and due to a novel migratory habit they migrate, in part, to the northwest and have established new wintering areas on the British Isles (section 4.1). It has been hypothesized that the rapid development of this wintering population is based on the inheritance of the novel migratory direction (Berthold and Terrill, 1988). When wintering birds from England were bred, their offspring indeed showed a mean WNW migratory direction in autumn, which differed significantly from the southern directions measured in other populations (Figure 2.18). When wintering birds from England were cross-bred with German birds showing an original southern directional preference, F1-hybrids were again intermediate, showing an average westerly direction of 253°, but still about one third of the hybrids tended to north from west (Helbig *et al.*, 1994).

Innate directional information in blackcaps is expressed in captive hand-raised birds at least until the second autumn migration (Helbig, 1990, 1992). Whether migrants use this information beyond the first autumn migratory period is open to debate.

### 2.4.5 Other features

As mentioned above, a possible genetic basis for the pattern of fat deposition in migrants has not yet been demonstrated. When, however, blackcaps from Central Europe, which are relatively heavier, and those from Africa (Canary Islands), which are about 2 g lighter, were

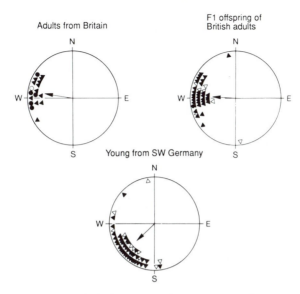

**Figure 2.18** Orientation of blackcaps caught in winter in Britain and tested the following autumn ('adults'), their $F_1$ offspring ('$F_1$') and a control group of hand-raised birds from SW Germany. Each symbol gives the mean direction of one bird during 15–20 tests. Among adults triangles show birds that are parents of $F_1$ offspring in upper right diagram. Arrows, group mean vectors. (Source: Berthold *et al.*, 1992a.)

cross-bred, body masses of F1-hybrids were intermediate, indicating that at least the largely fat-free premigratory body mass is under genetic control (Berthold and Querner, 1982b). A corresponding result was obtained in the same experiment for wing length, which is an adaptive feature of migratory performance (section 1.6). Further, a direct genetic control could be demonstrated for the time course (Berthold and Querner, 1982b) and the temporal pattern (Gwinner and Neusser, 1985) of the juvenile moult (in the blackcap and the stonechat, respectively) which is adaptive to the migratory season and performance (section 2.22). Differences in scheduling of the post-nuptial moult is also most likely under genetic control in different populations of garden warblers (Berthold, 1985b; section 2.22) and in two subspecies of orioles (Rohwer and Johnson, 1992).

## 2.5 THE ROLE OF PHOTOPERIOD

Photoperiodicity was assumed to play an important role in the control of bird migration at the latest during the last century, by both scientists and poets (Berthold, 1975). Its importance was experimentally demonstrated first by Rowan (1925; section 1.2). Since seasonal changes in day length appear with the greatest regularity from year to year among all known environmental factors, photoperiod was long thought to provide the most reliable environmental cue for the control of annual cycles, including migration. This view has been supported by many hundreds of experiments since Rowan's pioneering investigations. Photoperiod controls bird migration in at most four different ways, and at least in two. Its role as a synchronizer of endogenous annual rhythms is well established, as its role in accelerating or inhibiting individual migratory events or processes related to them. Its function as an ultimate factor for migratory movements and its previously hypo-thesized primary stimulatory role for migratory events is still specula-tive. First, the two uncertain roles will be treated briefly and then the two well established functions will be considered in more detail.

In higher latitudes day length decreases dramatically from summer to winter and falls, for instance, below 6 hours in regions north of the polar circle which, during the summer period, are inhabited by many migratory species. Although food availability in higher latitudes during winter is largely controlled by low temperatures, snow cover, etc., it appears likely that a number of migrants would have a chance to winter successfully in such areas if day length remained longer and thus allowed for longer diurnal feeding periods. This appears likely for granivorous and omnivorous species. This possible ultimate role of photoperiod could be examined to an extent in experiments where maximum ingestion rate would be tested in relation to temperature-dependent energy balance. It has been proposed that decreasing day length in autumn and increasing day length in spring, combined with parallel changes in temperature and other environmental factors, may create an unfavourable or improved energetic situation. These situations by themselves could lead directly to migratory movements (for details see Berthold, 1975). This view has only been supported to some extent by studies of typical migrants (Zwarts, 1990), for others see section 2.19. In this context, King (1961a) used a fractionated photoperiod to demonstrate in the white-crowned sparrow that day length does not directly influence the metabolic status of this species. Increased feeding time appears to be insignificant. The most likely role of photoperiod in this case is that of a synchronizer (see below). But the situation may be different in other cases (Owen *et al.*, 1992), and in quite a number of species metabolic preparations for migration appear to depend on increased feeding times and in part on increasing day length, for examples see section 2.9. A permissive or even stimulatory role of photoperiod in this context should be studied in more detail.

There is another respect in which photoperiod could act as an ultimate factor for bird migration. When experimental birds are transferred from long to short days, or vice versa, rather dramatic accelerating or inhibitory effects can be observed on subsequent annual processes depending on the season as well as on the differences in day length (see below). These differential effects would be minimized when migrants living in extremely long days during the breeding season avoided wintering in extremely short days, e.g. around the equator, or would try to winter in comparably long days. The fairly widespread phenomenon that migratory populations breeding in high latitudes also tend to winter in rather high latitudes on the other hemisphere, as it is expressed to an extreme in the Arctic tern (section 1.1), has been discussed in terms of preferences of suitable photoperiodic conditions (e.g. Seibert, 1949). However, whether an ultimate selection force in this respect exists remains open to debate.

Rowan's observations that increasing day length during the winter can evoke vernal migratory behaviour long before the normal migratory season (section 2.6), and the subsequent general finding that changes in day length can exert considerable effects on the course of annual processes and the annual cycle in experimental birds, has led to a comprehensive photoperiodic model of avian annual cycling. In this model photoperiod is considered to be a primary stimulatory factor. It was supported most by King (1963) and Farner (1966), following mechanistic considerations. In the spring migratory period, fat deposition and migratory activity are thought to be directly induced by increasing vernal day length (as direct long-day effects). In the autumn migratory period both phenomena are believed to occur as indirect long-day effects, again induced by an increasing vernal photoperiod that acts as a remote timer. This exclusive photoperiodic model was developed mainly on the basis of data obtained from the white-crowned sparrow, which was believed to be under the control of its annual cycle, i.e. an entirely 'photoperiodic' species. It was also later applied to other species like the chaffinch (Dolnik, 1975). The latter conclusions were criticized by Evans (1970) who stated that a remote timer set in spring could hardly be a trigger for autumn migration of young birds hatched during the summer. Furthermore, King (1968, 1970) found that in the white-crowned sparrow cycles of migratory fat deposition could persist for at least one year (after the experiment had ended) in conditions previously assumed to maintain an indefinite photorefractory state. These conditions were regarded as non-stimulatory. The observations indicate that the white-crowned sparrow may have endogenous circannual rhythms (as has the American dark-eyed junco; see section 2.2) and in this case photoperiod could act as a synchronizer. At present, it is not clear whether species exist in which photoperiod represents an exclusive stimulus for migration. Since endogenous circannual rhythms have been found in both typical and also in less typical migratory species, like partial migrants and species performing irregular irruptive movements (crossbills), it is likely that these rhythms are widespread (section 2.2). In this case photoperiod would generally act as a synchronizer.

As shown in section 2.2, endogenous circannual rhythms deviate regularly, and often considerably, from the calendar year. Even a slight deviation could make it impossible for most of the migrants to rely solely on that rhythmicity. Thus, for a proper timing of migration which is highly adaptive in long-distance migrants ('calendar birds', section 1.1) additional cues are necessary. Since the bird's physiological rhythms are obviously not capable of providing this, environmental cues are required. For reasons mentioned above photoperiod has become the crucial factor in adjusting the endogenous circa-rhythms to the biological seasons.

The role of environmental cycles, such as a synchronizer or 'zeitgeber', can easily be demonstrated using four types of experiments (Gwinner, 1986). (1) Exposure of a free-running rhythm with period $\tau_n$ to an environmental cycle, with period $T$. If the latter is effective as a zeitgeber, the endogenous rhythm should assume its period so that $\tau = T$. (2) Varying the period of the environmental cycle. In this case the period of the biological rhythm should follow changes in the period of the environmental cycle within certain limits. (3) Phase shifting the environmental cycle. If it is a zeitgeber, the biological rhythm should follow that phase shift within some cycles. (4) Exposure of animals kept under constant conditions to pulsatile or stepwise changes of an environmental variable. If it is a zeitgeber, the pulse or step should induce a phase shift, the size and direction of which should depend on the phase exposed to the stimulus. A systematic study of this type would lead to a phase-response curve, for details see Gwinner (1986).

Relevant experiments have been carried out in three migratory or partially migratory species, the European starling, the garden warbler and the Sardinian warbler, in which circannual rhythms have been demonstrated (Gwinner, 1986). But, migratory events have only been tested in the *Sylvia* warblers. An example is given in Figure 2.19. When the normal annual photoperiodic cycle of day lengths was shortened to a period length of six months, i.e. allowing for two complete photo-periodic cycles per year, the patterns of annual processes changed accordingly. Garden warblers went through four instead of two moult periods within a calendar year and showed four distinct periods of migratory restlessness. The latter also occurred in the Sardinian warblers. In this species the frequency of moult periods increased from one to two. These results clearly show that the annual photoperiodic cycle is a zeitgeber of the warblers' circannual rhythms of moult and migratory restlessness (as well as of gonadal size; Gwinner, 1987).

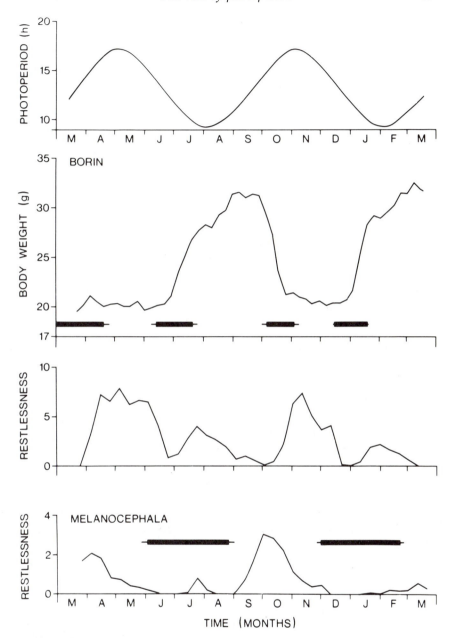

**Figure 2.19** Photoperiodic synchronization of moult (black bars) and migratory activity (restlessness) in the garden warbler (top) and Sardinian warbler (bottom) with two photoperiodic cycles within a calendar year (uppermost curve). (Source: Berthold, 1988c.)

The situation was very different, however, for the body mass cycle. Although garden warblers held in constant experimental conditions regularly showed two body mass cycles per year, and hence behaved like free-living conspecifics, the birds exposed to two photoperiodic cycles per year only increased their body mass once per year (Figure 2.19; and similar experiments in Gwinner, 1987). This could be explained by assuming that body mass is controlled by a different circannual oscillator than the other functions. Such a body mass oscillator could have a smaller range of entrainment and, therefore, the body mass rhythm would be synchronized by frequency demultiplication at a 2:1 ratio within the secondary range of entrainment. Rhythms of moult and migratory restlessness, however, with their larger ranges of entrainment, would be synchronized at a 1:1 ratio in the primary range of entrainment (Berthold, 1979a; Gwinner, 1987). This interpretation appears, however, to be incorrect, since garden warblers exposed to a photoperiodic cycle of the same general shape but with a period of one year did not go through two body mass cycles as the interpretation would have required (Gwinner, 1987). At present it remains open why body mass differs from the rhythms of moult and migratory restlessness as described. A possible explanation could be that in the synchronization experiments vernal fattening has been inhibited by exceptionally long photoperiods at a critical time. These long photoperiods may have rapidly activated the hypothalamo–hypophyseal system and the hypothalamo–pituitary–gonadal axis, and initiated gonadal growth to an extent that vernal migratory obesity was prevented (for details see Gwinner, 1987).

The ranges of entrainment for circannual rhythms are obviously extremely large and comparably much larger than those of circadian rhythms. Although not yet established for migratory processes an example will be given. In the European starling, photoperiodic cycles as short as 1.7 or even 1.5 months (i.e. eight cycles per year) are apparently still capable of synchronizing the circannual rhythms of testes size and moult (Figure 2.20). However, under these very short photoperiods the normal sequence of moult was often disturbed and incomplete and the testes often neither reached maximum size nor regressed to the normal minimum level (Gwinner, 1986). This surprisingly large range of entrainment for circannual rhythms, which appears to differ for various annual functions, is so far not understood in terms of its biological significance. Its adaptive value may be to enhance the ability of the system to cope with and to compensate for extreme conditions.

By the concurrence of endogenous annual rhythms as a framework for seasonal activities and photoperiod as a synchronizer, a functional entity for the control of annual periodicity has evolved with respect to

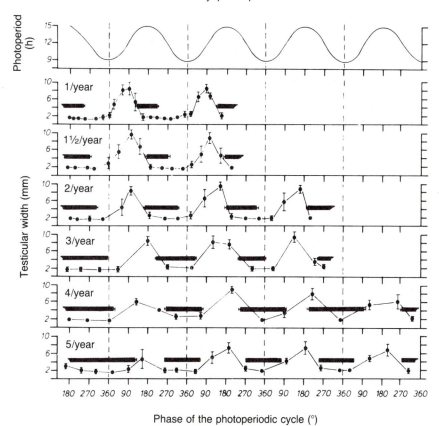

**Figure 2.20** Rhythms of testicular width and moult (bars) in six groups of European starlings exposed to sinusoidal changes of photoperiod as shown in upper panel. Amplitude and general shape of these photoperiodic cycles were the same in all groups but their duration varied from one cycle per year (cycle duration 365 days; second panel) to five cycles per year (cycle duration 73 days; lowest panel). Duration of photoperiodic cycles is normalized to 360° and data are plotted relative to the phase of photoperiodic cycles (0° = phase of shortest photoperiod; 180° = phase of longest photoperiod). Vertical lines at curve points and horizontal lines at bars, standard error. (Source: Gwinner, 1986.)

reliability, precision and flexibility. Depending entirely on the photoperiod would make an organism vulnerable to interferences from exceptional photoperiodic conditions, e.g. during an extended migratory journey. Depending exclusively on endogenous rhythms would bear the risk that any slight deviation of these rhythms from the calendar year would uncouple an animal's rhythm from the geophysical seasons. The dual system of endogenous circa-rhythms and photoperiod as a precise zeitgeber, however, provides many organisms with a viable basis for proper seasonal timing. The circa-rhythms start to prepare an organism physiologically for each of the subsequent annual processes in time often long before such processes are due. This adaptive preparedness is especially provided by one of the main characteristics of these rhythms, i.e. a period length shorter than the calendar year (section 2.2). Furthermore, the circa-rhythms have a certain degree of momentum to buffer short-term environmental influences. On the other hand, their large range of entrainment allows for considerable advance or delay shifts. These properties allow strong zeitgebers like photoperiod to effect acceleration when a delay in preceding processes had occurred or to slow them down when appropriate.

Although photoperiod is the predominant zeitgeber for circannual rhythms, and also appears to play a role in tropical areas, other environmental factors can synchronize endogenous annual rhythms as well. Probable candidates with respect to the timing of migratory events would be temperature, food supply, precipitation and others (e.g. Gwinner and Dittami, 1990). A few examples will be treated below.

In addition to synchronization effects on circannual rhythms, photoperiod also exerts rather dramatic effects on the course of juvenile development in migrants which are highly adaptive. Individuals hatched late in the season in higher latitudes regularly differ from earlier hatched conspecifics as follows: individual processes of juvenile development start at an earlier age, proceed more rapidly, end earlier and thus are of shorter duration; in effect, the processes are 'compressed'. This accelerated juvenile development means that even extremely late broods can prepare for autumn migration on time. In the example given in Figure 2.21 juvenile development of blackcaps from a German population is compared between broods from the end of May and late broods from August. Despite the difference in hatching dates of 72 days the late birds developed migratory activity just 18 days later than early conspecifics. Hence, both broods would have been able to leave their breeding area by September. Without this compensation late birds would have remained until the second half of November (a time that would in many cases be dangerous). This acceleration was primarily due to the extremely short duration of juvenile moult, about two-thirds the duration of early birds. Despite the rapid course the moult of body feathers was complete. It differed simply in its temporal organization (section 2.22). This amazing highly adaptive acceleration in the juvenile development of late-hatched birds, which is commonly observed in many species (e.g. Berthold, 1971), is at least predominantly photoperiodically controlled – it is an effect of short or decreasing day

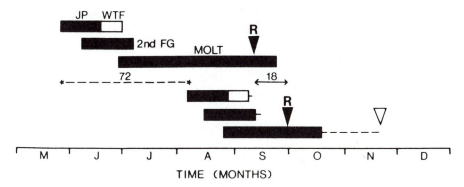

**Figure 2.21** Juvenile development of two groups of hand-raised German blackcaps hatched in May and August, respectively. JP, juvenile plumage; WTF, wing and tail feathers; 2nd FG, second set of body feathers; R, onset of migratory activity; 72 and 18, time differences in the dates of hatching and in the onset of migratory activity, respectively; open triangle, theoretical onset of migratory activity in the late-born birds in case of no 'calendar effect'. (Source: Berthold, 1988a.)

length. This can be easily demonstrated experimentally. When early broods are raised and kept on simulated autumnal photoperiods, they behave like birds hatched late in the season (Berthold *et al.*, 1970; Berthold, 1988a).

The accelerating effect of photoperiod on juvenile development in late-hatched birds is also a synchronizing effect. In blackcaps and stonechats it has been shown by cross-breeding experiments that the pattern of juvenile development is population-specific and genetically preprogrammed. The same conclusion appears likely for the garden warbler (section 2.22).

Short day lengths do not only accelerate juvenile development in late-hatched birds, and advance their onset of autumn migration, but there is also some evidence from field studies and experimental investigations that relatively short day lengths in late summer and autumn accelerate migratory fat deposition (Gifford and Odum, 1965; Berthold *et al.*, 1972b). Such an accelerating effect may also occur during a few overcast days compared with cloudless days. Possibly, a somewhat shorter activity period on overcast days combined with a somewhat longer resting period, based on basal metabolic rates, could create a favourable energetic situation to accelerate fattening, provided that sufficient food is available. This relationship, for which suggestive evidence was obtained during field studies in a warbler program (Anon., 1968), urgently needs further investigations.

According to Pittendrigh (1981) it is a functional necessity that a zeitgeber acts by differentially effecting the phase of the rhythm which it synchronizes. In agreement with this prediction it is found that, in autumn and winter, long and short photoperiods have opposite effects on the speed of the circannual system and on individual annual processes. Accelerating effects of short day lengths occur until the so-called photorefractory period is broken. The end of this period is then characterized by accelerating effects exerted by long day lengths. When that period is terminated annual processes, and the entire circannual system, may be accelerated by long days to an extent that spring migratory events are evoked during winter as, for example, has been demonstrated by Rowan in his classical illumination experiments (section 2.6). In some cases, a day length may be found that is critical to act first in the sense of accelerating short days and later of accelerating long days. This was found in European robins when kept in a constant daily light–dark ratio of 12:12 hours. Some individuals of an experimental group kept under such conditions produced migratory activity in the autumn migratory period which was immediately followed by a 'spring' migratory period and egg laying around Christmas (Berthold, unpublished).

Although the role of photoperiod in the control of avian migration is fairly well understood as a zeitgeber for circannual rhythms and an accelerator of juvenile processes in late-hatched individuals many aspects remain to be clarified. As mentioned above, the question of whether photoperiod can also act as a primary trigger for migration is unanswered. With respect to synchronization of circannual systems it is largely unsolved as to whether the relevant stimulus of a photoperiodic zeitgeber is the change in photoperiod in a distinct direction during a critical period or the exact day length *per se* before and after photoperiodic changes. This, at first sight, simple question is not easy to investigate since some circannual systems obviously require extremely specific photoperiodic conditions for their expression (Gwinner, 1986). Also, the exact timing for homeward migration of Eurasian migrants in Africa appears to depend on close relationships between circannual systems and specific photoperiodic schedules experienced (section 2.22). Other unanswered questions concern, for example, light intensity. What the effective threshold stimuli in twilight periods are and, thus, what the effective day length depends on is still not exactly known, neither is the problem whether slight seasonal changes in the duration of twilight in the tropics are capable of exerting zeitgeber effects on migratory events (e.g. Pohl, 1992). Moonlight and the intensity of artificial illumination of experimental rooms have been shown to influence the amount of migratory activity in a number of species to some extent (Berthold, 1975). There is also evidence that the intensity of nocturnal bird migration is higher in moonlit than in overcast nights (section 3.3). Most likely, a certain light intensity is a necessary prerequisite for any expression of nocturnal migratory activity (section 2.3), but precise relationships to actual migration are not known. Finally, a hypothesis has been proposed that the summation of day lengths could be important in the control of migration in transequatorial migrants. But, as earlier works have shown (see Berthold, 1975), there is no convincing evidence for such a view.

Kok *et al.* (1991) noted, for a continuous period of 36 years, the annual arrival and departure dates of the spotted flycatcher in the Orange Free State. They found that day length is by far the most important environmental factor influencing the date of departure. Most likely, also in this impressive example, photoperiod acts as a synchronizer of an endogenous annual rhythm. Photoperiod (possibly again as a synchronizer) is also thought to control the timing of migration in Bewick's swans (*Cygnus bewickii*; Rees, 1989). Consistency from year to year in the migratory pattern of individual swans exceeds a level referable to chance. Moreover, swans that arrived early for wintering were more likely to leave late in spring. It is suggested that the individuals' response threshold to changes in day length regulates the onset of migration. Heritability estimates for the migratory thresholds for full-grown offspring were low, accounting only for 10–20%, but may have been underestimated due to insufficient data.

In other species, food availability may be an important synchronizing factor. According to Zwarts (1990), whimbrels could not leave their winter quarters in Mauritania before the end of April. Before this time, prey species for fattening and day length for feeding are insufficient. Knots may be triggered to depart from stopover areas during autumn migration before a serious decline in the harvestable prey biomass takes place (Zwarts *et al.*, 1992). In the above mentioned Bewick swans, social factors also influence migration times. Males, which predominate in determining the wintering site for the pair, maintained their normal time of arrival in the wintering grounds when accompanied by a new mate. They also tended to lead the movements of the pair in autumn. After the turn of the year, flight initiation by females proved more common (Rees, 1987).

It is clear that in migrants conditions anticipated on breeding grounds must ultimately play a role in setting arrival times, initiating homeward migration and in controlling the course of migration (e.g. Morse, 1989). At present, we do not know in how many species this initiation is mainly triggered by endogenous factors or by the concurrence of endogenous and synchronizing external factors where photoperiod appears to play a major role.

## 2.6 ENDOCRINE AND NEURONAL MECHANISMS

Our knowledge of endocrine mechanisms in avian migration has been summarized by Wingfield *et al.* (1990). They conclude, among others 'Although physiologists have been studying migration in birds for almost 70 years, much of the data is contradictory and confusing . . . it is not always clear whether endocrine changes are a cause or an effect of migratory processes' and 'There is no reason to expect that any one combination of hormones is going to regulate all migratory processes'. Rankin (1991), in her general review on 'endocrine effects on migration', summarized 'In birds there is evidence that prolactin, cortical steroids, thyroid hormones, gonadotropics and gonadal steroids can all influence migration; considerable interspecies variation exists'. These conclusions reflect the high degree of uncertainty in our present knowledge of how endocrine mechanisms are involved in the exact control of avian migration. Reasons for these deficiencies are numerous, like former applications of hormone dosages outside the normal physiological range, ignorance of feedback mechanisms, etc. (Berthold, 1993). Above all, concept and search for specific 'migration hormones' (sexual hormones, Rowan, 1925; thyroid hormones, Putzig, 1938; corticosterone and prolactin, Meier *et al.*, 1980) is perhaps doomed to failure. It has led to an overemphasis of experimental approaches (Berthold, 1975). None the less, the study of endocrine mechanisms has now received a fresh impetus since the quantitative determination of hormones in small blood samples from living birds became practicable in the past two decades. In the following, the various types of hormones will shortly be treated with respect to their postulated roles in the control of migration whereas highly speculative data will largely be neglected.

### 2.6.1 Gonadal hormones

The possible regulatory role in migration of this group of hormones has been studied most intensively since the pioneering experiments of Rowan (1925). When Rowan exposed dark-eyed juncos and American crows to long days in midwinter their gonads developed earlier; released individuals showed a premature movement northwards, whereas castrates and non-photostimulated controls remained in the vicinity or even moved southwards. These results supported Jenner's (1824) earlier view that the gonads are involved in the control of migration and allowed for the more precise conclusion that gonadal development and gonadal hormones may trigger spring migration, whereas autumnal migration is independent thereof. The general parallelism in many bird populations between gonadal development and vernal migration, on the one hand, and gonadal regression and autumnal migration, on the other, made such a hypothesis widely attractive. Many subsequent castration experiments and studies based on injections of gonadal hormones to stimulate or depress migratory events remained contradictory (Berthold, 1975; Wingfield *et al.*, 1990). However, if gonad removal is performed carefully before the spring migration, then vernal premigratory fattening is abolished and migratory restlessness greatly diminished in both sexes (Weise, 1967; Gwinner, 1971; Stetson and Erickson, 1972; Mattocks, 1976; Schwabl *et al.*, 1988). Further, implants of small quantities of testosterone can reinstate vernal migratory hyperphagia, fattening and restlessness in gonadectomized male and female white-crowned sparrows, indicating that testosterone may be important in the regulation of spring migration in both sexes (Figure 2.22; Mattocks, 1976; Schwabl *et al.*, 1988). Schwabl and Farner (1989a) demonstrated, however, that implants of 5α-dihydrotestosterone or estradiol were ineffective in triggering vernal migratory fattening and restlessness in ovariectomized females. The effect of testosterone was suppressed by an androgenic inhibitor (ATD) that blocks aromatization of testosterone to estradiol. The result suggests that there may be a synergy of testosterone metabolites acting at the target organ level (Wingfield *et al.*, 1990). According to Yokoyama (1976, 1977) testosterone acts on spring migratory fattening by increasing the release of prolactin. A similar idea was earlier developed by Weise (1967). Finally, Wingfield and Farner (1978) were able to show that blood circulating levels of androgenic steroids increase prior to and during the spring migratory period. Measurements of sex steroid hormones in free-living white-crowned sparrows demonstrated elevations of testosterone in males and DHT in females but not of estradiol. Deviche (1991) found in male dark-eyed juncos that administration of testosterone increased food consumption in the spring migratory period. From the presently available evidence, Wingfield *et al.* (1990) conclude that androgens (testosterone and perhaps its metabolites) have a role in the development of premigratory events (hyperphagia, fattening and, to some extent, restlessness) for spring migration. They emphasize that, on the other hand, however, gonadal hormones have no influence over autumnal migration since gonadectomized birds show autumnal migratory events identical to intact controls (for reviews see Berthold, 1975; Wingfield *et al.* 1990). Thus, we are left with the present view that spring and autumn migration, although their phenomena are largely identical, have a very different hormonal basis and that the endocrine control of autumn migration remains particularly obscure. Under the circumstances the old question of how 'primary' the role of gonadal hormones in the control of avian migration might be (Berthold, 1975) is revived.

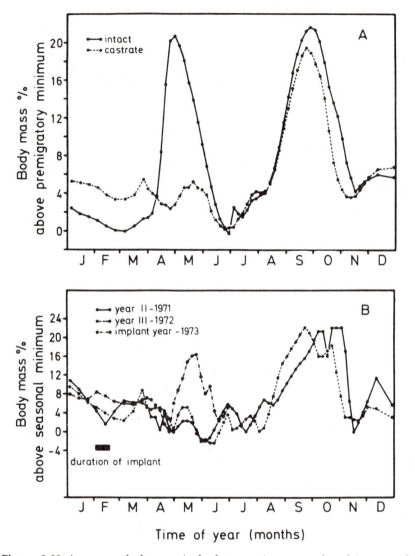

**Figure 2.22** A, seasonal changes in body mass in castrated and intact male white-crowned sparrows; B, in castrated males given an implant of testosterone for two weeks in February. (Source: Wingfield *et al.*, 1990.)

The role of gonadal hormones has also been discussed and investigated with respect to the termination of homeward migration (by high levels of testosterone associated with territorial behaviour; e.g. Wagner, 1961) and with regard to the control of partial migration (testosterone inhibiting departure in autumn; e.g. Lack, 1954). Schwabl and Farner (1989b) tested the first hypothesis by implanting testosterone in white-crowned sparrows. This failed to disrupt fat deposition and migratory restlessness towards the end of the premigratory period. It should, however, be recalled that the termination of homeward migration, at least by the so-called summer restlessness, is related to reproductive activities (section 2.1).

Changes of levels of testosterone have been investigated with respect to the proposed endocrine control of partial migration in two species, the obligate partial migrant European blackbird and the more facultative partial migrant willow tit (*Parus montanus*). In the blackbird, neither the decline of migratory urge with age in males nor sedentary behaviour within age and sex classes are associated with elevated androgen levels that may inhibit migration. Levels of 5α-dihydrotestosterone are even higher in migrant than in resident first-year females (Schwabl *et al.*, 1984a,b; Figure 2.23). In the willow tit, elevated levels of testosterone prior to migration occur, but they are not different between individuals that migrate and members of non-migratory wintering flocks. Implants of testosterone do not increase the probability of sedentary behaviour (Silverin *et al.*, 1989). Thus, Schwabl and Silverin (1990) conclude that testosterone secretion is apparently not involved in regulating partial migration in either the European blackbird or the willow tit.

There is a hypothesis that nocturnal migratory activity (restlessness) may possibly be based on a 'splitting' of locomotor activity into two components (section 2.2). Of two circadian oscillators, one controlling morning activity and the other that of the late afternoon and evening, the latter could be shifted into the night where it produces nocturnal migratory activity. This splitting is thought to be based on testosterone, since in European starlings testosterone injections induced splitting and a diurnal pattern reminiscent of that typical for the migratory season (Gwinner, 1974; Turek and Gwinner, 1982; Wingfield *et al.*, 1990). However, nocturnal migratory restlessness also develops in the autumn migratory period when gonadal hormones are unimportant and thus an androgenic regulation of migratory activity also remains open in this particular case.

## 2.6.2 Thyroid hormones

Studies of thyroid activity by histological measures, estimates of colloidal material, as well as of secretory activity, and investigations applying or, more recently, measuring circulating plasma levels of thyroid hormones, were about as numerous as those on gonadal hormones. A detailed summary of all these attempts would easily fill a booklet. Unfortunately, most of the results obtained, above all the earlier ones, are extremely contradictory, partly due to methodological

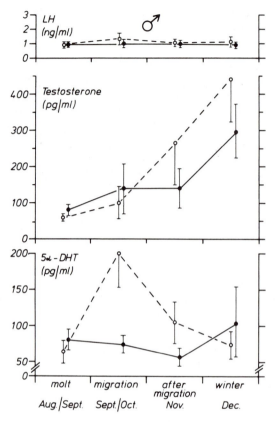

**Figure 2.23** Plasma levels of luteinizing hormone (LH) and androgens in first-year male European blackbirds that had developed migratory activity (closed symbols) or had failed to develop migratory activity (open symbols) during autumn in captivity. (Source: Schwabl and Silverin, 1990.)

insufficiencies (e.g. Berthold, 1975). But, even in the most recent review on these hormones, evaluating new sets of data Wingfield *et al.* (1990) conclude 'It is entirely possible, indeed probable, that although thyroid hormones may have a role in migratory processes, other factors are involved'. When one examines the data only a few examples of hard facts can be shown. In the red-headed bunting (*Emberiza bruniceps*) thyroidectomy reduced nocturnal restlessness and inhibited body mass gain, and injections of T3 and T4 increased locomotor activity and body mass (Figure 2.24, Pathak and Chandola, 1982a,b). Similarly, methyl-thiouracil, a thyroid blocker, resulted in decreased migratory restless-ness in European robins (Ieromnimon, 1977), and iopanoic acid, a potent inhibitor of peripheral conversion of T4 into T3, prevented

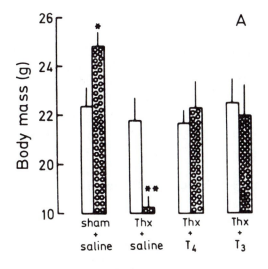

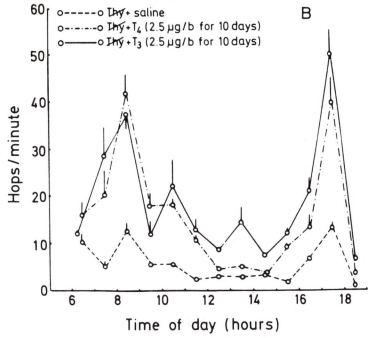

**Figure 2.24** Changes in body mass (A) and zugunruhe (B; perch hopping) in thyroidectomized and thyroxine and triiodothyronine injected male red-headed buntings. Vertical bars, standard errors. (Source: Wingfield *et al.*, 1990.)

migratory fattening in red-headed buntings (Pathak and Binosana, 1989). Some T3 or T4 plasma level elevation was found in several migratory species, in part contradictory. There is evidence that the increase in the T3 level in spring could be due to increased T4–T3 conversion rather than to an increase in secretion (for a review see Wingfield *et al.*, 1990). So far, it remains open to what extent thyroid hormones may be involved in any primary control of migratory events or whether they, more likely, exert fine-tuning or simply general metabolic effects.

### 2.6.3 Hypothalamic and hypophysial hormones

From this heterogenous group, prolactin has been most intensively studied. Meier *et al.* (1980) proposed that, in the white-throated sparrow, premigratory hyperphagia, fattening, migratory activity and even migratory direction are all regulated by changes in the phase relationship between the two circadian rhythms of prolactin and corticosterone plasma levels. These data are generally not considered to be conclusive (Wingfield *et al.*, 1990). What has been fairly well established, however, are the effects of prolactin on migratory fattening. For instance, circulating plasma levels are elevated during migratory periods in some species. Prolactin administration can also enhance lipogenic activity considerably and induce or accelerate fat deposition. Migratory activity (restlessness), on the other hand, may be induced or inhibited (Berthold, 1975; Wingfield *et al.*, 1990). As Wingfield *et al.* (1990) conclude, the actions of prolactin (like those of growth hormone and corticosteroids) on fattening are complex, many investigations are contradictory, and future studies would benefit by bearing in mind the effects of time of day which may be essential in mechanisms underlying prolactin action.

Schwabl *et al.* (1984a,b) found some evidence in the European blackbird that the reduced secretion of LH may be necessary for the development of zugunruhe, thus explaining the commonly observed influence of the migratory disposition on the recrudescence of the reproductive system (section 2.22). Bluhm *et al.* (1991) demonstrated that concentrations of pituitary and plasma LH can spontaneously change under constant photoperiods.

The role of growth hormone is contradictory with respect to lipolysis and lipogenic activity but may be associated with thyroid hormones. In migratory ducks, thyrotropin releasing factor increased not only T4 levels but also resulted in an elevation of growth hormone, and in the green-winged teal (*Anas crecca*) a significant correlation between T4 and growth hormone was found, with highest levels of both hormones just before onset of autumnal migration (Scanes *et al.*, 1980; Campbell *et al.*,

1981; for a review see Wingfield *et al.*, 1990). Furthermore, growth hormone is suggested to play a role in the hypertrophy and atrophy of flight muscles associated with migration (John *et al.*, 1983; section 2.7.2). In the willow tit, a facultative partial migrant, low body mass and high levels of growth hormone were found in migrating individuals, but possible hormonal regulatory mechanisms remain to be elucidated (Schwabl and Silverin, 1990).

Small injections of TSH enhanced nocturnal activity in the migratory period in European robins (Ieromnimon, 1977) and may possibly be involved in the expression of migration. Its precise role as that of the related thyroid hormones remains open.

The neurohypophysial hormones have actually received little attention. Vasotocin was found to release free fatty acids from adipose tissue in the domestic pigeon (John and George, 1986). This can only serve as a preliminary indication of a mechanism which needs to be elucidated.

### 2.6.4 Glucocorticosteroids, catecholamines and pancreatic hormones

Interrenal tissue investigations in a number of migratory species have not shown clear-cut trends for the involvement of adrenocortical secretions on migration. Changes in circulating corticosterone levels are also found not to be consistently correlated with migratory events (Wingfield *et al.*, 1990). Experimental evidence, however, suggests that adrenocorticosteroids may be involved in the regulation of migratory fattening and activity. Injections or implants of corticosteroids can induce migratory events, like hyperphagia, and blockers caused a reduction, but not in all species (Gray *et al.*, 1990). Still, Wingfield *et al.* (1990) conclude that these effects of corticosteroids on fattening may primarily affect fine-tuning and not the primary regulation. Schwabl *et al.* (1991) found, in garden warblers showing nocturnal migratory activity, elevated levels of corticosterone at the end of the dark phase and low levels during daytime. When migratory activity was disrupted this rhythm was absent. These results suggest the existence of diet changes in adrenocortical hormonal activity that could be involved in regulation of migration. Péczely (1976) has actually pointed out that any increased activity of the adrenal during migration may possibly be a result of intense activity rather than its cause, a hypothesis supported by other data (Wingfield *et al.*, 1990). High circulating levels of glucocorticosteroids, as a classical stress indicator, may play an important role in survival or accommodation during severe conditions (Mench, 1991). Hormonal data of such situations are scarce. Schwabl *et al.* (1991) reported low levels of corticosterone for garden warblers stopping over in the Sahara desert. They concluded that these birds, carrying large fat deposits, are not stressed by prolonged flight or lack of appropriate feeding areas. Corticosterone, the major avian corticosteroid, is known not only to be associated with stress but also with conflict, social status and food availability. It has therefore been thought to play an important role in the control of partial migration. However, in European black-birds, the levels were not different in sedentary and migratory individuals before departure. Therefore, it is unlikely that stress (social or food) induces migration in blackbirds nor that the pituitary–adrenal axis is proximately involved in the control of wintering strategy in this species (Schwabl and Silverin, 1990). Ramenofsky *et al.* (1990) reported fourfold higher levels of corticosterone in spring than in autumn in the western sandpiper (*Calidris mauri*). This may possibly be related to increased energetic requirements due to more severe weather. Gwinner *et al.* (1992b) found rather low corticosterone levels in a number of passerines when crossing the Alps during autumn migration. They suggested that birds during this first part of their migratory journey are not normally stressed during the actual migratory flight. However, a lean pied flycatcher showed a very high value, suggesting that migratory performances may become stressful if fat reserves decline. The authors predict increased adrenocortical activity under circumstances in which birds are urged to make metabolic adjustments (e.g. gluconeogenesis from protein) or behavioural changes (e.g. interruption or acceleration of migration).

To study baseline levels of corticosterone may be misleading. In future studies the ability to increase corticosterone levels in response to stress as a measure of adrenocortical activity should be looked at. Investigations of free-living pine siskins (*Carduelis pinus*) and Lapland longspurs (captured during snowstorms) showed that stress-elevated levels of corticosterone were higher in pine siskins, suggesting that mechanisms underlying irruptive migration may be different from those of regular migration (Astheimer *et al.*, 1992).

A few studies on epinephrine, norephrine and reserpine indicated some parallels which may effect migratory events. Wingfield *et al.* (1990) concluded that cause and effect in these correlations have not been established. Finally, according to the same authors, the roles of pancreatic hormones in migratory processes do deserve more study. For example, insulin was found to depress activity during the migratory period. A decline in glucagon during the migratory period may allow for an accelerated lipid synthesis necessary for fat deposition. Precise relationships, however, are unknown.

### 2.6.5 Melatonin

In a recent detailed analysis John and George (1989) found high secretory activity of the pineal during the migratory periods in the Canada goose (*Branta canadensis*), but the role of the pineal in migration is unclear. However, melatonin is involved in the control of the circadian system. Since circadian rhythms and their splitting may be the basis for the development of nocturnal migratory activity (section 2.2), melatonin could be related to its regulation. Although, any relationship appears to be extremely complex and far from being understood. At present it has been hypothesized that, in the control of circadian rhythms, the pineal and the suprachiasmatic nuclei (SCN) of the anterior hypothalamus are components of an inhibitory feedback system. It is thought to be based on a neuroendocrine loop in which the pineal inhibits SCN activity at night via its nocturnal secretion of melatonin, whereas the SCN inhibit pineal activity during the daytime via their multisynaptic connection. The eyes are possibly also incorporated in this system, since the retinae of at least some species show rhythms of melatonin and/or N-acetyltransferase which may reach the circulatory system. Furthermore, there is some evidence that locomotor activity and feeding are controlled by different suboscillators of that complex pacemaker system which might have implications for the detailed control of migratory activity (sections 2.1 and 2.7) and hyperphagia (sections 2.1 and 2.9). For review of these complex relationships see Gwinner (1989) and Cassone (1990).

## 2.6.6 Neuronal mechanisms

The sum of factors, both exogenous and endogenous, that are involved in the regulation of migration are perceived by the central nervous system and the information obtained is transduced directly into behaviour or neuroendocrine and endocrine factors that can also regulate migratory processes. Our knowledge of the relevant neuronal and neuroendocrine mechanisms is, unfortunately, even more limited than that of the endocrine processes (Wingfield *et al.*, 1990).

In the control of migration and its related endogenous rhythms (section 2.2), light and photoperiod act as an important external synchronizer (and possibly, in addition, in a primarily stimulatory way; section 2.5). Photoreceptors in the eyes and extraretinal photoreceptors in the pineal and hypothalamus are likely transducers of the photic information. In addition to neural connections, humoral signals, such as melatonin, may be involved in, for example, synchronizing the output of complex circadian and circannual pacemakers (e.g. Norgren, 1990; Gwinner, 1986; also see above). A number of neurosurgical and neuroendocrine treatments have provided evidence for the control of various behavioural traits in the infundibular nucleus and median eminence, as well as the ventral medial hypothalamus. Lesions, hormonal implants, injections and photostimulation through fibre optics during short-day exposure result in a variety of effects on migratory fattening and the expression of migratory activity (restlessness). For example, lesions of the posterior or entire median eminence of the white-crowned sparrow decreased fattening and restlessness. Those in the basal infundibular nucleus abolished restlessness but not fattening (Yokoyama, 1976). Lesions of the ventral medial hypothalamus in the white-throated sparrow, on the other hand, increased fattening (Kuenzel and Helms, 1967; for more details see Wingfield *et al.*, 1990). These areas are known to be involved in the regulation of a variety of behaviours and in the processing of sensory information. Hence, the results of lesioning studies here are perhaps neither surprising nor conclusive. However, the lesions mentioned may probably disrupt secretion of hypothalamic hormones rather than abolish neural mechanisms. The knowledge of the neuronal mechanisms involved in avian migration is still in its infancy. Deviche (1992) recently used the dark-eyed junco as a model for studies on the endorphinergic regulation of food intake. Results obtained suggested that opioid mechanisms participate in the control of feeding in this species.

The avian hippocampus (hippocampal region or dorsomedial cortex) is known to play a role in processing and/or storing memory. A study by Healy and Krebs (1991) did not show an enlargement of its volume in migratory birds parallel to that of food-storing species. However, the study did not rule out the possibility of a more subtle relationship between migratory behaviour and development of the hippocampus.

Richardson *et al.* (1992) have started to study the roles of central and peripheral regulators of food intake in the white-crowned sparrow. So far it remains open whether altered sensitivity to neuropeptides underlies this seasonal process which is presently being tested. Future investigations of the humoral control of migration may greatly benefit if they are properly adjusted to define premigratory and migratory, and autumnal and vernal migratory seasons. In addition, clear distinction should be made between basic, supplementary and modifying factors. For instance, Wingfield *et al.* (1990) point out that it seems likely that behavioural interactions that synchronize the migration of groups of individuals might act through the endocrine system, but this aspect of hormonal control has obviously not been explored.

## 2.7 METABOLIC ADAPTATIONS

Lipids are the major substrate fueling long flights in birds (Blem, 1990) and, thus, it is not surprising that these energetic resources are the 'miracle drugs' of avian (and other animal) migration. They are the currency for premigratory acquisition of energy stores, for expenditure of those stores during flight and for repleneshing energy stores in the course of migration (Walsberg, 1990). Therefore, basic processes of lipid digestion, anabolic functions of lipogenesis and fat deposition, as well as catabolic functions of mobilization and utilization of fat, are of prime importance in understanding metabolic adaptations of birds to migration and energy production of migrants. In addition, processes regulating adaptive body composition in order to avoid unnecessary ballast, and to provide water balance and protein supply, are also of great importance. These processes providing flight energy will be treated here and, to an extent, in subsequent sections. In elucidating metabolic processes, the study of blood chemistry has recently enabled the use of subtle methods which provide a very promising approach and will also briefly be considered.

### 2.7.1 Fat metabolism, muscle physiology and blood chemistry

The reasons that fats have evolved as major fuel for animal migration are: (1) they have the highest concentration of metabolic energy stores of all body chemicals (section 2.1); (2) hydrophobic and non-polar in nature, they can be stored without water or protein (section 2.1); (3) they might be handled more efficiently than proteins or carbohydrates in intermediary metabolism (less loss of molecular enthalpy as heat during digestion and storage; Leibel, 1992) and their oxidation does not affect body glucose and muscle protein; (4) most tissues in the body, including kidney, liver and muscle, are able to oxidize fatty acids (FA; Ramenofsky, 1990); and (5) muscle fibres with fat deposits are exhausted relatively slowly (Wade *et al.*, 1990).

Fat metabolism of migrants is comprised of many different processes during the premigratory phase and migration. In the latter, one has to consider stopover periods during which fuel is absorbed and replenished, and in the flight phase that it is mobilized and used. Ramenofsky (1990) presented a review of these processes (Figure 2.25), which will be closely followed here. She emphasized that much of our present basic information results from studies on domestic birds and that many specific adaptations of migrants are most likely still unknown.

### (a) Lipogenesis and fat deposition

The characteristic increase in migrants food intake – hyperphagia or 'overeating' – in part accompanied by an increase in the efficiency of food utilization (section 2.9), are the main factors which lead to hyper-lipogenesis (Figure 2.26). Lipogenesis, the *de novo* synthesis of long chain, unsaturated FA, occurs in birds primarily in the liver. Here, acetyl-CoA carboxylase (ACC) and fatty acid synthease (FAS) are the two major enzymes involved. The specific role of these enzymes during

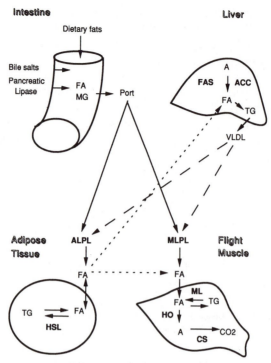

**Figure 2.25** Fate of fats ingested in the diet and synthesized in the liver in birds. Arrows depict directional movement of metabolites. Abbreviations: A, acetyl-CoA; ACC, acetyl-Co carboxylase; ALPL, adipose tissue lipoprotein lipase; CS, citrate synthase; FA, fatty acid; FAS, fatty acid synthetase; HO, β-hydroxyacyl-CoA dehydrogenase; HSL, hormone-sensitive lipase; MG, monoacylglycerol; ML, muscle lipase; MLPL, muscle lipoprotein lipase; Port, portomicron; TG, triacylglycerol; VLDL, very-low density lipoprotein. (Source: Ramenofsky, 1990.)

migration is not yet fully understood. Hyperphagia and increased nutrient utilization may also lead to an increase in the supply of substrates for portomicrons and low density lipoproteins (VLDL), i.e. lipoproteins that deliver triaglycerol (TG – FA combined with glycerol) to tissues. As an example, there is a vernal increase in the activity of hepatic FAS in the dark-eyed junco prior to the onset of premigratory hyperphagia. Therefore, FAS may initiate the increase in hepatic lipogenesis and later ACC may regulate FA synthesis. These possible internal feedback systems deserve further study. In the dark-eyed junco it was also shown that the increased levels of liver FAS activity persisted through the whole spring migratory period. Lastly, an

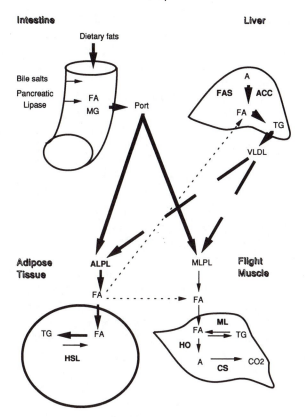

**Figure 2.26** Fate of fats ingested in the diet and synthesized in the liver in birds during the premigratory or absorptive phase of migration. Arrows depict directional movement of metabolites. The thickness of the arrows indicates the relative flux of metabolites through that portion of the pathway. For abbreviations, see Figure 2.25. (Source: Ramenofsky, 1990.)

increase in malic enzyme in the liver was found during fat formation, although its role is uncertain.

Before transported FA can be deposited in, or used by, target tissues, hydrolysis of TG contained in the lipoproteins must take place. Lipoprotein lipase (LPL) and adipose tissue lipoprotein lipase (ALPL) are the essential enzymes involved. Their functions are fairly well known, but research is needed to define the endocrine regulation of ALPL, especially in migratory birds. With the huge fat deposits found in many migrants one would predict an increase in ALPL during the premigratory period. So far, there is an indication of this in the European quail (*Coturnix coturnix*), but not in the dark-eyed junco. Food restriction and refeeding experiments with garden warblers have, however, shown a peak of ALPL during the early stages of refeeding. These data suggest then that ALPL activity is refractory to changes in fat deposition and body mass, except when refeeding follows depletion of adipose depots (Ramenofsky, 1990).

Lipid deposits in migratory birds consist mostly (often greater than 80%) of triglycerides in the form of unsaturated FA (section 2.7.2). The low melting points of these FA may influence ease of mobilization and use of lipid reserves. Thus, this kind of triglyceride composition may be an important factor in adaptive strategies of lipid storage (Blem, 1990). There may even be different compositions in warmer abdominal and cooler subcutaneous deposits. To this, perhaps the increased body temperature of birds during flight (section 2.10) increases flight muscle efficiency and energy transport due to these low melting points (Blem, 1980). In a detailed study of the distribution of lipids and their FA in the semipalmated sandpiper (*Calidris pusilla*), Napolitano and Ackman (1990) found that triacylglycerols were the dominant lipid class in all tissues analysed, accounting for between 49% of the total lipids in liver and 95% in adipose tissue. Squalene, apparently endogenous, was the only hydrocarbon detected. For details regarding FA see the next section. For a possible role of G3PD in the control of obligate partial migration see section 2.4.

### (b) Fat deposition rate

Under favourable environmental conditions and a specifically programmed migratory disposition, migrants can attain a daily body mass increase of up to about 10% or, in passerines, of up to about 1.5 g (Berthold, 1975; Ramenofsky, 1990; Lindström, 1991; Bairlein and Totzke, 1992; Kvist *et al.*, 1993). Alerstam and Lindström (1990) and Lindström (1991) recently summarized normal and maximum fat deposition rates ($FDR_{max}$) in migrating birds. During initial and subsequent fattening they found, in 58 different species or populations, an average (24 hour) gain in mass relative to lean body mass of 2.4% for passerines ($n = 31$) and 1.3% for waders ($n = 27$) with a range from less than 1 up to 7%. $FDR_{max}$-values obtained from field studies were 4.3–5.4% $d^{-1}$ for 10–20 g passerines and 2.6–4.3% for 20–100 g shorebirds, where median values were 2.4% ($n = 31$) and 1.3% ($n = 22$). In addition, they found a significant positive correlation between fat load at departure and fat deposition rate in both waders and passerines, indicating that there is an important element of time reduction involved in bird migration. Still, all mass gain prior to migration was interpreted as fat so that values may have to be reconsidered to an extent when protein accumulation (section 2.7.2) is more carefully investigated. Further, not all values have been sampled close to the crossing of ecological barriers and thus may not necessarily represent true maximum values. There was general decline in the $FDR_{max}$ with body mass, both for individuals and populations. Thus, the $FDR_{max}$ in birds of different sizes has been predicted, on a theoretical basis, to be proportional to $M^{-0.27}$ ($M$ = lean body mass). Field data indicate that some bird species accumulate fat to an extent that the maximum daily metabolizable energy intake ($DME_{max}$) is reached. Here the restricted capacity of the digestive system may affect the metabolic capacity and the migratory performance of those birds ('metabolically limited migrants'). In other cases, however, migrants limited by foraging time will not reach $FDR_{max}$ due to various kinds of environmental influences. When, for instance, migrating bluethroats (*Luscinia svecica*) at a stopover site were supplied with mealworms they showed a much higher rate of body mass increase than under natural conditions (Lindström, 1991). Zwarts *et al.* (1990) in a review of studies on waders preparing for migration showed that the total migratory reserve adds 20–80% to winter mass, the rate of mass gain is 0.1–4% per day and the period of mass increase lasts four weeks on average.

The beginning of fat deposition depends very much on the type and course of migration. Also, there may be some age-dependent programs or even specific feeding conditions to which deposition has been adapted. Many young passerines depart on their initial autumn migration slowly with almost no fat, or at least without distinct premigratory fattening for migration (Berthold *et al.*, 1974; Norman, 1987; Ormerod, 1990a; Bairlein, 1991a; Kaiser, 1992). This may be adaptive with respect to ballast and migration speed (section 2.7.2). Others, preparing for long non-stop flights, may start to accumulate fat up to 1–1.5 months before departure (Johnson *et al.*, 1989; Piersma and Jukema, 1990). Kaiser (1992) found no difference in the mean fat content stored between long- and short-distance migrants in a staging area in S Germany where long-distance migrants mainly stop during initial parts of their autumn migration. In a number of species the fattening rate is higher in adults than in juveniles (Spina and Massi, 1992). At present it remains open as to whether this is due to increased foraging efficiency, dominance, changes in fattening programs in adult birds or a mixture of several factors. In fat-depleted birds mass gain can occur more rapidly than in fat conspecifics due to adjusted foraging behaviour (Moore, 1991).

*(c) Costs of fat deposition*

Fat deposition results, by itself, in energetic costs. According to Blem (1990) they have seldom been quantified but may account for about 1%

of an adult's annual energy budget, as in herring gulls. Kersten and Piersma (1987) estimated the efficiency of transforming metabolized energy to fat at about 88%. It is not known to what extent different nutritional and metabolic adaptations (section 2.9) may minimize energetic costs of fat deposition in migrants. Lindström (1991) estimated the absolute minimum level of daily energy expenditure in free-living, fat depositing birds as $1.5 \times$ BMR and the reasonable upper level as about $3 \times$ BMR.

*(d) Lipolysis and blood chemistry*

In the regulation of lipolysis (Figure 2.27), i.e. the mobilization of FA from stored TG, hormone-sensitive lipase (HSL) within the adipocytes

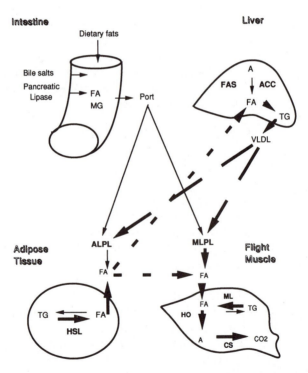

**Figure 2.27** Mobilization and utilization of fats during the flight or catabolic phase of migrations. Arrows depict directional movement of metabolites. The thickness of the arrows indicates the relative flux of the metabolites through that portion of the pathway. For abbreviations, see Figure 2.25. (Source: Ramenofsky, 1990.)

plays a critical role. Lipolytic activity of HSL in birds has been shown to be stimulated by epinephrine, norepinephrine, glucagon, growth hormone, mesotocin and arginine vasotocin (Ramenofsky, 1990). The exact temporal pattern of mobilization of stored fat in migrants is still speculative. In part it may occur during the overnight fast, as shown in a few species in winter (Ramenofsky, 1990). Vallyathan and George (1969) demonstrated that during simulated flight in domestic pigeons the levels of FA in the adipose tissue decreased while those in the plasma, liver and muscle increased. Recent studies with captive dark-eyed juncos suggest that mobilization of FA from adipose tissue increases throughout the period of zugunruhe. This would mean that there is an adaptive increase to supply sufficient levels of energy for migratory flights (Ramenofsky, 1990). Savard *et al.* (1991) found that, in the same species, fat-cell lipolysis was lower at the beginning than at the end of nocturnal migratory activity, while the opposite was observed for muscle lipoprotein lipase (MLPL). These results suggest that the amount of work of one night of nocturnal physical activity modifies both MLPL activity and fat-cell lipolysis in an interrelated fashion. This phenomenon could act to direct a steady supply of FA to the site of energy utilization, i.e. flight muscles. Jenni-Eiermann and Jenni (1991) investigated plasma metabolites in free-living migrants during actual migration. Individuals of three songbird species were mist-netted on an alpine pass during their nocturnal flight. It was confirmed that the birds utilized fat during migratory flight along with an increased protein metabolism. One surprising result was the extremely high triglyceride level in these exercising animals. The highest levels of fat metabolites were found in the garden warbler as the most pronounced migrant in that study. In addition, these small birds also had an elevated level of the very low density lipoprotein (VLDL) fraction (Jenni-Eiermann and Jenni, 1992). The authors hypothesized that free FA re-esterification in the liver and delivery of triglyceride-VLDL to the flight muscles helps to circumvent constraints in energy supply during endurance locomotion in small migratory birds with very high mass-specific metabolic rates. Bairlein and Totzke (1992) made a similar comparative study of grounded migrants in the Sahara desert and a Central European staging area. They found that in garden warblers the variation in blood chemistry was closely related to the extent of fat deposits. In birds with high fat stores there were high levels of glucose, triglycerides, cholesterol, $Mg^{2+}$ and haematocrit. In specimens with almost depleted fat reserves relatively high levels of urea nitrogen, aspartate aminotransferase and creatine kinase were found, indicating the use of body proteins. The elevated levels of $Na^+$ and $HCO_2^+$ found were postulated to reflect dehydration.

*(e) Fat utilization in the flight muscles*

The flight muscles of migratory birds are especially adapted to utilizing FA for energy. Both the morphological (section 1.6) and biochemical efficiency of these muscles increase during the premigratory period. Detailed studies in a number of species have shown that an increase in enzymatic activity enhances the capacity of migrating birds to utilize fat, and when preparing for migration the aerobic capacity of the flight muscles dominates glycolytic activity. The flight muscles are thus morphologically and biochemically adjusted to complete oxidation of FA (Ramenofsky, 1990; Butler and Woakes, 1990). An increased oxidative fat capacity was also demonstrated in the migratory fraction of obligate partial migrants. This indicated that migratory strategy and premigratory fat deposition may influence the enzymatic pattern in the pectoralis muscle (Lundgren, 1988).

An open but most interesting question is whether, in birds, training effects can also enhance flight muscle capacity to utilize FA, as has been shown in humans (Wade *et al.*, 1990). Endurance training can increase muscle capillarity and mass specific activity of citrate synthase (CS – an indicator of the capacity for oxidation of acetyl-CoA in the citric acid cycle), and cause a transformation of muscle fibres from FG to FOG (section 1.6) in the leg muscles of tufted ducks (*Aythya fuligula*; Butler and Woakes, 1990). Increased oxidative capacity in migratory reed warblers may also be a training effect (Lundgren and Kiessling, 1986). Further, 'conditioning' of the pectoral muscles after periods of relative inactivity, as occur during moulting and incubation, is important in improving muscle efficiency (Butler and Woakes, 1990).

FA from deposits are delivered to the flight muscles through the circulation, and MLPL releases FA from lipoproteins into the capillary beds from where they diffuse into the muscle fibre. There, they may then undergo β-oxidation for the production of acetyl-CoA that then enters the Krebs cycle and electron transport system for complete oxidation, to $CO_2$, $H_2O$ and energy (Ramenofsky, 1990). With respect to fat transport, MLPL appears to be a key factor in providing an increased flow of FA to working muscles during migration. In dark-eyed juncos, it reaches peak activity when zugunruhe is exhibited. It appears to be increased by corticosterone and/or by flight activity, which obviously acts as a stimulus for secretion of corticosterone and other hormones (Ramenofsky, 1990; section 2.6). In the grey catbird (*Dumetella carolinensis*) there is, associated with premigratory fattening, no change in the specific activity of CS but an increase in activity of HAD (3-hydroxyacyl-CoA-dehydrogenase), a key enzyme in β-oxidation of FA (Marsh, 1981). In other species there is, however, an increase in CS and COX (cytochrome oxidase; Butler and Woakes, 1990; Lundgren and Kiessling, 1985, 1986).

## 2.7.2 Adaptive body composition

It has already been shown in sections 2.1 and 2.7.1 that fat deposition and hyperlipogenesis are major characteristics of migratory birds. Many migrants have virtually no lipid reserves during summer (except for some 'tissue fat', about 0.5–1 g in passerines), or very small reserves of only about 3–5% (Berthold, 1975; Blem, 1980, 1990), but may depart a few weeks later with fat deposits which can increase their body mass by more than 100% (section 2.8). In addition to fat deposition, migrants are characterized by further changes in body composition which appear to be adaptive in several aspects and will be treated here.

### (a) Fat and protein

Fifteen to 20 years ago it was not clear whether a reduction of glycogen, protein and/or water content in body mass during the premigratory or migratory period could be interpreted as an adaptive reduction in ballast that reduces the energetic costs of migratory flight (Berthold, 1975). There was a mixed set of results concerning changes in body mass and individual components related to migration (Blem, 1980). One possible explanation here would be that, in many cases, final processes of the moulting period were not properly separated from those of the migratory period and this may partly be the reason for contradictory results. Today, the situation is still similarly unclear but the approach has changed. Piersma (1990) recently described what is a reasonable working hypothesis. Current studies have shown that fat deposition in geese, waders and songbirds can be accompanied by a substantial increase in fat-free body mass, primarily due to protein storage. In waders and songbirds 20–50% of the body mass gains before spring and autumn migration were found to be attributable to increases in fat-free mass (Piersma, 1990; Drent and Piersma, 1990; Lindström and Piersma, 1993), and up to 40% wet protein storage was estimated from studies on garden warblers (Biebach, 1992). On the contrary, in adult sandhill cranes (*Grus canadensis*) fat-free mass remained constant during migratory fattening (Krapu and Johnson, 1990).

The commonly observed increase in protein during migratory fattening is often expressed, as has been known for a long time, in breast muscle hypertrophy. In waders, about two-thirds of the increase in flight muscle mass can result from myofibril changes and a quarter from mitochondria (Evans and Davidson, 1990). In an ultrastructural study in several wader species (Evans *et al.*, 1992) it was found that, just before spring migration, fibre diameters had increased but the proportion of myofibrils decreased alongside a compensatory increase in mitochondria. The proportion of sarcoplasma did not change. George *et al.* (1987) have analysed in detail the seasonal degradative, reparative and regenerative ultrastructural changes in the breast muscle of the migratory Canada goose. None the less, it is known that at least in the bar-tailed godwit (*Limosa lapponica*) only 15–40% of the fat-free mass gain could be accounted for by flight muscle changes. Thus, hypertrophy and protein storage go on in other parts of the body as well. This is in agreement with some earlier findings (Berthold, 1993). On the other side of the wing, the loss of fat-free body mass during migratory flight in godwits could only be accounted for by a 20% decrease in the mass of the breast muscles (Piersma, 1990). Driedzic and Crowe (1993) found heart size and protein content increase in semipalmated sandpipers, as had earlier been assumed by Kuroda (1964) for heart and lung in migratory thrushes. A number of details suggest that in the semipalmated sandpipers the heart increase in total protein occurs without an increase in mitochondrial proteins. Biebach (pers. comm.) observed substantial changes in intestine mass in the garden warbler.

Piersma (1990) has suggested that the premigratory protein storage and the loss of fat-free tissue during long-distance flights can be explained by the minimum requirement for protein for repair. From a physiological point of view, this would characterize long-distance migration as a 'very rapid starvation process' for which protein storage represents a supply of fresh parts 'to keep the engine going'. This view seems very plausible. Similarly, Bairlein and Totzke (1992) suggested that premigratory increases in body proteins may not only be related to the power requirements for migratory flights but may be necessary to maintain appropriate protein turnover rates, or breakdown of proteins may even facilitate the oxidation of free FA. Whether premigratory and migratory protein storage is as basic as fat deposition needs to be carefully investigated. So far, there are contradictory findings. For instance, Kaiser (1992) found increases and decreases in fat-free body mass in staging songbirds of Central Europe. Still, this variability may be attributable to different stages of the migratory disposition or to differences in the ratios of flight distances and stopover periods. Other contradictions may be cleared up, as Piersma (1990) pointed out, when more appropriate statistical analyses are applied or a restricted 'scientific search image' is given up. Swain (1992) found in the horned lark (*Eremophila alpestris*) that overnight fasting caused significant depletion of protein in flight muscle and liver, but not in other muscle groups. The disproportionate catabolism of flight muscle sarcoplasmic protein may be due to a greater susceptibility to proteolysis.

## (b) Carbohydrates

Carbohydrates are primarily stored in the form of glycogen in liver and muscle tissues in birds. They are the most important energy source in birds not involved in intensive activity or exposed to low ambient temperatures (Blem, 1990). This also appears to hold for migrants during non-migratory seasons. However, during the migratory period glycogen stores are low compared with fat deposits. For instance, in small birds they normally appear to amount to less than 3% of wet liver mass (Blem, 1990). In grey catbirds, Marsh (1983) reported values which are among the highest known for birds. Maximum concentrations during migration were 77 mg/g liver glycogen and of 20 mg/g muscle glycogen. A number of investigators have suggested that carbohydrate metabolism and storage are adaptively suppressed during the migratory season in order to facilitate fat metabolism or to reduce ballast. Indeed, there is clear evidence for such a reduction in a number of species, including the catbird (Marsh, 1983) but not in others (Berthold, 1975; Blem, 1990). It is still unclear whether these conflicting results are due to species differences or to different stages of the migratory disposition (Berthold, 1975).

## (c) Body water

Adaptive ballast reduction in the migratory season has also been intensively discussed in the so-called adaptive 'premigratory dehydration hypothesis'. According to a recent review by Piersma (1990), it is 'empirically poorly supported and theoretically unlikely'. Piersma (1990) substantiated his view by reviewing a number of investigations, including his own, that more or less support the idea of constancy of body water rather than adaptive dehydration. Child and Marshall (1970) found water ratios of 68.7 ± 0.11% in a large number of small and medium-sized birds. They found no consistent variation due to sex, season, family or general body size, as had been suggested in other studies (Blem, 1980; Åkesson *et al.*, 1992). Even under severe dry conditions when crossing the Sahara desert migrants were found to have body water ratios of the same magnitude. Only lean birds showed more or less pronounced dehydration (section 2.10). However, some authors, like Johnson *et al.* (1989), have reported decreases in body water pacific. In Pacific golden plovers (*Pluvialis fulva*) and sanderlings they went from about 67–62% during the premigratory period to about 60–57% in the migratory period. This may represent an adaptive premigratory dehydration. Finally, Kaiser (1992) did find significant decreases in body water content in four out of 11 songbird species in a

Central European staging area. The average water content dropped from 72.1 ± 0.38% during the premigratory period to 69.8 ± 0.17% during migration.

At present it remains unclear whether slight premigratory dehydration may be adaptive as a strategy to increase flight range or energy storage capacity. According to Piersma (1990) many small water losses which have been detected could also have been due to methodological errors. Kaiser (1992) discussed four possibilities to explain observed water content changes in migrants: (1) adaptive dehydration; (2) artefacts of analysis; (3) side-effects of metabolic shifts; and (4) effects of moult. It is not unlikely that at least part of the observed dehydration reflects a post-moult return to 'normal' body water levels (often increased during the main moult) that has erroneously been interpreted as adaptive to migration.

*(d) Size and composition of fat deposits*

Two more points are often discussed with respect to body composition in relation to migration: the detailed composition of body fat by FA and the size of stores in relation to different migrations and destinations, like autumn or spring migration.

In birds, most deposits of FA are $C_{16}$ and $C_{18}$ molecules, in some cases more than 90% of the total (Blem, 1980, 1990). Oleic or linoleic acid may account for 30–40% of the total (Table 2.1; Berthold, 1975; Blem, 1990). There are, however, many significant differences in FA composition among different species, sex and age groups. In addition, effects of geographical region, season, diet, ambient temperature and day length can also be found. Since energy content and melting point of FA increase with chain length, and decline with unsaturation, and the melting point possibly effects the ability to mobilize TG (Blem, 1990), adaptive FA composition changes could well be a strategy for different migratory requirements. Some results, for instance, have shown northward and overwintering increases in FA with low melting points, like in resident house sparrows. Still, no specific adaptations in migrants have been found (Blem, 1980, 1990). However, Pruitt *et al.* (1990) found differences in membrane lipid composition of leg tissues in different wader populations which may be adaptations to various overwintering sites. The phosphoglycerides of the dunlin subspecies (*Calidris alpina alpina*) that overwinters on the cold water coasts of Great Britain, are significantly more unsaturated than those of the subspecies (*Calidris alpina schinzii*) that overwinters on the warmer water coasts of northern Africa or southern Europe. Since FA unsaturation is largely responsible for the viscosity of phosphoglyceride bilayers it is suggested that the accumulation of monoenes in the cold-water wader may be an

**Table 2.1** Fatty acid composition of adipose tissue from four species of birds (after Blem, 1990)

| Fatty acid $(c:d)$[a] | Emperor Penguin[b] | Herring Gull[c] | Ruby-throated Hummingbird[d] | Lapland Longspur[e] |
|---|---|---|---|---|
| 14:0 Myristic | 8.3 | 3.3 | 1.0 | 1.1 |
| 16:0 Palmitic | 7.3 | 18.5 | 17.6 | 22.3 |
| 16:1 Palmitoleic | 2.0 | 4.0 | 5.2 | 5.2 |
| 18:0 Stearic | 2.1 | 6.2 | 20.9 | 6.7 |
| 18:1 Oleic | 9.7 | – | 33.4 | 39.7 |
| 18:2 Linoleic | 0.7 | 30.5 | 1.8 | 19.9 |
| 18:3 Linolenic | 0.1 | – | 3.7 | – |
| 20:1 Eicosenoic | 32.2 | – | – | 2.8 |
| 20:4 Arachidonic | 35.2 | 20.3 | 10.6 | 0.9 |
| 22:4 Docosatetraenoic | – | – | 4.5 | – |
| 22:5 Docosapentaenoic | 0.4 | 16.5 | – | – |
| 22:6 Docosahexaenoic | 0.4 | – | – | Trace |

[a] Number of carbon atoms to number of double bonds.
[b] *Aptenodytes forsteri.*
[c] *Larus argentatus.*
[d] *Archilochus colubris.*
[e] *Calcarius lapponicus.*

adaptation to conserve the fluidity of cellular membranes in exposed extremities. These problems need be treated in future in more detail.

It appears that body composition can vary adaptively in both fat and fat-free mass in relation to specific requirements of different types or routes of migration, or even depending on the environmental conditions of the goal areas. In species like the garden warbler, which has similar routes during autumn and spring migration, has to cross the same ecological barriers on both journeys, and arrives relatively late in temperate zone areas, maximum fat levels are similar in autumn and spring and the birds arrive with largely exhausted fat deposits (Bairlein, 1991a; Berthold, 1975). In a variety of species, especially songbirds migrating short and middle distances, e.g. American sparrows, European robins or chaffinches, higher energy reserves have been observed in the spring migratory period which are thought to provide a safety margin for more rapid migration. In addition, they may be an insurance against more inclement weather and poor feeding conditions both at staging areas and on the breeding grounds (Berthold, 1975; Evans, 1991). In waders like knots, sanderlings and turnstones (*Arenaria interpres*) very high fat deposits in spring are interpreted as adaptive 'optimal overloading' (Alerstam, 1991b; Gudmundsson *et al.*, 1991). Deposition of more fuel than is needed to cover the flight distance to the next destination indicates that by-passing of possible staging sites may well occur. Both longer direct flights and increased flight speed due to high energy loads appear to maximize the overall course of homeward migration in these Arctic waders (Gudmundsson *et al.*, 1991). This may also be true in other species. In addition, waders breeding in high latitudes often carry a considerable amount of fat to the breeding grounds which may be interpreted as a safety margin. Increases in muscle mass in the wintering area or at staging areas might be used as a protein store to enhance survival or accelerate egg production on the breeding grounds. Such a protein transport for egg production has been reported for geese and shorebirds (Davidson and Evans, 1988; Johnson *et al.*, 1989; Evans and Davidson, 1990). As opposed to this, McNeil (1969) showed that waders can also be fatter in autumn than spring, possibly related to relatively longer migration routes. There are similar reports for other species (Berthold, 1975). Overall, however, relatively higher fat deposits and fat-free body mass in spring appears to be more common. For European robins, Karlsson *et al.* (1988) and Åkesson *et al.* (1992) have shown that in southern Sweden 'short-stage migrants' occur, travelling over land with small fat reserves, in contrast to 'long-stage migrants' which cross the Baltic Sea and therefore are equipped with larger fat reserves. It would be very interesting to elucidate how these differences are controlled, possibly by different metabolic programs (section 2.3). For further examples see the next section.

## 2.8 FLIGHT RANGES, LONG-DISTANCE FLIGHTS AND ENERGY BALANCE

Along with the elucidation of the physiological basis of migration, considerable efforts have been made to calculate theoretical flight ranges of migrants and to discuss the course of migration with respect to non-stop flight capacity and the necessity for stopover periods. Overall, then, an effort has been made to calculate the total energy expenditure for migration and the all-year energy budget of migrants. Despite considerable progress in understanding basic physiological processes we are far from reliable calculations due to several shortcomings. First, predictive equations for total body fat estimates may not be valid even when applied to different populations of the same species due to structural differences (Castro and Myers, 1990). Other methods, with the exception of lipid extraction, have also met with limited success (Blem, 1990; section 1.3). Another primary factor restricting analyses of energetics of migration is the difficulty in accurate estimations of power consumption during flight. Predictions from extant models are largely untestable at present (Walsberg, 1990). Also, the ratio of mechanical energy output to total chemical energy input, assumed normally to be rather constant, at about 20–25%, may vary depending on flight speed, wake gait or body size (Rayner, 1990; Alerstam, 1990, 1991b; also see below). Further, because flight conditions and the effects of wind, flocking, formation flight and cruising altitude are rarely known, flight range calculations have to be taken with caution. They can only be considered as relative measures of flight range capabilities of different species or populations (Gudmundsson *et al.*, 1991). Finally, if part of the body mass increases are due to protein storage instead of fattening, as outlined in the last section, then all previous flight range estimates have to be reconsidered (Biebach, 1992). Regardless of these restrictions, calculations of flight range capacities, and especially the development of adequate equations, have recently made substantial progress, as will be shown in the following.

### 2.8.1 Models and equations for flight range estimates

The following is a short list of various approaches which have recently been taken to estimate ranges. Then, some direct calculations will be presented. As Castro and Myers (1989) and Kaiser (1992) stated, many models have been proposed and many equations exist from several decades of work (Raveling and Lefebre, 1967), however, overall they were originally derived either from aerodynamic theory (Pennycuick, 1975; Greenewalt, 1975), or more empirically from a combination of fat deposits of migrants and their flight costs (McNeil and Caideux, 1972). Most recent calculations show some different approaches. For instance, Johnson *et al.* (1989) have used both the Davidson (1984) and Summers and Waltner (1979) formulae to predict flight ranges in Pacific golden plovers. These are modified versions of the McNeil and Caideux (1972) equation which accounts for weight loss during flight and for the specific flight metabolism of nonpasserines. Biebach (1992) based his flight range calculations of passerine trans-Sahara migrants on Rayner's (1990) equations, which, in his opinion, currently seems to be the most reliable model to predict power input during cruising flight. Kaiser (1992) used five different formulae to estimate flight ranges in Central European passerines and made average values of the results for final considerations. Castro and Myers (1989) developed a new model of their 1988 equation to estimate shorebirds' flight ranges, which now also considers body mass decrease during the migratory flight. The equation: $R = 26.88 \times S \times L^{1.614} \times (M_1^{-0.464} - M_2^{-0.464})$, gives the flight range $R$ in km using $M_1$ = body mass at the end of the flight (g), $M_2$ = body mass at the start (g), $S$ = flight speed (km/hour), and $L$ = wing length (cm). This model predicts similar flight ranges for large and small species.

### 2.8.2 Some current flight range estimates

From the above-mentioned equations and models the following estimates have recently been made. Johnson *et al.* (1989) predicted maximum flight ranges of up to 12300–14098 km for 111 lesser golden plovers, with an average of 6290–7280 km, and an estimated flight speed of 104.7 km/hour according to measurements. These estimates easily explain lengthy non-stop migrations across the western Pacific where, however, the maximum distances traversed by this species are unknown. For homeward migration from Hawaii, for the heaviest individual, a 176 g bird containing 70.7 g of lipids, 6879–7953 km was estimated. With this flight capacity, that bird could easily migrate to several places in Alaska with considerable remaining fat reserves. Biebach (1992) (Figure 2.28) calculated flight ranges of 1380 and 1070 km for garden warblers and willow warblers, respectively. Surprisingly, even the fattest of these potential non-stop flyers did not seem to have sufficient energy reserves to get across the Sahara in calm air and appeared to need the aid of tail winds, regardless of whether they perform a non-stop or intermittent migration. This prediction conflicts with many earlier assumptions of an easy non-stop crossing of the Sahara (Moreau, 1972; Biebach, 1988). Piersma and Jukema (1990) concluded, from a detailed energy-budget study, that bar-tailed godwits are also not able to cover the distance between west Africa and the Wadden Sea in one flight without making good use of favourable high-altitude winds. The same applied for estimates for the knot made by Piersma *et al.* (1991). Castro and Myers (1989) (Figure 2.29) predicted flight ranges for six shorebird species in the magnitude of 4000 km. In contrast to other models, they were similar for large and small species, and they were considerably lower than those predicted by other formulae (as shown in their paper) or by other authors (e.g. up to at least 8000 km, Barter and Hou, 1990). Corresponding predictions to Castro and Myers' (1989), have recently been made by Drent and Piersma (1990), Evans and Davidson (1990) and Gudmundsson *et al.* (1991) for godwits, knots and other waders, respectively.

### 2.8.3 Fat deposition, muscle efficiency and flight range in migrants with different body size

Schaefer (1968) found that the power input:weight ratio decreased very slowly from the smallest to the largest migratory species. He concluded that waders should be able to cross the Sahara, for example, with only about half the proportional fat deposit of the small passerines, while most ducks would find this trip even easier. Such relationships have been treated recently in more detail by Lindström (1991),

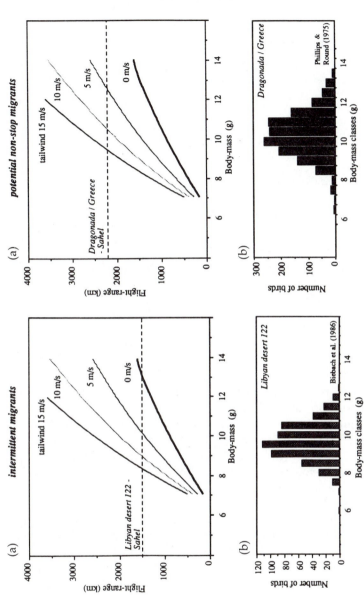

**Figure 2.28** (a) Flight-range for willow warblers grounded in the desert (intermittent migrants) and for willow warblers before the desert crossing (potential non-stop migrants). Calculations were based on various equations. Different lines are for different tail winds. The stippled, horizontal line represents the necessary flight-range for birds to reach the Sahel from the study sites. (b), Frequency distribution of body-mass of willow warblers. The flightrange for birds from the different mass classes can be read directly from (a). (Source: Biebach, 1992.)

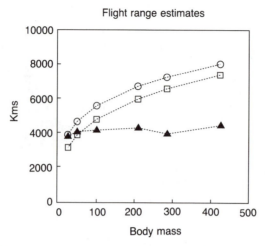

**Figure 2.29** Flight-range estimates (in km) using the equations of Summers and Waltner, 1979 (circles); Davidson, 1984 (squares); Castro and Myers, 1989 (triangles); for six species. It was assumed that individuals depart with a fattening level of 40% (fat:total body mass), and that flight speed is 75 km/hour. (Source: Castro and Myers, 1989.)

who made the following main conclusions. For birds travelling with similar fat loads, the overall speed of migration is directly proportional to the fat deposition rate and to the lift:drag ratio. If the lift:drag ratio is independent of body mass, then maximum speed should vary approximately in proportion to $M^{-0.27}$. However, there are at least three factors which can lead to a general increase in the lift:drag ratio with body mass in migrants: (1) larger birds generally have relatively longer wings, with wings span increasing proportionally, $M^{0.394}$; (2) due to their higher flight speeds, larger birds have lower drag coefficients; and (3) because larger birds have a lower basal metabolism relative to the power required to fly, their lift:drag ratio will be larger than that of smaller birds. Lindström (1991) found that the lift:drag ratio increased with body mass approximately in proportion to $M^{0.13}$, and hence the maximum speed of migration will, on average, be proportional to $M^{-0.27} \times M^{0.13} = M^{-0.14}$. The author concluded that by being more efficient flyers, larger migrants can partly compensate for their lower maximum fat deposition rates. It is assumed that only birds with less than 750 g lean body mass are able to double their mass by fat deposition and still fly at the maximum-range speed (Pennycuick, 1969; section 2.14).

Power output muscle efficiency in birds, as in other animals, is low because of the low efficiency of chemical to physical energy transformation, and much of the output is, therefore, in the form of heat (Butler and Woakes, 1990). Muscle efficiency for small passerines is estimated between 0.07 and 0.15. However, there are indications from muscle ultrastructure and physiology in flying birds that efficiency can be expected to increase with size (Rayner, 1990); Nachtigall (1990) mentioned 0.25 for the domestic pigeon. Hence, range could also be higher in larger birds if muscle efficiency increases with size as suggested. However, maximum gross muscle efficiency probably does not exceed 30% (Walsberg, 1990). Biewener *et al.* (1992) found, in the starling, that in the first direct measurement of muscle force during activity an overall flight efficiency of 13%. These interesting relationships between size of migrants, flight efficiency, fat deposition rate and range of migratory flights require more empirical data to be better understood.

### 2.8.4 Migratory episodes, type of migration and adaptive energy storage

Although the majority of passerines proceed slowly from the breeding areas in short hops, often interrupted by longer stopovers, the homeward migration is known to contain longer bouts (for details see section 2.15). Furthermore, as Piersma (1987), Johnson and Herter (1990) and others, have pointed out, shorebirds and waterfowl use different 'hop, skip, or jump' strategies to migrate over long-distances, depending on the availability of suitable resting places and time for migration. As for Arctic waders, the distances between breeding and wintering areas are covered in a few long-distance flights that may even by-pass possible staging sites. Reasons for these rapid direct flights are that appropriate feeding habitats are usually unevenly distributed along the migratory routes, and the need exists to arrive within a short period at the nesting sites when conditions there allow for breeding. In white-rumped sandpipers (*Calidris fuscicollis*), an extremely long-distance migrant between the Canadian Arctic and Patagonia, long, non-stop, short-distance and multiple-stop flights occur during migration (Harrington *et al.*, 1990). Insufficient fat deposition on a falsely chosen staging area may interfere with further migration, or even breeding success (the domino effect; Piersma, 1987), so that a multitude of migration strategies, even in closely related populations, may be advantageous.

For the generally more rapid homeward migration, or longer migratory jumps, larger fat deposits appear to be adaptive. In waders this aspect of fattening has been termed an adaptive, 'optimal overloading' (section 2.7.2). This is, however, only one of several relationships which have been proposed for the interdependency between migration type, migratory episodes and adaptive energy storage, and which are listed in the following. First, smaller long-distance migrants, like passerines, hummingbirds, waders and others, deposit large amounts of lipid, especially before long non-stop flights, in the range of 60 to slightly more than 100% of fat free body mass (maximum 102–115%; Kvist *et al.*, 1993). Short- and mid-range intracontinental migrants conversely accumulate smaller reserves, and a few forms deposit intermediate amounts (Berthold, 1975; Blem, 1990). Even in short- to medium-distance migrants fat deposits have been found to vary adaptively intraspecifically in relation to different migration distances. This was shown for various populations of red-billed queleas (*Quelea quelea*; Bruggers and Elliott, 1989). Some cross-species comparisons have been done by Alerstam and Lindström (1990). They calculated, in passerines migrating long distances, fat loads between 40 and 70% (median 50%, $n = 63$), approaching 90% in a few species, and 50–90% (median 66%, $n = 17$), with a maximum of about 100%, for waders. Partial migrants on the contrary had values of about 30%. One exception to these patterns was the spotted flycatcher (*Muscicapa striata*). This species is capable to hunt prey continously while crossing the Sahara. Thus, this in-flight hunter migrates with exceptionally low fat stores (Bairlein, 1992). Little is known about fat deposition in raptors during or before migration. Species like the Amur falcon (*Erythropus amurensis*), which makes long water crossings, Swainson's hawk (*Buteo swainsoni*), the longest-distance soaring migrant over land, or the honey buzzard (*Pernis apivorus*), presumably take on large fat deposits during migration. Many other raptors using gliding flight may need maximum fat deposits of 10–20% (Kerlinger, 1989).

Depending on the *en route* requirements of migration, and the food access, short-, medium- or long-distance migrants may begin migration either before or after peak lipid deposition. In some species migration begins even before significant fat deposition has occurred (Blem, 1980; Kaiser, 1992; section 2.7.1); reed and sedge warblers (*Acrocephalus schoenobaenus*) represent well-known examples for this. The first species migrates in short hops, fattening slowly during later migration, whereas the second species travels in longer flights and accordingly energy reserves are accumulated more quickly and are more pronounced (Bibby and Green, 1981; Yrjölä *et al.*, 1989). As an approximate value, the majority of passerine migrants depart with fat loads amounting to 20 or 30% of their lean body weight (Alerstam and Lindström, 1990). As outlined in section 2.7.2, in many long-distance migrants fat deposits appear to be of similar magnitude during autumn and spring migration. Energy reserves have been found to be higher for homeward migration, especially in a number of short- and middle-distance migrants. It is important for the considerations in this section to examine whether these higher reserves in spring are necessary for migration, or as a safety margin for poor conditions on the breeding grounds or as a reserve to ensure breeding. In Swedish willow warblers, Hedenström and Pettersson (1986) found that males which arrive early in a southern population are heavier than females. In a northern population with later arrival the opposite is true. Hence, the relationship between arrival and breeding may affect fattening patterns. Only a few species build up higher fat deposits for the migration from breeding areas than the homeward leg. This was found in several shorebird species, and it was proposed to be related to the relatively longer migration route (section 2.7.2).

In a number of species fat deposits have been shown to be larger in adults than in juveniles [e.g. songbirds: marsh warbler, Kelsey *et al.*, 1989; bluethroat, Ellegren, 1990b, 1991; raptors: steppe buzzard (*Buteo buteo vulpinus*), Gorney and Yom-Tov, 1990; cranes: sandhill crane, Krapu and Johnson, 1990; for more examples see Alerstam and Lindström, 1990]. These differences appear to be adaptive, allowing a more rapid migratory journey in adults as indicated by shorter stopover periods (Ellegren, 1991). For juveniles the slower migration may be related to constraints set by the correct course of their endogenous programs (Berthold, 1993; section 2.3). However, in a number of species, juveniles do migrate with somewhat larger fat reserves. This is the case in chaffinch, yellow-rumped warbler (*Dendroica coronata*) or western sandpiper. These differences may be related to either specifically time- or energy-selected migration (Alerstam and Lindström, 1990). Finally, fat deposits may differ even in populations of the same species in a given area. This could represent differing demands of their migration routes, as has been shown for European robins migrating mainly over land versus over sea, mentioned in the last section.

Certainly, these different relationships are far from being fully understood. Many more cost and benefit analyses (or considerations of 'calculated' and 'uncalculated' migration; Baker, 1978) will be needed before the how's and why's of energy storage in relation to migration have been elucidated. One important constraint should be kept in mind – flight costs increase with increasing body mass (Pennycuick, 1975). Migrants, therefore, in order to minimize energy consumption, should avoid large fat deposits as long as food resources along their route are sufficient (Gudmundsson *et al.*, 1991). Alerstam and Lindström (1990) estimated that set amounts of additional fat allow birds with fat reserves equal to lean body mass to fly an extra distance of only 35% over that attainable without any prior fat reserves. On the other hand, a bird accumulating fat at a certain rate will increase its potential speed of migration, which in many cases may be advantageous. However, this is only possible up to a maximum speed, with further increases in fat load speed will fall off again (Alerstam, 1990). Minimizing transport costs may be the reason why many migrants, including long-distance migrants, initiate a slow migration from breeding areas with low fat reserves. On the other hand, this may also be related to the seasonal or human-made deterioration of the resources available to migrants. Massive habitat destruction and reduction of insects by the application of biocides have been assumed to impair the possibility of building up fat stores according to preprogrammed levels (Kaiser, 1992). Similarly, human-induced disturbance may influence fattening during staging in migratory birds and may effect energetic consequences (e.g. Bélanger and Bédard, 1990).

All in all, adaptive fattening has to be considered under the selection pressure of at least four different constraints – time, energy, predation minimization (Alerstam and Lindström, 1990) and increasing human influences. On the one hand, energy load increases transport costs; on the other hand, it also increases flight speed and speed of migration up to a point (optimal fattening; section 2.14). High energy loads up to overloading may fuel long, non-stop flights and allow safety margins in goal areas. Fat birds are, however, more vulnerable to predation due to effects on flight characteristics. Therefore, fat load to maximize safety from predation during the migratory journey is clearly smaller than the corresponding fat level for maximizing migration speed (Alerstam and Lindström, 1990). According to McNamara and Houston (1990), there is an optimal level of reserves at which total mortality (starvation plus predation) is minimized. Lindström (1990) found some evidence that bramblings (*Fringilla montifringilla*) seem to prefer avoiding predation to attaining maximum speed of migration. However, in his study the importance of maximum migration speed cannot be ruled out. It will be a most interesting task to elucidate to what extent migrants deposit energy stores in relation to time, energy or minimization of predation, possibly taking into account other factors as well (Witter and Cuthill, 1993). In a recent test using data of migrating bluethroats and rufous hummingbirds (*Selasphorus rufus*), Lindström and Alerstam (1992) obtained qualitative, and to some extent quantitative, agreement with the time-minimization hypothesis.

### 2.8.5 Overall energy budgets for migration and all annual activities

The necessary information to work out an adequate energy budget for all annual activities, including migration, is so far only available in a few species. Drent and Piersma (1990) have presented a prominent example for the bar-tailed godwit, a species migrating in relatively few hops, which is relatively easy to grasp (Figures 2.30–2.32). In the population migrating within Europe, covering 2500 km one-way, the

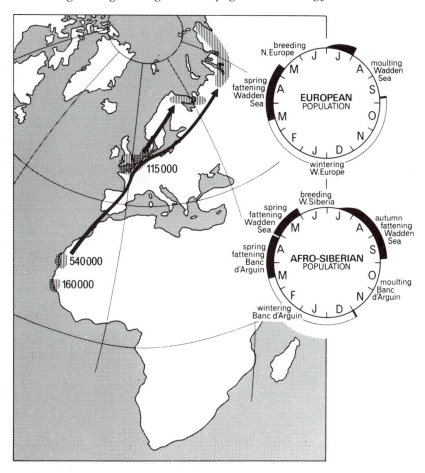

**Figure 2.30** Diagram of the main breeding and wintering areas of the two populations (European and Afro-Siberian) of bar-tailed godwits using the East Atlantic flyway. The annual clocks indicate the sequence of phases experienced by birds of the two populations in the course of the year. (Source: Drent and Piersma, 1990.)

energetic cost of migration accounts for about 22% of the annual energy expenditure as compared to 48% in birds migrating between Siberia and Africa, covering 8300 km one-way. These differences, however, are balanced by high- and low-cost wintering areas. Piersma *et al.* (1991), Piersma and Davidson (1992) and Piersma (1994) have added a further example of different subspecies of knots, where they included considerations of intraspecific differences in the basal metabolic rate and evaporative water loss during wintering.

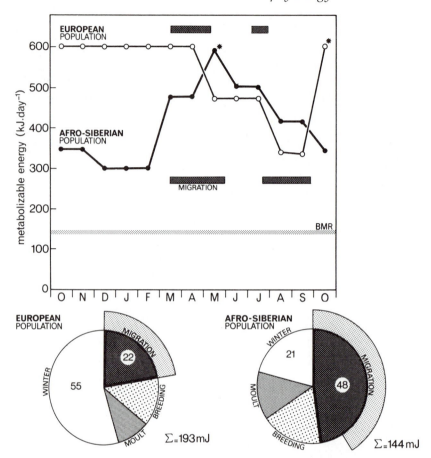

**Figure 2.31** Provisional energy budgets (ME, metabolizable energy) in the course of the annual cycle for the European (open circles) and Afro-Siberian (filled circles) populations of the bar-tailed godwit. The annual totals are collected in the pie diagrams below, where the percentages of annual ME are shown for the calendar periods concerned with migration, breeding, moult and winter. For migration the proportion of the annual time budget is indicated on the outer circle. (Source: Drent and Piersma, 1990.)

Klaasen *et al.* (1990) have measured daily food intake, assimilation efficiency and body mass in five species of waders kept in cages under tropical conditions. In their study, the so-called 'reduced endogenous heat production hypothesis' was supported by the findings that assimilation efficiencies and maintenance metabolism under such conditions were relatively low. This may be adaptation to avoid heat stress under tropical conditions.

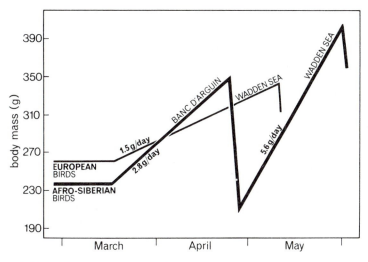

**Figure 2.32** Body mass changes during spring migration of males of two populations of bar-tailed godwits. (Source: Drent and Piersma, 1990.)

So far, comparable estimates of overall energy budgets for songbirds are not available. In these species, premigratory fattening is rarely sufficient to provide the necessary energy for the whole migratory journey, or even for a few hops. Thus, reliable calculations would be necessary for quite a number of episodes and this is presently extremely difficult and full of shortcomings, as outlined in the introduction.

## 2.9 FEEDING, NUTRITIVE AND ASSIMILATION STRATEGIES

Premigratory and migratory fattening requires considerable additional energy supplies that often need to be collected in periods of high-energy demands during actual migration, post-nuptial moult or the final stages of juvenile development. Several nutritional strategies and adaptations have been developed that provide the necessary surplus energy. These include: (1) hyperphagia ('overeating') as probably the most important mechanism (section 2.1); (2) dietary changes to easily accessible food; (3) selection of energy- or nutrient-rich food or food stuffs that specifically promote fat deposition; (4) increased food absorption and assimilation efficiency; (5) decreased rates of standard metabolism, (6) movements to more favourable feeding grounds, as observed in European starlings in intermediate migratory movements between juvenile development and autumn migration (Fliege, 1984; section 1.1); and (7) devoting more time to actual feeding at the expense of other activities such as vigilance, as observed in turnstones

(Lindström, 1991). These feeding and assimilation strategies may be enhanced by some energy-saving measures, like reduced locomotor activity during the daytime (section 2.13) or simply by avoiding temporal overlap between fattening and other seasonal activities (section 2.22). Fattening may further be favoured by high ambient temperatures which are present during the vernal premigratory period in migrants wintering in the tropics (Blem, 1980), or even by short-day effects before autumnal migration (section 2.5).

### 2.9.1 Hyperphagia

Although many details of nutritive and assimilation strategies related to migratory fattening are not well understood, hyperphagia generally appears to be the most important (Berthold, 1975; Blem, 1980; Bruns and Ten Thoren, 1990). During hyperphagia, both food intake and food metabolized rise considerably. Metabolized energy levels at this time increased by 20% in white-crowned sparrows (King, 1961b), 25–30% in bobolinks (Gifford and Odum, 1965), and 40% in garden warblers (Bairlein, 1985b) (Figure 2.33). The temporal pattern of hyperphagia appears to depend on the species and nutritive conditions. In the white-crowned sparrow, hyperphagia did not result in important changes in the daily feeding pattern (King, 1961b). When birds are restricted to a very nutrient-poor vegetable diet, however, they may feed all day. These birds then metabolize food on the magnitude of two times of their own body mass, and food intake can reach short-term maxima of up to about 300% (Berthold, 1976). In waders preparing for spring migration some species, such as the knot, increased the total time they spent feeding and fed more at night, throughout neap tides and at high temperatures. In other species, however, such as the little stint (*Calidris minuta*), feeding time did not increase (Zwarts *et al.*, 1990). The turnstone exhibited decreasing vigilance (Evans and Davidson, 1990). In the sanderling, Castro *et al.* (1992) found that diurnal time budgets varied among locations, with feeding times ranging from 40 to 90% of daylight hours depending on different living conditions. The complex of hyperphagia and patterns of food intake needs much more investigation. The same holds true for relationships between hyperphagia, fat depletion and fat restoration. It is well known that adaptive hyperphagia normally results in rapid fat restoration, but details of feedback mechanisms and differences in postulated primary and secondary hyperphagia (Farner *et al.*, 1969) are barely understood (section 2.15). Hyperphagia may be favoured by increased preference of more variable food, as has been shown by feeding experiments with yellow-rumped warblers (Moore and Simm, 1986).

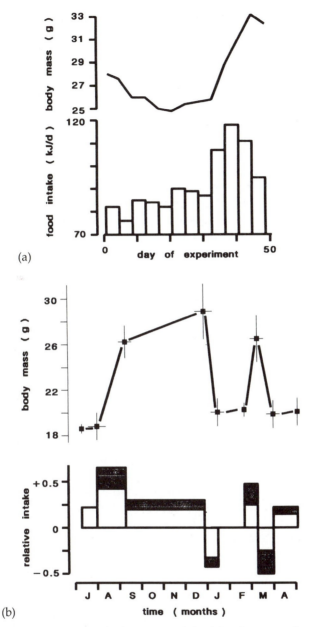

(a)

(b)

**Figure 2.33** (a) Changes in body mass and food intake rates of an individual captive white-crowned sparrow during the spring migratory period. (Source: King, 1961a.) (b) Average net food intake of 10 captive garden warblers (12:12 light–dark, hours) relative to the low body mass phase in 'winter' during various periods of their seasonal body mass cycle (above). The open bars show the changes in gross food intake (food eaten: dry matter), the dashed bars show additional changes in net food intake (food assimilated) due to changing assimilation efficiencies. (Source: Bairlein, 1990.)

### 2.9.2 Adaptive seasonal frugivory

During autumnal migratory fattening, birds are often confronted with the following difficult situation. On the one hand, fattening requires considerable hyperphagia and thus increased foraging, on the other hand, many food stuffs like insects decline seasonally before the migratory period. Further, decreasing ambient temperatures can increase the costs of thermoregulation, and extended nocturnal flights may require resting periods during the day rather than allow for extended foraging. Many species are able to get around this dilemma by seasonal shifts in their diet. One dramatic shift of this nature is found in many carnivorous/insectivorous species which, after the breeding period, shift to mixed animal/plant or possibly even to exclusively plant nutrition during part of migration. Such shifts are well known in many songbirds whose diets change from insects to berries and other fleshy fruits. They are, however, also common in many other taxa like cranes, gulls, ducks, geese, waders, and even bee-eaters and raptors (Berthold, 1993). They also occur in birds preparing for homeward migration, e.g. during dry periods in their wintering areas (e.g. Blake and Loiselle, 1992). For many migrants, fruits like berries are an 'easy prey', often available in almost unlimited quantities with a minimum foraging expenditure. Consumption of a large variety of fruits is greatly facilitated by an extreme tolerance in birds to substances like alkaloids which are poisonous to mammals, and accidental death due to ethanol intoxication from fermented fruits has only rarely been reported in birds (Fitzgerald *et al.*, 1990). There is an extremely wide range of vegetative food utilization in carnivorous/insectivorous migrants (Brensing, 1977; Snow and Snow, 1988), and even among frugivore species there is extreme variation with regard to food preference. The regularities here are just beginning to be understood. In the following, the main hypotheses and general results are briefly summarized.

Theoretically, seasonal frugivory could simply reflect seasonal changes in food availability for omnivorous bird species. Most berries, for instance, in higher latitudes ripen during the autumn migratory period of many bird species. Some authors (Dolnik and Blyumental, 1964) have proposed that there is a programmed, possibly complete, shift to vegetable diet in many carnivorous/insectivorous species in order to provide sufficient carbohydrate intake for fattening. Systematic field studies in various regions (e.g. Brensing, 1977; Jordano, 1985) have shown that species consuming large quantities of berries often supplement their diet with small quantities of animal food. Therefore, the question of complete shifts in nutrition is difficult to answer on the basis of field observations. The hypothesis of endogenously controlled shifts in seasonal food preferences was first tested by Berthold (1976).

Garden warblers, blackcaps, European blackbirds and European robins, kept under constant experimental conditions, were offered a mixed diet of insects and berries of alder buckthorn (*Frangula alnus*), which these species regularly consume in large quantities at stopover sites (Brensing, 1977). In the experiment, the species regularly ingested berries. Outside of the blackcap the berry consumption was rather limited. The garden warbler, the species with highest migratory fat deposits, clearly preferred the insect diet during fattening and showed only a short-term preference of berries during the period of winter moult and low body weight (Figure 2.34). Blackcaps consumed more berries, but the preference was not exclusively related to fattening and only present in the first experimental year. On the other hand, garden warblers showed normal body mass and fat cycles when kept for nine

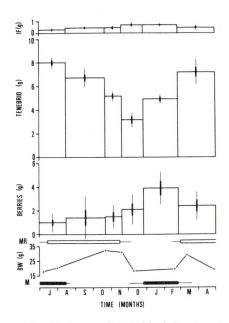

**Figure 2.34** Body weight (BW), moult (M, black bars), migratory restlessness (MR, white bars) and consumption of insect fodder (IF) as well as *Tenebrio* larvae and berries of *Frangula alnus*, for a group of eight hand-raised garden warblers. Mean values are presented with standard deviation, and for some with standard error of the mean. Food consumption is averaged for six periods: (1) for the duration of the juvenile moult; (2) for the period from the end of that moult to the mean maximum of body weight during the autumn migratory period; (3) from the mean maximum of body weight to the onset of continuous rapid body weight decline; (4) from the onset of weight decrease to the beginning of winter moult; (5) for the duration of the winter moult; and (6) for the period of spring migration. (Source: Berthold, 1976.)

months on a pure insect diet. Further, when the four experimental species were kept on an exclusively vegetable diet during the autumn migratory period they rapidly lost body mass, in a few days, to critical levels (Figure 2.35). This loss was probably due to protein and/or lipid deficiencies (Bairlein, 1991b). It also occurred in individuals which were trapped during the autumn migratory period and which, according to their coloured faeces and their stomach contents, were used to berry consumption of the various kinds offered. From these results, one can conclude that feeding on berries and other pulpy fruits by carnivorous/insectivorous migrants, mainly in higher latitudes, can be understood as simply providing a supplementary food resource. This does not exclude, however, that in critical situations, like during migration, plant food may greatly enhance fattening when supplemented by a minimum of animal diet for body maintenance (Berthold, 1976). Recently, Izhaki and Safriel (1990) reported on similar weight loss in experimental birds kept on a fruit diet and related it to poor nitrogen assimilation, possibly effected by secondary compounds.

In the meantime, both intensive field and laboratory studies have considerably increased our knowledge (Bairlein, 1990, 1991b). In these studies garden warblers were provided *ad libitum* with a semi-synthetic 'insect' diet and simultaneously with different fruits. They assimilated rather small or moderate amounts of energy by feeding on privet, honeysuckle and red elder, but more than half of the energy in case of black elder (Figure 2.36). Black elder was also clearly preferred in selection experiments to other berries (except white dogwood). When garden warblers were fed exclusively on black elder berries, the birds did not undergo any significant long-term loss of body mass, but instead were even able to regain their initial body mass on that diet after several days of food deprivation (Figure 2.36). Similar results have been obtained with figs. Finally, garden warblers fed *ad libitum* with both a semi-synthetic 'insect' diet and black elder berries increased body mass during migratory fattening faster than if fed only an animal diet (Figure 2.36). These results correspond well to field data on warblers and figs (Ferns, 1975). Also, Gardiazabal (1990) found a positive correlation between the amount of vegetables consumed, body mass and fat deposition in four passerine species in the field. Boddy (1991) studied fruit diets of *Sylvia* warblers and some thrushes during the autumn migratory period in England. Larger species commenced earlier and ate more fruit. Mean fruit contents of faecal samples exceeded 80% infrequently, indicating that even the most frugivorous species always took other food, normally insects, to supplement their diet. Elder berries became the preferred fruit for all *Sylvia* species. The author calculated that, for instance, the blackcap can obtain 75–90% of its daily energy requirement from (about 410) elderberries in less than

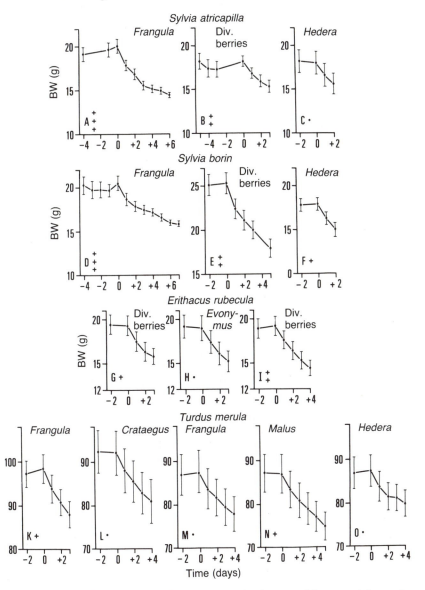

**Figure 2.35** Mean body weight (BW; with standard error of the mean) for each of eight hand-raised and/or trapped blackcaps, garden warblers, European robins and European blackbirds which were supplementally (until day 0) or exclusively (from day 0) fed on various fruits (*Frangula alnus*, diverse berries, *Hedera helix*, etc.). 0, Onset of the experiment, exclusive feeding of fruits; minus, days before; plus, days after the onset of the experiment; 1–3 plus, significance levels. (Source: Berthold, 1976.)

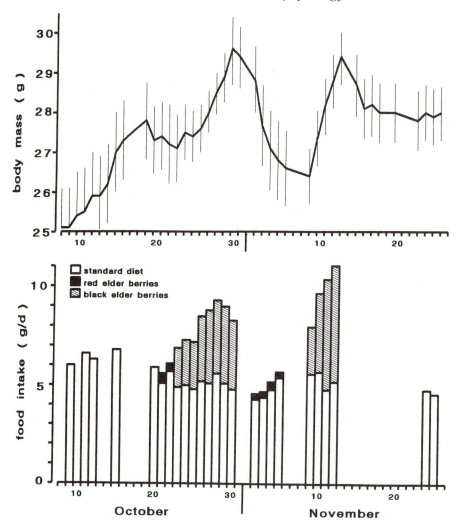

**Figure 2.36** Changes in average body mass (± standard error, above) and changes in food intake (dry matter, below) of eight captive garden warblers. The birds were continuously fed *ad libitum* on a standard lab. diet. On some days they additionally received fruits *ad libitum*. (Source: Bairlein, 1990.)

10% of the time available for feeding. However, blackcaps had trouble in maintaining body mass at a low level when fed exclusively on common elder berries for two and a half weeks during the migratory period, so that the experiment had to be broken off (Berthold, unpublished). During wintering, individual European robins showing different levels of fruit consumption did not show significant differences in body weight or fat accumulation (Jordano, 1989).

So far, no obvious correlation between black elder (and other fruit) preference and fruit characteristics, like colour, size, macronutrients, fibre content or handling time, has been found which could explain its specific importance for the garden warblers. Similarly, the concepts of the 'optimal foraging' theory cannot simply be applied to fruit consumption and nutrient content in the Mediterranean area to explain shifts (Gardiazabal, 1990). Most likely, complex interrelationships between basic fruit characteristics, secondary plant chemicals and as yet unknown physiological adjustment mechanisms need be described to explain the adaptive significance. For instance, fruits are, in general, rich in unsaturated fatty acids that are major fat deposit components in birds (section 2.7.2) and there is suggestive evidence that for a variety of bird species lipids are important factors in their selection of fruits (Stiles, 1993; for experimental evidence see below). On the other hand, they are generally low in protein and overall lipids. However, garden warblers have much lower daily protein requirements (2.7 mg/g/day) than previously supposed and, in addition, are able to compensate for low dietary protein levels by increased assimilation efficiency (below). Further, plants offer different kinds of sugars, and the preferences and physiological traits of avian seed dispersers are broadly correlated with the sugar composition of the nectar and fruit that they feed on, and appear to have influenced the evolution of the sugar composition. For instance, hummingbirds prefer sucrose whereas many passerines prefer glucose and fructose. Preference for hexoses in passerines appears to be restricted to the sturnid–muscicapid lineage where intestinal sucrase is lacking (del Rio *et al.*, 1992). del Rio and Karasov (1990) have presented a 'hummingbird model' and a 'frugivore model' for the digestion of sugars in birds and various physiological and evolutionary consequences.

At present, it can be stated that facultative frugivory in many carnivorous/insectivorous migrants is a strategy that provides an 'easy prey' as an energy supplement to fattening with generally low fibre content which reduces digestion costs associated with mechanical breakdown. Special fruits, like black elder or figs, may be of prime importance either for specific species or in general (Gautier-Hion and Michaloud, 1989). In lower latitudes, many fruits appear to be generally nutrient-richer than in the temperate zone and may there be especially valuable for pronounced fattening (Herrera and Jordano, 1981; Snow and Snow, 1988). However, a series of questions remain to be tackled. For instance, to what extent is seasonal frugivory preprogrammed by endogenous components? Or is it primarily caused by changes in food availability? Here Gardiazabal (1990) emphasized that one should not exclude the possibility that temporary frugivory may result from the lack of suitable animal prey. Secondly, it is not clear how far variations in intestine morphology and physiology might increase efficiency of seasonal frugivory (Bairlein, 1990), or to what extent enlargement of the digestive tract accompanying seasonal frugivory (Kuroda, 1964) may constrain the premigratory reduction of fat-free body mass. The question of why some migrants, like leaf and reed warblers, do not shift to seasonal frugivory cannot be answered (Brensing, 1977). Recently, differences in intestine structure were discussed in this context (Gardiazabal, 1990), which may be related to predominant feeding on aphids, as reported for sedge warblers, including appropriate habitat selection (Ormerod, 1990a). Levey and Duke (1992) suggested that possibly functional adaptations to frugivory (i.e. how fruit is processed) are more likely to be important than structural adaptations. They found in cedar waxwings (*Bombycilla cedrorum*) antiperistalsis mixing digesta in the intraluminal zone adjacent to the rectal mucosa which may increase nutrient absorption. And indeed, active uptake of D-glucose was as high in the rectum as elsewhere in the gut.

It also remains open to what extent widespread 'exceptional' food consumption as, for instance, seed-eating in waders (Baldassarre and Fischer, 1984) or fruit-eating in raptors (Berthold, 1976), may also contribute to favour a migratory state under certain conditions. Extremely specific adaptations may occur. For instance, yellow-rumped warblers and tree swallows (*Tachycineta bicolor*) are among a small group of birds in North America that regularly eat waxy fruits. Place and Stiles (1992) found that these species have specific gastrointestinal traits which permit efficient assimilation of saturated fatty acids, the main component of the waxy material, which in most animals are poorly assimilated. Due to this adaptation both species can feed extensively on fruits of the bayberry (*Myrica* sp.) which may allow these small passerines to maintain more northerly wintering ranges than closely related species. Lastly, the problem of plant coloration and its attractiveness to frugivorous birds is still under debate. On the one hand, Stiles (1982) hypothesized that early autumnal changes to conspicuous leaf coloration in fruit-bearing trees serves to attract frugivorous birds. On the other hand, bright colours have been proposed to repel birds in other cases (James, 1990). Willson *et al.* (1990) found in American migrants little tendency to favour red and black, and a weak tendency to reject yellow. The authors suggested that avian colour preferences may not provide strong selection, favouring the evolution of the common fruit colours, and that the frequency distribution of fruit hues is best explained in other ways. These and many other problems, like considerable individual variation in fruit choice (Jung, 1992), require an enormous effort in a scarcely-treated field of nutritional ecology, and co-evolution of plants and birds where various aspects of individual species and their potential animal and plant nutrition have to be considered. Also in this context, interactions between seed dispersers, fruiting plants and microbes that infect fruits appear to play an important role and exert effects on co-evolutionary processes (Buchholz and Levey, 1990).

### 2.9.3 Increased assimilation efficiency and digestive modulation

Relatively few studies have been carried out to test whether the efficiency of food digestion, absorption and assimilation in birds changes seasonally (Berthold, 1993). Theoretically, it would be possible that birds with their generally high metabolic rates would permanently live with a maximum assimilation rate. King (1961a) found in white-crowned sparrows a largely unchanged food utilization coefficient throughout the year. Place (1990) compared the properties of the avian digestive system, i.e. high throughput rate, high extraction efficiency and minimal volumes, with those of an optimal chemical reactor. He recommended that chemical reactor theory be used to gain new insights into the 'optimization constraints' confronting avian digestion. In contrast to this approach, Karasov and Levey (1990) investigated *in vivo* digestive efficiency of radioactively labelled sugars in four passerine species, and found that the anatomy and physiology of fruit eaters apparently result in less than complete digestion and absorption of sugars. The studies carried out, summarized by Bairlein (1990), favour the opinion that there are possibly two types of birds: granivorous passerines which do not have any significant variation in food assimilation during migratory fattening, and insectivorous/frugivorous songbirds which have temporary increases in assimilation efficiency. A detailed analysis in this respect has been made in the garden warbler by Bairlein (1990). This species has a significant increase in the efficiency of food utilization or assimilation efficiency (i.e. proportion of food ingested to digested food) during migratory fattening. This increases both metabolizable energy and nutrient intake. The changes in assimilation efficiency account for about one-third of the entire increase in food assimilation during fattening. The other two-thirds are based primarily on hyperphagia (Figure 2.33). Most of the increase in assimilation efficiency is due to higher efficiency of fat utilization, while minor changes are also found in carbohydrate and protein utilization. The mechanisms of these changes are so far unknown. Endogenous factors may be involved, but slowly progressing adaptations to, for instance, a low-protein diet over several weeks implies that more immediate responses to food characteristics may be important. In any case, an increased assimilation efficiency must somehow be a costly inherited strategy, otherwise one would expect birds to rely on it permanently. Levey and Karasov (1992) tested three non-exclusive hypotheses with respect to digestive modulation in the seasonal frugivore American robin: differences in intestinal absorption of sugars and amino acids, retention time of digesta and changes in gut morphology. The results obtained suggest that short retention time strongly increased over the range of short winter days to continuous light in the Arctic breeding area in relation to a changing food spectrum (Prop and Vulnik, 1992). In the phainopepla (*Phainopepla nitens*) a major seasonal change in gizzard size was found, correlatable to a dietary shift in fruit consumption. Similar alterations in the avian digestive tract have also been reported for other species (Walsberg and Thompson, 1990). In another study, where faeces were analysed for protein, crude fibre and energy contents, in dark-bellied brent geese (*Branta b. bernicla*; Bruns and Ten Thoren, 1990) improved physiological utilization of nutritional components was not indicated, and hyperphagia may therefore be the only factor leading to fat deposition. In the same species it was found that symbiotic cellulose decomposition by bacteria does not play a role in body mass increase (Pankoke and Holländer, 1991). Certainly, the problem of adaptations in assimilation efficiency to migration requires much more investigation. So far, at least two distinct carriers for sugars and at least four distinct carriers for amino acids are known in birds. The whole bird's integrated uptake capacity may change depending on many factors (Obst, 1990), thus providing quite a puzzle to be clarified. And there may be considerable differences among species, as in gastrointestinal transit and lipid assimilation efficiency experiments, as was found between petrels, albatrosses and penguins (Jackson and Place, 1990).

### 2.9.4 Selection for energy and nutrients

If, for instance, seasonal frugivory of carnivorous/insectivorous migrants is mainly due to economical reasons, the 'easy prey' hypothesis raises the question: to what extent are migrants capable of selecting optimal food within food classes? As mentioned above, no obvious general correlations between fruit choice and characteristics, like macro-nutrients, have been found, and thus the question is open to debate (Gardiazabal, 1990). On the other hand, it is well known that geese, for instance, are capable of feeding selectively on various plant species during migration, depending upon their particular nutrient require-ments (Drent *et al.*, 1981). Again, Bairlein (1990) has carried out a crucial experiment. He held garden warblers on a semi-synthetic diet where all food variables were kept constant and only nutritional composition was varied (Figure 2.37). The results showed that birds

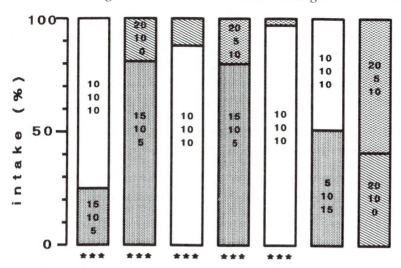

**Figure 2.37** Average food selection by six captive garden warblers feeding on isocaloric diets differing in nutritional composition. The figures within the bars show (from top) % protein, % fat and % nonstructural carbohydrates, respectively, in the wet diets (50% water content). Stars below the bars indicate statistically different selection. (Source: Bairlein, 1990.)

can sense the energy resources on the nutrient level. In general, the birds preferred the diets with high fat levels. Among diets similar in fat content choice was determined more by sugar content than protein content. Trials with isocaloric diets also showed that garden warblers are able to feed selectively on foods differing in dietary nutrients independent of the caloric value. Similarly, captive dark-eyed juncos, induced into premigratory disposition by long photoperiods, chose a diet higher in fats and proteins than did conspecifics exposed to short photoperiods, or house sparrows as non-migrants (Fitzgerald, 1990). Moreover, garden warblers provided with two diets simultaneously, identical in the gross lipid content but differing in the fatty acid composition of the lipids, exhibited obvious preferences with a tendency to choose foods rich in C-18 unsaturated fatty acids. These are the predominant fatty acids in the birds' lipid deposits (section 2.7.2; Bairlein, 1991b). Mallards (*Anas platyrhynchos*) may optimize fat storage for wintering and migration by eating foods such as acorns and crustaceans that yield large quantities in C-18 and C-16 fatty acids (Heitmeyer and Fredrickson, 1990). However, it has to be borne in mind that other species may behave differently (e.g. the white-throated sparrow; Borowicz, 1988). In this species Odum and Major (1956) found that lipid levels of the diet had little effect on the rate or final level of fat deposition in caged individuals. Thus, finding generalizations may be complicated due to influences of a variety of factors (for detailed discussions see Berthold, 1976; Bairlein, 1990).

There is little evidence for decreased rates of standard metabolism in several bird species tested (Blem, 1980). However, Pohl (1971) found a lower rate of oxygen consumption and thus obviously a reduced metabolic rate in bramblings caged outdoors during the migratory period. Finally, lowering body temperature and becoming torpid may be used to save migratory lipid stores in some birds. Carpenter and Hixon (1988) found that migrant hummingbirds may go into torpor even when very fat and may also improve their energetic basis for migration in this way. Correspondingly, torpor could also protect migratory fat depots in swallows which can become torpid in critical situations (e.g. Schüz et al., 1971).

## 2.10 THERMOREGULATION AND WATER BALANCE

So far, records of body temperature of actually migrating birds are not available. However, it is well known for many bird species that there is a substantial increase in body temperature during flight in the magnitude of 1–4°C. Most of the temperatures of birds measured during active flight are greater than 41°C (Gessaman, 1990). Hyperthermia during flapping flight is even found at low ambient temperatures, which may improve muscle efficiency and increase maximal work output (Butler and Woakes, 1990). In detailed studies with domestic pigeons during long wind-tunnel flights, Nachtigall and Hirth (1990) found that mean core temperatures increased in the first minutes of flights by 1.5–3.0°C and remained at this higher level. Temperature further increased with flight speed, body mass and ambient temperature. In flying domestic pigeons and white-necked ravens (*Corvus cryptoleucus*) body temperatures of up to 44.5–44.8°C have been measured (Gessaman, 1990), and in small birds up to 46–47°C. The latter values are higher than the mean levels for highly active birds (43.85 ± 0.94°C, *n* = 74; Prinzinger et al., 1991) and are close to the lower limit of lethal values which range from about 46–48°C (Prinzinger, 1990; Martin, 1987). Gessaman (1990) has reported the first measurements of body temperatures in birds trapped during migration. Sharp-shinned and Cooper's hawks (*Accipiter striatus* and *A. cooperi*) netted on a mountain ridgetop had average body temperatures of between 40.2 and 40.6°C. In these high-altitude soaring migrants, however, a relatively low ambient temperature and low body temperature would have been expected. Taking all evidence on body temperature measurements during bird flight together, overheating may theoretically become a problem for migrants at least during extended continuous flights with high fat loads under high ambient temperatures, e.g. when crossing deserts. Birds during normal sustained flight must be able to dissipate more than eight times as much heat as during rest in order not to be overheated (Giardina *et al.*, 1990). Thus, heat dissipation may be of crucial importance for migrants under certain circumstances. Heat dissipation, however, is closely linked to water metabolism so that the problem of overheating may also be one of dehydration. On the other hand, migrants flying at heights above 5000 m (section 2.12) may spend hours in ambient temperatures in the region of −50°C (Elkins, 1988a) and may then have to counterbalance cold stress and hypothermia.

Birds aloft are able to dissipate large proportions of metabolic heat by radiation and convection. Herring gulls, for instance, can lose up to 80% of their total heat production during flight through their feet, and at successively higher ambient temperatures; for example, flying domestic pigeons let their legs hang down using them as coolers (Butler and Woakes, 1990; Nachtigall, 1990). In the domestic pigeon, heat loss from legs and feet could account for 50–65% of the total heat produced during flight. Thus, the legs and feet can play a major role in whole-body thermoregulation, both at rest and during flight (Martineau and Larochelle, 1988). According to recent studies on domestic pigeons by Craig and Larochelle (1991), the contribution of the wings to heat dissipation during flight may not be nearly as important as has been formerly supposed. Among other mechanisms, a peripheral thermo-receptor in the head, which monitors the temperature of the inspired air, may cause metabolic rate compensations in the domestic pigeon (Bech *et al.*, 1988). According to experiments in a few species, by Giardina *et al.* (1990) and Clementi *et al.* (1991), there is evidence for the existence of a molecular mechanism by which haemoglobin is not only used for various ways of oxygen transport (section 2.11) but also for heat dissipation. At higher ambient temperatures, however, the proportion of the heat produced that can be dissipated by non-evaporative mechanisms (convection and radiation) declines according to simple physical laws so that evaporative mechanisms (mainly water content of the air exhaled) must be employed to a great extent (Nachtigall, 1990). The rate of evaporation can be increased by raising breathing rate and by panting or fluttering the throat pouch. In the white-necked raven, for instance, preflight breathing rates of about 40 breaths $min^{-1}$ gave way to rates of about 180 breaths $min^{-1}$ nearly at the instant of take-off (Martin, 1987; Bernstein, 1989). However, despite these mechanisms, even at an ambient temperature of 30°C, respiratory evaporation has been found to account only for about 20–30% of the total heat loss in the species investigated (Butler and Woakes, 1990). Evaporation thus rarely dissipates more than half of the total heat production. None the less, excessive evaporation may easily lead to dehydration and thus become dangerous. Evaporative water loss, however, can be balanced, at least to some extent, by metabolic water production when fat is used as a fuel, so that under ideal environmental conditions during flight, metabolic water production may equal water loss due to evaporative cooling and water homeostasis can be maintained. The use of fat during flight in birds produces a metabolic energy of about 38–39 kJ (section 2.1), and slightly more than 1 g metabolic water per gram. Nachtigall (1990) found that domestic pigeons early in flight used carbohydrates as fuel, but over the first hour they gradually switched to fat fuel and were then able to use that high amount of metabolic water. However, at moderate to high ambient temperatures water is lost at a greater rate than its metabolic production. In European starlings, Torre-Bueno (1978) found that water loss and production is balanced only at ambient temperatures below 7°C, and for the domestic pigeon, Nachtigall (1990) found the same result at ≤5°C. During migration, birds could actually attain water homeostasis by ascending to altitudes where the air is cool enough to keep evaporative heat dissipation at the required level, or by flying at night using lower temperatures. To some extent, evaporative water loss may be reduced by countercurrent heat exchangers within the nasal passages of some bird species (Blem, 1980). In order to attain water homeostasis, migrants should seek fairly ideal conditions according to the following calculation by Bairlein (1995): per g of body water 2.4 kJ of heat can be dissipated, i.e. 2.54 kJ through 1.06 g water obtained per g of metabolized fat. Because burning 1 g of fat results in 29 kJ of heat, heat dissipation through the endogenously produced body water accounts only for about 9% of the metabolic heat production.

Biebach (1990, 1991) and Bairlein and Totzke (1992) gathered relevant data in the Sahara desert and analysed dehydration versus fuel exhaustion in passerines in detail (Figure 2.38). Their main results are that water content of small migratory birds is in general about 67% of the fat-free weight (see also section 2.7.2), and the tolerance to dehydration is about 5–10% below this value (Bairlein, 1995). Measurements of individuals trapped in the Sahara yielded that, in general, water content of migrants resting in the desert is in the range of hydrated birds, and circulating levels of blood components like $Na^+$ and UREA as indicators for the state of body hydration increased only in lean birds. Even dying small passerines, like willow warblers, investigated by Biebach (1991) in the Libyan desert showed normal water budgets but had run out of fat. Thus, crossing the Sahara even by a non-stop flight should be possible with a balanced water budget, if an altitude with an air temperature of about 10°C is obtained. This, however, may be difficult during the day in the autumn migratory period. According to most recent flight-range estimates (Biebach, 1991), many passerines

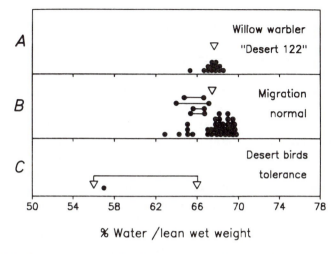

**Figure 2.38** Water content and tolerance to dehydration. Triangles are means. A, circles represent individual values of willow warblers resting in the Libyan desert; B, circles represent mean values of small migratory bird species. Connected circles are water contents of lean (left) and fat (right) birds of the same species; C, circle represents the mean value of reed warblers preparing to cross the desert, arrows indicate normal water content (right arrow) and lower level of tolerance against dehydration (left arrow) in resident desert birds. (Source: Biebach, 1990.)

may need tail winds for a successful crossing. In autumn northerly (tail) winds prevail over the Sahara from ground level up to about 1000 m, whereas winds higher up are more westerly to southerly. Thus, unfavourable wind conditions might be the reason for the interruption of flight during the day and the evolution of an intermittent strategy to cross the desert (section 2.16). Carmi *et al.* (1992) concluded, from a computer-simulation model, that dehydration rather than energy will limit flight duration in migrants. The model predicted that small passerines cannot cross the Sahara in a non-stop flight, but should confine flying to the cooler hours at night and rest during the day.

Birds stopping in the desert during the day are, however, confronted with high air temperatures and low humidity and thus may also be faced with problems of water balance. According to detailed measurements and estimates on willow warblers, Biebach (1990) proposed that assuming a balanced water budget during flight in the night, the water content would remain above critical values of dehydration even if two stopover days and three nights of flight were necessary to cross the Sahara. This also assumed, however, sufficient fat deposits. Carmi *et al.* (1993) found, in a number of bird species, that dehydrated individuals were excellent plasma volume conservers, comparable to the most xeric-adapted mammal species. This capability may greatly enhance their dehydration tolerance during both flight and stopovers in arid areas.

In summary, migrants appear to be capable of avoiding both over-heating and dehydration, even in critical environmental conditions, via a number of behavioural adaptations. Hence, the main limiting factor for long-distance migration appears to be fuel. Because wind significantly increases heat loss from birds, as experiments with dark-eyed juncos by Bakken *et al.* (1991) have shown, the question arises whether migrants may favour specific wind conditions also with respect to thermoregulation. For specific mechanisms to conserve heat during flight at extreme altitudes, and to avoid hypothermia, see the next section. Another potential limiting factor of long-distance migration, which has not so far been considered seriously, is muscle fatigue – it may be worth taking this into account.

## 2.11 ADAPTATIONS TO HIGH ALTITUDE

As mentioned in section 1.1, birds like a Rüppell's griffon vulture have been observed up to altitudes of 11 300 m, and griffon vultures (*Gyps fulvus*) are able to soar to altitudes of more than 10 000 m (Bögel, 1990). Barheaded geese (*Eulabaia indica*) routinely fly over the Himalayan mountains at 8000–9000 m, and swans have been recorded during a transoceanic flight (off the west coast of Scotland) at an altitude of 8000–8500 m (Stewart, 1978). These high-altitude capacities necessitate certain physiological adaptations. Tests have shown that bar-headed geese can remain conscious and stand erect in hypobaric chambers under simulated high-altitude conditions of slightly over 12 000 m (12 190 m) (Black and Tenney, 1980). This altitude far exceeds the limits of tolerance of mammals (perhaps with the exception of bats; Fedde, 1990). House sparrows are alert and active and exhibit normal behaviour at an altitude of 6100 m, whereas mice, in contrast, exposed to the same altitude are comatose. Four species of buntings and finches, tested in conditions simulating altitudes up to 10 000 m, also showed high tolerance to hypoxia (Novoa *et al.*, 1991). Although humans have reached the summit of Mt Everest without supplemental oxygen, it is clear that they do so with great difficulty and are at the limit of human performance. The demands on birds are, however, not only extreme, they also occur with only short periods of adaptation. Bar-headed geese, for instance, may begin their migration near sea level and reach altitudes of about 9000 m in a few hours, allowing little time for acclimatization (Faraci, 1991). Under such conditions these birds

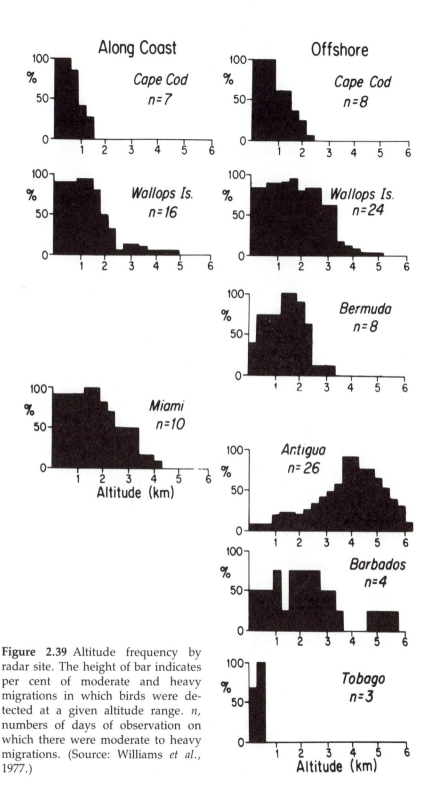

**Figure 2.39** Altitude frequency by radar site. The height of bar indicates per cent of moderate and heavy migrations in which birds were detected at a given altitude range. *n*, numbers of days of observation on which there were moderate to heavy migrations. (Source: Williams *et al.*, 1977.)

must be equipped with special mechanisms that prevent hypoxia and altitudal sickness. These mechanisms may not be specific migratory traits, as shown in data on Rüppell's griffon vulture and the house sparrow. A recent comparative study of four passerines from low and high altitudes, by Novoa *et al.* (1991), actually concluded that high resistance to hypoxia may be a common feature shared by all flying birds. It is presumably related to their very efficient respiratory system, and not an adaptive trait restricted to high-altitude species. However, as will be shown, there are specific adaptations in species residing permanently at high altitudes or performing extreme high-altitude flights.

To begin with one should look at the constraints of high altitude. At 9000 m, which is regularly reached by migrating bar-headed geese, the partial pressure of oxygen in air entering the lung is only about one-third of that at sea level. On the other hand, it is known from studies of freely flying birds in wind-tunnels that oxygen consumption may increase 15-fold during flight. An important question then is: how species, such as bar-headed geese, maintain adequate oxygen transport while performing such vigorous exercise during severe hypoxia? (Faraci, 1991.)

Over the past 10 years, the phenomenon of unusual hypoxic tolerance of birds has been investigated in considerable detail. In effect, all of the organ systems involved in the pathway of oxygen transfer from atmosphere to the mitochondria have been studied (Hiebl and Braunitzer, 1988; Butler and Woakes, 1990; Fedde, 1990; Faraci, 1991). Although there have not been physiological studies on birds flying at high altitudes (real or simulated; Butler and Woakes, 1990; Faraci, 1991), it is now fairly well understood which adaptations birds would allow to cope with such a hostile environment. According to Faraci (1991), a major adaptation in birds is their ability to tolerate extreme hypocapnia (a reduction of arterial carbon dioxide partial pressure due to ventilation; see below). This ability appears to be related to unique aspects of regulation of the cerebral circulation and oxygen delivery to the brain. Other important adaptive mechanisms involve gas exchange efficiency of the lung, oxygen-binding properties of haemoglobin, enhanced cardiovascular performance and cellular adaptations. Part of these adaptations have already been treated in section 1.6, but they will be considered in more detail.

### 2.11.1 Pulmonary adaptations

The avian parabronchial lung has a higher efficiency for gas exchange than the mammalian alveolar lung because of the cross-current arrangement of the convective gas flow through the parabronchi and the returning mixed venous blood (Fedde, 1990). When exposed to severe hypoxia, in response to hypoxemia, domestic chicks and bar-headed geese react with pronounced hyperventilation. Under these circumstances gas exchange in the parabronchial lung is nearly perfect. Pulmonary $O_2$ extraction is improved by an increased $O_2$ flow, but hyperventilation also shifts blood acidity, favouring $O_2$ loading, and compensates for metabolic acid production (Bernstein, 1989). It is true that increased ventilation reduces the oxygen partial pressure difference between the inspired gas and arterial blood, but this is partly made up for by the fact that the diffusing capacity of the avian lung membrane is quite pronounced, even compared with mammals (Fedde, 1990; Faraci, 1991). Observations on hummingbirds during hovering, and on other birds during running and moderate hypoxia, suggest that exercise increases the oxygen-diffusing capacity. The mechanisms that produce this apparent increase in gas exchange efficiency are unknown, but they may involve a decrease in inequality of ventilation-perfusion. Temperature may also play a role. Extreme cold, as experienced by birds migrating in high altitudes (e.g. $-30°C$ at about 7000 m; Bernstein, 1989), increases the efficiency of gas exchange in the avian lung by a mechanism that may involve a decrease in ventilatory shunt, or an increase in oxygen-diffusing capacity (Faraci, 1991). At the low temperatures found at high altitudes it is probably in part advantageous for migratory birds to conserve heat, even during flight. Therefore, hyperventilation is probably accompanied by a reduction in exhaled-air temperature by specific mechanisms (Bernstein, 1989).

While birds perform high ventilation during hypoxemia, they actually lose large amounts of carbon dioxide from the body and become hypocapnic and alkalotic. In humans, such a loss results in disorientation and incapacitation, whereas in birds mental and physical activities appear to be uneffected (for possible mechanisms see below). Repeatedly, the question of synchrony of wing-beats, and inhaling and exhaling of flying birds has been treated (section 2.13). In European starlings it was found that they do not inhale and exhale with each wing-beat. Rather, wing-beat and breathing in this species are essentially mechanically independent (Banzett *et al.*, 1992). Cineradiographs, however, have established that furcular and sternal movements are synchronized 1:1 with wing-beat. Goslow *et al.* (1990) hypothesize that the sternum may act as a pump between the air sacs and the lungs that might serve the increased metabolic demands of flight.

### 2.11.2 Cardiovascular adaptations

The cardiovascular system and blood also have very important adaptations for the conditions of high altitude. Generally, birds have relatively large hearts, a large stroke volume and thus a large cardiac output, particularly when compared to similar-sized mammals. The size of the avian heart suggests that it can produce large increases in cardiac output during hypoxia. This has been shown during rest periods but remains to be examined in flying birds. During hypoxia, in a number of species, blood flow is distributed away from several organs, like the gastrointestinal tract. In the bar-headed goose, however, which maintains a comparably high arterial oxygen content during severe hypoxia, blood flow remains more or less unchanged to these organs, and regional oxygen delivery to tissues is maintained at normoxic levels (Fedde, 1990; Faraci, 1991). In contrast to other birds (below) and mammals, bar-headed geese do not increase red blood cells nor change haematocrit or viscosity of their blood under hypoxic high-altitude conditions. Thus, their heart is not subjected to a viscosity-induced workload increase under these conditions. It is still, however, unknown wether the lack of polycythemia (increase of erythrocytes, haematocrit, etc.) in this goose is attributable to a blunted erythropoietin response to hypoxia or some other mechanisms. Presently, it is known that the lack of pulmonary hypertension and polycythemia during hypoxia allows the heart to pump blood more efficiently during flight (Fedde, 1990). In mammals, hypocapnia constricts cerebral blood vessels even during normoxia and causes a multitude of behavioural disturbances. This is not the case in birds. Moreover, hypoxia causes a greater increase in cerebral blood flow in birds than in mammals. Thus, even during severe hypocapnic hypoxia, cerebral oxygen delivery in birds is maintained at normal levels or even increases. One sees here a unique regulatory mechanism that may essentially contribute to the exceptional tolerance of birds to extreme altitude which, however, has not yet been analysed in detail. The rete opthalmicum as a site of countercurrent heat exchange may possibly also function as a site of gas exchange, and may essentially be involved in providing enhanced oxygen delivery to the brain during hypoxia (Bernstein, 1989; Faraci, 1991). In mammals and several bird species, the pulmonary arteries normally constrict in low-oxygen environments. This is not the case in the bar-headed goose. The mechanism preventing hypoxic pulmonary vasoconstriction is not known, but this further adaptation helps the species to alleviate workload on the heart (Fedde, 1990).

### 2.11.3 Blood adaptations

The oxygen affinity characteristics of the haemoglobin molecules represent another very important mechanism enabling birds to perform at high altitudes. Species that are well-adapted to high altitudes as permanent residents, like the Andean goose (*Chleophaga melatoptera*), or as occasional migrants possess haemoglobin with a relatively high oxygen affinity. In the Andean goose, the affinity of whole blood is elevated. In the bar-headed goose and Rüppell's griffon vulture, haemoglobin polymorphism provides a two-stage and a three-stage cascade of haemoglobins with graded oxygen affinities. These different types of haemoglobin molecules show different oxygen-binding capacities, which are adaptations for differing partial pressures of oxygen at different altitudes. When exposed to transitory hypoxic stress, the species mentioned are thus equipped with a portion of specific high-altitude blood. The different oxygen affinities of the single haemoglobins within the cascades are due to amino-acid substitutions. Differences in the amino acids at key locations on both the $\alpha$- and the $\beta$-chains of haemoglobin molecules enhance the number of binding sites that can be filled (Hiebl and Braunitzer, 1988; Braunitzer and Hiebl, 1989; Fedde, 1990; Faraci, 1991). In addition to the intrinsic oxygen affinity characteristics of the haemoglobin molecule, other factors that modulate the position of the oxyhaemoglobin dissociation curve also improve the affinity for oxygen. This is achieved by a marked decrease in arterial $P_{CO_2}$, an increase in arterial pH and respiratory alkalosis and perhaps a blood temperature decrease (which remains to be demonstrated) (Faraci, 1991).

In a number of studies elevated haematocrit and haemoglobin values during the migratory period were reported (Bruns and Ten Thoren, 1990; Bairlein and Totzke, 1992). They may indicate increased oxygen carrying capacity, as in winter-acclimatized dark-eyed juncos (Swanson, 1990). However, more studies are necessary to clarify the significance of haematocrit values, and also the rapid coagulation of avian blood has to be carefully considered (Bairlein and Totzke, 1992).

### 2.11.4 Muscle and cellular adaptations

As already outlined in section 1.6, avian muscles and cells have a number of specific capacities that facilitate migratory flights in general and high-altitude flights in particular. Generally, avian muscle has a high oxidative capacity. This is achieved by a high blood capillary density (relative to mammals) in skeletal and cardiac muscle as well as in brain tissue. This decreases the diffusion distance for oxygen. Further, the pectoralis as the principal flight muscle consists primarily of small fibres with comparatively high oxidative enzyme activities and a high density of large mitochondria. Avian cardiac muscle also has a high density of mitochondria. Finally, high myoglobin concentrations facilitate the diffusion of oxygen since the majority of oxygen that is consumed in cardiac and red skeleton muscle is transported to mitochondria by myoglobin. In the bar-headed goose, the myoglobin concentration in the pectoralis is elevated compared to species adapted to lower altitudes (Fedde, 1990; Faraci, 1991). To conclude, however, it is not clear to what extent genetic and environmental factors interact in these adaptations. It is well known, for instance, that capillary density in skeletal muscle, concentration of myoglobin and the activity of the oxidative enzyme citrate synthease can increase in response to exercise, physical conditioning or hypoxia. Lastly, apart from the specializations mentioned there may be others that also facilitate the oxygen transfer within tissues, and may be of importance in high-altitude flight adaptation (Fedde, 1990). More essential details may become available when satellite-telemetry will produce physiological measurements during natural free flights at high altitudes which so far are lacking (Bernstein, 1989). This section can be concluded with an 'approximation to a definition' of a high-altitude bird (or mammal) recently given by Monge and León-Velarde (1991): a high $O_2$-haemoglobin affinity, a moderate or absent polycythemic response, a low venous $O_2$ pressure, a thin-walled pulmonary vascular tree that responds to hypoxia with moderate increase in the pulmonary arterial pressure, and the absence of chronic mountain (altitude) sickness. For the many more physiological details already known see the reviews of Hiebl and Braunitzer (1988), Bernstein (1989), Butler and Woakes (1990), Fedde (1990), Monge and León-Velarde (1991) and Faraci (1991).

## 2.12 CONTROL OF FLIGHT ALTITUDES

Bird migration normally occurs at all altitudes from sea level to about 8000–9000 m and at times perhaps even greater heights (sections 1.1 and 2.10). At the lower end, migrants crossing rough seas often fly in wave troughs almost in contact with the water surface in order to minimize wind effects. At the other end, migrants which have to overcome mountains like the Himalayas are forced to ascend to altitudes of up to 8000–9000 m. One can also find these extremes in the absence of such obvious environmental barriers. For instance, as mentioned in section 1.1, swans have been recorded at about 8200 m on their route from Iceland to Ireland (Stewart, 1978; Elkins, 1988a), or as Williams *et al.* (1977) reported migratory altitudes of over 6000 m (Figure 2.39) in the North Atlantic. Also, passerines have been recorded as high as 6800 m (Martin, 1990).

Clearly principal bird migration occurs between these extremes. Systematic sight observations and, above all, radar studies have given fairly good information on the altitudinal distribution of migrants over continental and coastal America and Europe and some other areas. Although still fairly little is known about the altitudinal distribution of single species, especially in nocturnal migrants, a number of regular patterns have emerged. For continental migration, examples are treated from Central Europe since there are no large waterbodies comparable to the Great Lakes or the Gulf of Mexico in North America to complicate interpretation. For transoceanic migration, examples from the North Atlantic and the Gulf of Mexico are treated. These are by far the most intensively investigated areas. Unfortunately, almost nothing is known at what altitude the world's largest desert – the Sahara – is overcome, although preferred altitudes there would be of greatest interest for ecophysiological considerations (section 2.16).

In Central Europe, altitudinal distribution of migrants has been systematically studied in prealpine and alpine areas by Bruderer (1971), Bruderer and Jenni (1990) and Liechti (1992), in The Netherlands by Buurma *et al.* (1986), in northern Germany by Jellmann (1989), and additional data are available from recent studies in southern and central Germany (Bruderer and Liechti, 1990). These studies demonstrate, like many others (Elkins, 1988b), that most migratory bird flight occurs below 3000 m. In detail, Bruderer (1971) found median altitudes of 700 m for nocturnal migration and of 400 m for diurnal migration during spring migration. The upper 90% confidence interval was approximately 2000 m for both periods (Figure 2.40). The observed maximum heights were about 4000 m in spring and 4500 m in autumn. Values obtained by Jellmann (1989) were similar. In nocturnal spring migration the median height was 910 m, and in early autumn migration 430 m; with regard to the distribution, 33 and 14% of all tracks were found above 1000 m, 16 and 6.5% above 2000 m, and 3.5 and 1.5% above 3000 m, respectively. Differences between the prealpine area and northern Germany, as well as between periods in the same area,

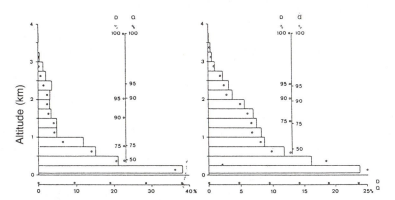

**Figure 2.40** Altitude frequency of migrants during spring migration in Switzerland. Left, diurnal migration; right, nocturnal migration. (Source: Bruderer, 1971.)

may be partly due to differences in species composition, for example different proportions of waders. There are several indications that larger birds, and especially fast, powered flyers like waterfowl and shorebirds, are more likely to fly at higher altitudes where winds are stronger, temperatures are cooler and less turbulence may be present (Bruderer, 1971; Kerlinger and Moore, 1989; Gauthreaux, 1991). In the study of altitudinal aspects of daytime (broad-front migration over East Netherlands), Buurma *et al.* (1986) found that only 17% of 359 radar records of flocks of migrants occurred above 200 m. Although this low level migration is typical for diurnal migration over bare country the proportion of birds over thrush size still increased with altitude. From these data it was proposed that larger birds were hindered less by head winds at higher altitudes. However, in case of unfavourable winds, even waders may fly low over the sea surface when departing from Iceland, whereas under favourable conditions they climb steeply up to altitudes of 600–2000 m above sea level (Alerstam *et al.*, 1990).

Gauthreaux (1991) has studied the altitudinal distributions of migrants arriving on the north coast of the Gulf of Mexico in detail. In a series of radar studies median values varied between altitudes of 1000 m increasing up to 1500–2000 m. The nocturnal migration of most song-birds usually occurred at altitudes below 500 m. Williams *et al.* (1977) used a network of radar along the eastern coast of North America, Bermuda, the Caribbean Islands and on ships, to investigate movements of birds over the North Atlantic Ocean during the autumn migratory period (Figure 2.39). The average altitude of migration increased *en route* to a maximum of 4000–6000 m over Antigua and then decreased abruptly near the South American coast. High-altitude flight over the Caribbean was assumed to offer clear advantages in terms of reduced head winds and avoidance of low-level turbulence (Elkins, 1988b) compared with flight at lower altitudes.

Comparably little is known about the altitudinal strategies of migrants when crossing large deserts like the Sahara. Dolnik and Bolshakov (1985) used moon-watching techniques in order to estimate the altitudinal distribution of migrants over the Karakum and the Kizilkum deserts in Asia. The ranges of altitudes with the greatest density of migration were 300–950 m above ground level. This relatively low height was partly biased by descending migrants and the inclusion of birds just beginning migration. In the Tien Shan, and especially in the neighbouring Pamir, much higher flight levels were found according to moon-watching (Figure 2.41; Dolnik, 1990). Schaefer (in Moreau, 1972) found that the bulk of migrants in the area of the western Sahara were moving between about 1500 and 2500 m. Some were found higher, over 3000 m, where they possibly avoid lethal sandstorms (Elkins, 1988b). For some recent observations see section 2.16.

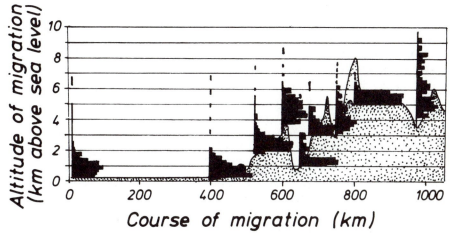

**Figure 2.41** Frequency distributions (filled histograms) of the altitude of nocturnal migration above the Tien Shan and Pamirs deserts in a section along latitude 72° E. (Source: Dolnik, 1990.)

Gauthreaux (1991) found considerable day-to-day variance in the altitudinal distributions of migrants arriving in the Gulf of Mexico area. It was possible to demonstrate from these studies that the variance was essentially due to differing flight tactics in response to changing wind fields. A significant correlation existed between the mean migratory altitude and the altitude of most favourable winds. Lower altitudes, however, were still favoured over slightly better conditions at much higher altitudes. Cochran and Kjos (1985) applied radio-telemetry to migrating thrushes (*Catharus*) in North America with the following results. After take-off, thrushes ascend until they find suitable winds. With moderate wind, they minimize lateral drift by adjusting altitude to conditions with favourable winds. With less wind they accept lateral drift. If winds at all altitudes above about 75 m have unfavourable head or side components the birds simply do not fly. This indicates that long-distance migratory strategy often involves a constant-heading behaviour where wind drift is mitigated by altitudinal change rather than by compensation for wind. When an obstacle like the Alps has to be overcome, Liechti and Bruderer (1986) suggested that the flight strategy consists of optimizing between minimum climb and minimum deviation in direction. Liechti (1992) found that high-flying birds tend to maintain their heading, which in fact often results in drifted tracks, whereas lower-flying birds tend to turn their headings in order to compensate for drift.

As Elkins (1988b) and Gauthreaux (1991) pointed out, there can be little doubt that the variance in the altitudes of migratory flights is related to atmospheric dynamics, especially wind speed and direction, but also cloud thickness, moisture content and others. Elkins (1988b) summarized that most species tend to fly higher in favourable winds and clear skies, but at low levels in strong head winds, low cloud base and heavy precipitation – if indeed they fly at all (section 2.17). A migrant may also seek a level at which its heat and water budget are in equilibrium (section 2.10), and perhaps at which it can avoid icing. Flying below the freezing level may be especially appropriate when in cloud (Elkins, 1988b). However, little is known about how birds are able to find the optimal flight level. The method by which a bird detects its altitude is generally assumed to be by the use of the ear in sensing atmospheric pressure changes. This question was addressed in more detail by Elkins (1988b). During the take-off climb, a migrant will normally be able to monitor changes in wind velocity with references to the topography beneath. If it is necessary to continue climbing, he proposed that high-altitude migrants could recognize favourable wind flows by various means: (1) sensing vertical wind shear (VWS), i.e. the change in wind velocity and direction with height; (2) recognizing the distinctive cloud formations associated with VWS and jet streams; (3) detecting the infra-sounds made by jet streams (as proposed by Kreithen, 1978, as a means of orientation; see also section 3.4); (4) using an 'accelerometer' to sense accelerating air flows. Such an accelerometer in the avian body has not been demonstrated but has been discussed with respect to sensing wind parameters as well as to inertial navigation (Elkins, 1988b; section 3.4).

What migrants do over extended deserts is still unclear and may show us other fascinating adaptations. It may well be that high altitudes are selected in order to overcome the obstacle quickly and safely in addition to intermittent migration strategies now established (section 2.16).

## 2.13 GENERAL BEHAVIOURAL ASPECTS

Attaining a migratory disposition and the specific activities during migratory period are not only characterized by pronounced physiological adaptations but also by behavioural changes. The most obvious patterns will be treated in some detail.

### 2.13.1 Diurnal patterns of locomotor activity and foraging behaviour

The most obvious change in diurnal patterns of locomotor activity in migrants is the development of nocturnal migratory activity. This occurs in many bird species which are exclusively daytime-active outside the migratory season (section 2.1). Diurnal patterns of locomotor activity are also changed in several other ways. In general, many migrants tend to reduce the amount of daytime activity during periods of migratory flights. For instance, Bergman (1941) described the appearance of a characteristic pause ('einschlafpause') in the evening before nights with nocturnal migratory restlessness in captive migrants. Later this was shown to be accompanied by a cessation of feeding (anorexia, spontaneous reduction of food intake; Berthold, 1975; le Maho, 1990). Such a 'flightless period' before nocturnal migrants typically depart, between 30 and 45 min after sunset, is thought to be used for decision making as to whether to migrate and in which direction (Martin, 1990). It is this period when sunset cues may be used for initial orientation (section 3.3). Farner (1955) found a general tendency to sleep more lightly in birds in a migratory disposition. Gardiazabal (1990) observed that individual garden warblers and pied flycatchers slept during daytime in resting areas in Spain. For a considerable number of bird species it has been shown that caged individuals reduce their daytime locomotor activity during periods of pronounced nocturnal restlessness (Merkel, 1958; Berthold *et al.*, 1990b; Figure 2.42). In the field, birds crossing the Sahara desert with sufficiently large fat deposits for further stages regularly rest motionless for the whole day in the shade (section 2.16). In experiments with garden warblers, Gwinner *et al.* (1988) found that when nocturnal activity after food deprivation increased, the duration of diurnal activity decreased, whereas its intensity increased.

Another typical characteristic for many migrants, along with the development of nocturnal migratory activity, is the disappearance of a second peak of locomotor activity in the late afternoon. This shift from a bigeminus to a single-peaked diurnal activity pattern has been

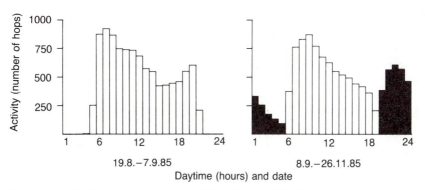

**Figure 2.42** Daily activity patterns of hand-raised blackcaps. Left, before the autumn migratory period; right, during the autumn migratory period, with nocturnal migratory activity (black columns). (Source: Brensing, 1989.)

proposed as a mechanism to allow for the development of nocturnal migratory activity by uncoupling several circadian oscillators on a hormonal basis (section 2.2). This characteristic change to a one-peaked pattern is regularly found both in the laboratory and in the field (Brensing, 1989; Figures 2.42–2.44). It is typical for individuals with sufficient fat deposits, whereas lean birds may behave differently. They often maintain the second peak of activity or even show pronounced hyperactivity, which under extreme conditions may lead to a box-shaped or multi-peaked activity pattern with high activity levels all day long. Such differences become quite apparent when one compares 'trapping patterns' of songbirds trapped in staging areas. Migratory individuals with high or normal fat deposits trapped only once show essentially one-peaked patterns; retraps, however, on average, less-migratory individuals with lower fat deposits which stay longer, show two-peaked patterns (Figure 2.43). While studying red-eyed vireos (*Vireo olivaceus*) during stopovers, Loria and Moore (1990) and Moore (1991) found the following characteristics in lean birds. They stayed longer (as known for many species; section 2.15). In addition, they moved at higher mean velocities while averaging the same frequency of foraging movements, and they increased their degree of turning after feeding attempts. In effect this broadened their use of microhabitat and expanded their feeding repertoire. The authors suggested that lean individuals adjusted their foraging behaviour to mediate compensatory mass increase. Similarly, mobile groups of wintering dunlins may obtain higher amounts of body fat than resident groups (Ruiz *et al.*, 1989). For relationships between energy demands, feeding and activity patterns see also sections 2.1 and 2.9.

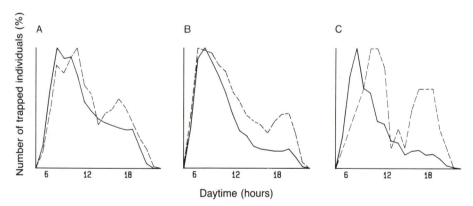

**Figure 2.43** Daily trapping patterns of blackcaps (A), reed warblers (B) and redstarts (C) during the autumn migration period in S Germany. Solid lines, first traps; broken lines, retraps within the season. (Source: Brensing, 1989.)

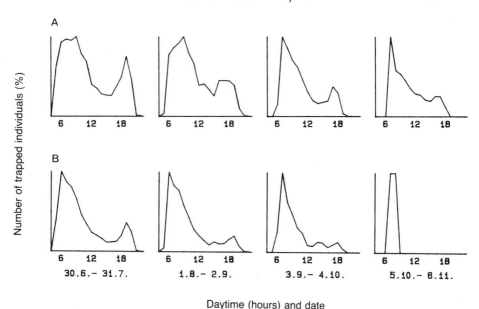

**Figure 2.44** Changes of daily trapping patterns of chiffchaffs (A) and willow warblers (B) during four successive intervals of the autumn migratory period in S Germany. (Source: Brensing, 1989.)

Brensing (1989) showed that species with different breeding habitats but the same diurnal activity patterns, under similar ecological conditions, can exhibit differences in diurnal activity patterns when they use the same habitats for stopover periods during migration. This chronological niche segregation is probably one of the mechanisms which allow co-existence during staging between species (for others, see section 2.21). For wintering shorebirds, day and night habitat segregation has been reported (McNeil, 1991). Gatter and Behrndt (1985) reported that first-year swallows, with 5% shorter wings than adults, have longer daily migration times. This extension is thought to compensate for less favourable flight performance and experience. In contrast, Tsvelykh and Goroshko (1991) assume that the more pointed and narrow wings in young swallows are more adapted to long flights with less energy expenditure and that the flying apparatus of adults is more adapted to manoeuvrability (see, however, also section 1.6). Finally, in diurnal migrants there is considerable interspecific variation in the periods used for actual migration. Many species migrate solely in the early morning hours, others exhibit increased activity before sunrise, migrate late in the morning, extend migration throughout the day or migrate again in the afternoon or evening. These differences have been interpreted as adaptations to feeding ecology, lowering risk of predation or as result of fuel conditions and environmental influences (Dorka, 1966; Dolnik, 1975; Flousek and Smrcek, 1984). Many mainly at night migrating species often show characteristic morning flights within 2–3 hours after sunrise. Wiedner *et al.* (1992) studied these flights in detail in North America and hypothesized three main functions: (1) returning to land from the ocean; (2) compensation for lateral drift incurred during the previous night's migration; and (3) seeking habitat in which to rest and forage.

### 2.13.2 Flocking behaviour

Flocks are a widespread phenomena after the breeding season among birds which are largely solitary or live only in pairs during the breeding period. Such flocking behaviour is also typical for many diverse migrants from small tits or finches to large cranes or swans. Flocks form occasionally when, for instance, birds which exploit thermal or ridge lift use favourable sites at the same time. Normally, however, flocks are formed intentionally, regardless of whether they consist of one species or as mixed flocks of several species [e.g. chaffinches and meadow pipits (*Anthus pratensis*), or Canada geese and sandhill cranes, or different wader species; Orr, 1986; Piersma *et al.*, 1990]. Many species like storks, raptors or, among nocturnal migrants, many shorebirds and waterfowl, tend to always migrate in flocks. In other cases flocks are extremely ephemeral, and many species appear to migrate exclusively singly (Svazas, 1990). Often, flocks may remain together for a whole migratory journey as in many larger species (see below). To what extent groups of passerines migrating at night remain cohesive, as has been suggested for the prothonotary warbler (*Protonotaria citrea*; Moore, 1990), is largely unknown. Svazas (1990) suggested that there is considerable contact between passerines migrating at night, whereas some groups like *Sylvia* and relative species migrate only singly. Piersma *et al.* (1990) described how wader flocks, before long-distance flights, usually assembled from clusters into V-formations or echelons, in most cases intensely vocalized, and usually headed into the wind with climb rates of about $0.6 \, \mathrm{m \, s^{-1}}$.

Kerlinger (1989) discussed the three most reasonable explanations for flocking in migrating hawks. These and several other aspects (Alerstam, 1990) will briefly be considered below.

### (a) The energy saving explanation

A common advantage has been assumed to be energy saving due to specific flight formations in flocks. Many large bird species, like geese, gulls, cranes or cormorants migrate in typical formation flight, e.g. V-formations, wedge, double-V, echelons, simple slant formations, billowing bow-shaped or dense 'clumped' flocks. In all these formations the individuals constantly shift positions, and the leading birds periodically swing out from their position and rejoin the line further back (see below). The birds in a flock can obtain a certain amount of lift for nothing by flying close to each other in the upward-directed air flow around each other's wings (Alerstam, 1990). Hummel (1973) and Hummel and Beukenberg (1989) have analysed hundreds of such formations with respect to reduction in flight power demand (Figure 2.45) and have also carried out relevant flight experiments with airplanes. In general, they found that in formation flight each wing is situated in an upwash field generated by other wings in the formation. This leads to a considerable reduction in flight power demand. The total flight power reduction of the whole formation depends on the

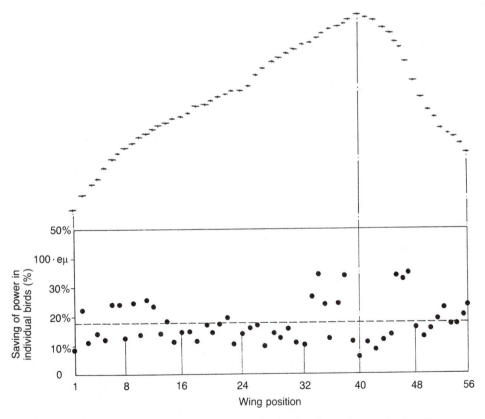

**Figure 2.45** Saving in power per bird in a flock of 56 brent geese. Broken line, mean saving 17.6%. (Source: Hummel, 1973.)

total number of birds and on their lateral distance. The distribution of this power reduction on the individuals involved is a function of the actual shape of the formation. Power reduction reaches a maximum by a certain overlap of the wing-tips of two adjacent birds and by a slight stagger in flight level. It appears that birds 'feel' the energetic benefits and, without disturbances, arrange formations with respect to optimal power reduction for each individual. Since the individual at the apex of a formation has the smallest reduction in flight power demand it regularly changes its position – a phenomenon which was already known to Emperor Frederick II (1964). According to Hummel (1973), the overall reduction in flight power demand in typical formations is on the magnitude of 20%. Theoretically, reductions of more than 50% would, however, be possible in ideal formations, which presumably do not occur. In migrating Canada geese, Hainsworth (1989) has measured details like distance between adjacent birds, distance between wing-tips of adjacent birds, wing-beat frequencies and extreme relative wing positions to test for use of variation in trailing wing-tip vorteces which requires a similarity in wing-beat frequency (Figure 2.46). Only 48% of the individuals were found with frequencies similar to those of the bird ahead. Thus, induced power saving may be limited by unpredictable moves of birds ahead and by an inability to track trailing vortex positions. Kshatriya and Blake (1992) have developed theoretical models of the optimum flock size of birds flying in formation. It turned out that within a myriad of possible flock sizes there is one that is optimal for maximizing energetic efficiency for a given maximum range speed under given conditions. Under certain conditions, such as long non-stop migration, solo flight is an optimum migratory strategy. Alerstam (1990) obtained evidence that waders may 'extract' the aerodynamic gain of flying in flocks in the form of increased speed instead of in the form of reduced transport costs.

Some association during migration may also occur when klepto-parasitic species like Arctic skuas (*Stercocarius parasiticus*) exploit and follow terns during autumn migration as suggested by Wuorinen (1992).

*(b) A cultural transmission explanation*

In many large bird species, like cranes, gannets, geese or storks, families and often much larger groups migrate together. Such birds regularly use traditionally very specific staging areas *en route* and very tiny specific wintering areas. Inexperienced migrants need then be guided by experienced conspecifics. In this way, for instance, young Siberian white cranes (*Grus leucogeranus*) are regularly 'brought' to the Bharatpur Bird Sanctuary in northern India, or whooping cranes (*Grus*

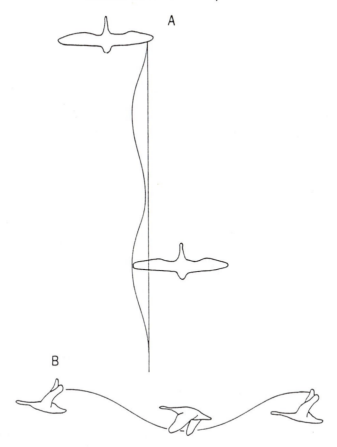

**Figure 2.46** Schematic representation of predicted wing positions relative to horizontal (A) and vertical (B) variation in trailing wing-tip vortex positions. To keep the inboard wing-tip at the centre of a vortex with variable position requires a greater degree of overlap in wing-tip spacing (A), while tracking vertical position variation requires wing phase variation with distance behind the bird ahead (B). (Source: Hainsworth, 1989.)

*americana*) to their wintering place near Corpus Christi in the Gulf of Mexico. Often, such regularly used areas change, as has been reported for the greylag goose (*Anser anser*; Rutschke, 1990). In such cases cultural transmission can rapidly lead to a new tradition in such long-lived species (see also chapter 5). Hashmi (1990) has discussed to what extent the geographical distribution during migration is influenced by tradition in the gannet (*Morus bassanus*), in which five ageclasses can be visually distinguished. Although tradition may play an important role in determining migration routes, staging areas and wintering places in many large gregarious species, it has to be kept in mind that they still may have genetically preprogrammed migratory components, as has been suggested in white storks (section 3.1).

## (c) The orientation/navigation hypothesis

There is a strong theoretical basis and a number of lines of evidence for the idea that migration in flocks may improve the accuracy of orientation. For instance, flock size of several species tends to be larger under overcast than under clear skies, and contact calling becomes more frequent. For improvement, flock direction would either represent the mean migratory direction of the individuals' directional tendencies, or flock orientation might be determined by the most competent leaders (Kerlinger, 1989; Alerstam, 1990). For nocturnal migrants, it has been proposed that they improve their orientation by responding to flight calls (see below), and possibly the associated Doppler shifts from surrounding conspecifics (Alerstam, 1990). In spite of its attractiveness, this hypothesis has only been experimentally tested in two studies of domestic pigeons, where results were positive (Tamm, 1980) and negative (Keeton, 1970). Badgerow (1988) analysed formations of Canada geese and the results obtained supported, to some extent, the orientation communication hypothesis. More work is certainly required.

## (d) The foraging efficiency hypothesis

This hypothesis has been discussed with respect to birds in detail by Kerlinger (1989). Foraging by groups of organisms, in this case migrants, is assumed to facilitate the location of patchy prey and may, through several degrees of cooperation, improve foraging efficiency. In some species, like small falcons and vultures, roosting flocks are suggested to serve as information centres. Ekman and Hake (1988) have examined how flocking in the greenfinch (*Carduelis chloris*) may reduce starvation risk. In principle, foraging in groups increases the chance to detect patchy, ephemeral and often superabundant food resources. Although the individual that finds the patch will have to share the food, it will, however, benefit in future situations. Hence, migration in flocks may guarantee a more consistent food supply and be an important basis for refuelling. More details on foraging efficiency and 'reduced variance benefit' are given by Szekely *et al.* (1991) in their dynamic model of flocking behaviour in passerines for the non-reproductive season. Increased foraging efficiency may also strongly promote flocking in migrants during wintering, which is often reported (Curry-Lindahl, 1981; Ewert and Askins, 1991). In barnacle geese family goslings were found to grow fatter than unattached goslings because they could feed without interruption for longer periods (Black and Owen, 1989). In lesser snow geese (*Chen caerulescens*) large families were found to dominate small families, which in turn dominated pairs and lone individuals (Gregoire and Davison Ankney, 1990).

### (e) The lowering predation risk hypothesis

Several models predict that flocking can be an effective mechanism to lower predation pressure. The main aspects are summarized by Milinski (1986) as follows. Prey animals in a group have the advantage over singles in that they are able to detect an approaching predator earlier. The information about the predator may then spread quickly across the swarm ('Trafalgar effect'). Individuals in dense groups of prey animals have a lower risk of being detected than individuals spread out widely. Within a group, each member has a decreasing risk of becoming the victim with increasing group size ('dilution effect'). Further, predators often have difficulties in attacking a swarm member successfully because of the 'confusion effect'. Trying to overcome the confusion may increase the predator's own risk of predation. Finally, flocks often have a chance to attack a predator jointly and to drive it away. Lindström (1989) found, in the study of migratory finches, that both attack frequency and hunting success of raptors (which may kill up to about 10% of finches during autumn migration in S Sweden) increased with flock size. The risk for an individual finch showed no correlation with flock size. Thus, patchy distribution of the preferred food rather than predation seemed to be the primary cause for formation of large flocks in finches.

### (f) The thermal location and utilization hypothesis

This hypothesis concerns species that are able to move on by gliding and soaring during migration. It has been discussed in detail by Kerlinger (1989). Migrants which are able to use thermals should avoid resorting to powered flight because it is so costly compared with gliding and soaring (section 2.14). Instead, they should locate thermals. Hypothetically this could be done more efficiently by flocks than by individuals. Kerlinger presents three models: (1) local enhancement; (2) random encounter; and (3) climb-rate feedback. Local enhancement simply means that if a migrant sees another migrant using a thermal, it flies to that thermal. The random encounter model predicts that members of flocks spacing themselves perpendicular to the migratory pathway increase their efficiency to detect useful thermals. The climb-rate feedback model suggests that flock members within thermals may easily obtain information about areas with optimum updraft strength.

The degree of importance of various factors leading to flocking in migrants certainly varies greatly among species and, unfortunately, it has not been clarified for any one species. Hence, there may also be other factors promoting gregarious migration. In swallows, for instance, migration in flocks during periods of inclement weather may facilitate common roosting in torpidity (Prinzinger, 1990) and may in this way favour survival.

### 2.13.3 Characteristic call notes

Some nocturnal migrants have characteristic call notes during migration. Species which regularly form flocks during the day have night migration call notes similar or equal to daytime notes. Other species, however, have very specific calls apparently only used during nocturnal migration (Hamilton, 1962; Dorka, 1966). Hamilton (1962) has tested the significance of such a nocturnal call in caged bobolinks. These calls are also uttered during intensive migratory restlessness. From play-back experiments the author suggests that calls of migrants aloft may induce grounded migrants in appropriate migratory condition to take-off. In many species nocturnal call notes may help to maintain contact in flocks and also to improve orientation, as discussed above and in chapter 3 regarding 'echosounding'. Schwanke and Rutschke (1990) noted structural differences in call notes in bean geese (*Anser fabalis*) under clear skies and mist, respectively. This wide field of largely mysterious behaviours remains a challenge for ecophysiologists.

### 2.13.4 Territoriality *en route*

The establishment of territories during the breeding season is a characteristic for most non-colonial bird species. In addition, many species occupy territories in the wintering areas whereas others are more vagrant during wintering (Debussche and Isenmann, 1984; Schwabl *et al.*, 1991). There is little information on how many species establish temporary territories *en route* during migration. Rappole and Warner (1976) found that northern waterthrushes (*Seiurus noveboracensis*) were strongly territorial in spring migration, and individuals known to be territorial gained weight during stopover periods. To some extent, American warblers also appear to temporarily defend territories where they gain weight (Morse, 1989). Carpenter *et al.* (1983) found that rufous hummingbirds periodically establish and defend territories along their summer southward migration route, very likely in order to optimize body mass gain. Wassmann (1989) reported such feeding and resting territories in the golden oriole. They were established in adult males, can be used for several days and were defended intensively.

The pronounced site faithfulness of many passerines in staging areas (within a migratory period; Bairlein, 1981; Streif, 1991) suggests that many individuals rest in restricted areas which may be considered as temporal territories. This matter needs much more careful consideration. For territoriality during wintering, where migrants may stay in temporary pairs, like stonechats, even females may sing, as in European robins, see sections 2.20 and 2.21. For further behavioural adaptations, e.g. related to feeding, resting or thermoregulation, see sections 2.9, 2.10 and 2.15.

### 2.13.5 Emergency measures

Migrants have developed numerous behavioural adaptations and reactions to cope with the various situations experienced during migration. Often, migrants allow closer or less close approach of humans depending on previous experience, above all the degree of persecution in their breeding and staging areas (e.g. Burger and Gochfeld, 1991). Individuals in migratory disposition can switch from being risk-prone to risk-aversion when maximum fat reserves are deposited (Moore and Simm, 1986). In case of untimely cold snaps or other adverse conditions, birds are astoundingly flexible in foraging. Foliage-gleaning species may then be seen hunting on the ground, insectivorous species may pluck grass seeds (Morse, 1989; section 2.9), etc. There are innumerable occasions in which birds have landed on ships (Elkins, 1988b), and even small birds hitching rides on the backs of larger species have been reported (Mead, 1983). These are probably cases of emergency landings of very tired individuals on such a 'flying island'. In ducks, hunting pressure may possibly lead to premature departure for autumn migration (Jakobsen, 1991). Certainly, to compile a comprehensive review of emergency measures developed by migratory birds would be a promising task.

## 2.14 CONTROL OF THE MODE AND SPEED OF LOCOMOTION

Besides flying, birds use all the other kinds of locomotion they are able to perform in order to cover migratory distances. The speed of migration is most likely shaped by natural selection. Favoured speeds should promote a reasonably fast, energy-efficient yet safe migration (Kerlinger, 1989).

### 2.14.1 Mode of locomotion

Not only the ancient *Hesperornis* perhaps migrated by swimming (section 1.1) but also recent forms, like some marine birds, swim for distances of over 1000 km. Penguins, for instance, are known to cover similar distances combining swimming, walking, hopping and sliding ('snow paddling', since wing-assisted). Otherwise, migration on foot, swimming or diving is only occasionally used to cover migratory distances (Salomonsen, 1967; Schüz *et al.*, 1971; Bastian, 1992; Berthold, 1993). In birds of similar mass, energy consumption during flight is only approximately 2.5 times greater than that when running or swimming at maximum speed (Butler, 1991). Because of its much higher speed, combined with reasonable transport costs, flight is the most attractive locomotion for migration (see below).

It has been repeatedly debated whether or not, and if so to what extent, resting nocturnal migrants may direct their diurnal foraging activity according to their migratory direction, and thus cover some migratory distances while feeding. 'Migration during resting' (Schmidt-Koenig, 1980), i.e. hopping in the migratory direction during foraging trips has been reported to occur regularly in nocturnal migrants on the Kurish point in eastern Europe (Dolnik, 1975) and occasionally for species elsewhere (Bastian, 1992). A systematic study carried out in the scope of the MRI-Program (section 1.3), however, showed that migration during resting either did not occur or was of no significant importance in the nocturnal migrants studied. It appeared, instead, to be advantageous to stay in an appropriate resting area. To leave an area during foraging in order to gain a relatively small distance seems to be too risky. It would be more easily covered during a short nocturnal flight (Bastian and Berthold, 1991; Bastian, 1992).

## 2.14.2 Flight – general aspects and powered flight

The world's heaviest flying birds are presently great bustards (*Otis tarda*) where old males may reach up to 21 kg. Swans have the second-highest average body masses among flying birds followed by other bustards, condors, albatrosses, pelicans and others. Great bustards are just capable of flying a few kilometres. Other very large birds are capable of performing sustained flights but are restricted in their optimal flight speeds. Swans, for instance, are obviously not capable of flying with the power required for maximum range speed (see below; Martin, 1987; Nachtigall, 1987; Alerstam, 1990). Other large birds are forced to glide over large distances (see below). Medium-sized and small birds have the highest flying ability and are capable of performing sustained powered flights for up to 100 hours and over thousands of kilometres (section 1.1). Flight, and especially powered flight (i.e. continuous flapping, bounding or undulating flight, Figure 2.47), is by far the most important method of locomotion and forward drive in migratory birds. Flapping flight, however, is also by far the most

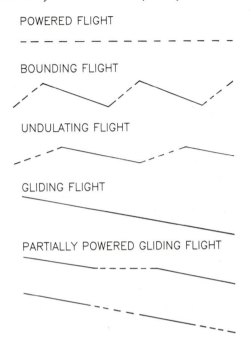

**Figure 2.47** Schematic of various types of avian flight. Solid line, no wing-beats, as in gliding or wings tucked into the body; dashed lines, wing-beats. The angle of a line to the horizontal represents the flight path relative to the air. (Source: Kerlinger, 1989.)

expensive method of locomotion, considering energy cost per unit of time (Alerstam, 1991b). Flying birds, in general, consume 4–20 BMR (basal metabolic rate; Masman and Klaassen, 1987). Minimum power input during flapping flight has been estimated to be generally about 10 times the resting level (Butler and Woakes, 1990), maximum field estimates range up to 30.8 BMR [black-necked grebes (*Podiceps nigricollis*), Jehl, 1994]. In raptors, energy costs of powered flight are 7–14 BMR. For migrating knots, Drent and Piersma (1990) have calculated 7–9 BMR. However, as Alerstam (1981) outlined, the normal speed of flight is often more than 10 times higher than that of sustained running or swimming for similar-sized animals. As a consequence, the cost of transport, i.e. energy cost per unit of distance travelled, is reasonably low or, at least, lower than for running, and only marginally higher than for swimming. Thus, the combination of relatively high speed and reasonable transport costs makes flight attractive as the most useful and practicable mode of locomotion for long-distance migrants. The average cost for the overall migration speeds in migratory birds is estimated to about 4 BMR (Alerstam, 1991b).

Flight enables birds to make astounding seasonal journeys. In general, the strategy of long-distance migration is largely restricted to flying. There are relatively few animals performing long-distance migration by swimming or walking as mentioned above for birds and others elsewhere (e.g. Berthold, 1993). They are normally above 1 kg in mass, since the cost of locomotion is an inverse function of body mass, and hence migration is relatively expensive in small organisms (Blem, 1980). As Houston (1990) pointed out, under the maximization of gross rate of energy delivery, it is always optimal for birds to fly, even during the parental feeding period.

Flight, in general, is not a uniform behaviour but rather a variable and adaptable process. Speed of flight, in particular, can be altered to a high degree on an individual basis according to circumstances and demands. A single species is also normally capable of more than one type of flight (Kerlinger, 1989; Figure 2.47). With respect to migration, the question arises concerning the range of speeds migrants use for their journeys and what the evolutionary forces and mechanisms are that control specific performances.

Bird flight as an integrated part of migration has been discussed extensively since 1969. It was Pennycuick who then developed a comprehensive theory of bird flight on the basis of classical aerodynamic theory and applied it to migration. The complicated theoretical background, and the small empirical basis, was put together in general books by Pennycuick (1989) and Norberg (1990), and in a special review on migrating hawks by Kerlinger (1989). In addition, there are other reviews on the subject by Tucker (1973), Butler and Woakes (1990), Nachtigall (1990), Rayner (1990) and Alerstam (1991). The following summary, which has to be brief, will closely follow Alerstam's (1991) report.

### (a) Flight speeds and optimization

Of key importance in Pennycuick's (1969) theory is the power curve for flapping flight. This shows the power required to fly as a function of speed. The power curve that can be estimated for a given bird on the basis of its body mass and wing span is characteristically U-shaped. This means that at very slow and very fast speeds the energy required for flight is greater than at intermediate speeds. Figure 2.48 represents such a power curve for the Arctic tern. The part of this curve showing very low flapping-flight speeds has been estimated from the vortex theory. In this approach (Rayner, 1990) power is calculated on the basis of assumptions about the circulation and geometry of vortex rings in the wake of a slow-flying bird. From the power curve three flight velocities of special interest for migrants can be predicted.

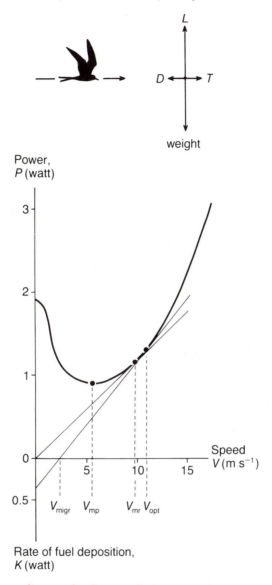

**Figure 2.48** Force diagram for flapping flight (L, lift; D, drag; T, thrust) and the power curve estimated for the Arctic tern (mass 0.11 kg, wing span 0.8 m, air density 1.11 kg/m³) according to the theory of bird flight mechanics. The mechanical power output indicated on the power axis should be divided by the conversion efficiency (normally assumed to be about 23%) to give the total chemical energy metabolized by a flying bird. The rate of fuel deposition is converted into its mechanical equivalents to be compatible with the flight power given. (Source: Alerstam, 1991b.)

1. $V_{mp}$ – the minimum power speed. It allows flight as long as possible on a given energy reserve and could be used for the purpose of minimizing energy costs per unit of time.
2. $V_{mr}$ – the maximum range speed. It will allow birds to cover the maximum range on a given fuel reserve. It is the optimal flight speed for minimizing the energy costs per unit of distance covered.
3. $V_{opt}$ – the optimal flight speed. It is that which minimizes the total duration of the migratory journey, incorporating time devoted to flight as well as to energy accumulation. It can be solved graphically by drawing a tangent from the relevant rate of energy deposition (marked on the extended ordinate in Figure 2.48) to the power curve. The speed at which this tangent intercepts the abscissa shows the maximum overall speed of migration, $V_{migr}$. Note that drawing a line to any other speed than $V_{opt}$ on the power curve would lead to a lower speed of migration than $V_{migr}$.

With the various predicted flight and migration speeds in mind, the previously-mentioned questions arise again: what velocities are selected by migrants and why?

A number of reviews and many case reports clearly demonstrate that migrants generally travel slower than the maximum flight speed and faster than with minimum power speed, i.e. in the range of $V_{mp}$ to $V_{mr}$ or $V_{opt}$ (Berthold, 1975; Kerlinger, 1989; Alerstam and Lindström, 1990; Alerstam, 1991b; Liechti, 1992). According to Alerstam (1990) and Evans and Davidson (1990), average flight speeds during migration range between about 30 and 60 km per hour in passerines and 50–80 km per hour in waders and other larger birds. The rule of thumb is that the speed roughly doubles when the mass of the bird increases 100 times (up to about 15–20 kg, when flying is no longer possible as mentioned above). Kerlinger (1989) summarized that flight speeds of migrating hawks are adjusted in a manner consistent with a distance maximization theory in three different flight situations: (1) flapping migrants fly faster with upwind than downwind; (2) ridge-gliding migrants fly faster with strong lift than with weak lift and faster with opposing winds; and (3) migrants gliding between thermals flew faster than those gliding in ridge lift. $V_{mr}$ is also dependent on the air density; the birds ought to raise their migration speed by approximately 5% per 1000 m of increasing altitude in order to save as much energy as possible. A few radar studies indeed suggest that migration speed increases at higher altitudes (Alerstam, 1990). However, despite the general agreement of theoretical expectations and empirical findings, many details with respect to the flight speeds preferred by migrants remain unclear even in the best studied groups like waders (Evans and Davidson, 1990). Alerstam (1991b) raised the question: 'Whether it will be possible to compare predictions and observations of flight speed accurately enough to reveal whether birds adjust their speed to minimize energy or time for migration, remains to be seen'. Thus, basic questions as to whether migratory flight speed and overall migration speed are selected with respect to time, energy or predation minimization, and possibly other factors like accomplishment of endogenous programs (section 2.3), cannot yet be readily answered. The number of details which have to be considered may be exemplified by the following points under discussion. Alerstam (1979) and Piersma (1987) argued that a series of short flights is always energetically cheaper than one long flight to cover the same distance. Drent and Piersma (1990) calculated flight costs (for knots) of 0.74 kJ/km in long non-stop flights and 0.5 kJ/km in shorter hops. Evans and Davidson (1990) pointed out that this may be true of level flight but that costs of climbing to the appropriate altitude at the start of each flight stage are also important. Since birds carrying heavy fat deposits have higher optimal speeds, the overall flight time for the total journey should be shorter if migration proceeds as a single long flight. Further, either strategy would be favoured if it produced better tail-wind assistance. It appears that many relevant details here will remain obscure until tools like telemetry and long-range satellite-telemetry have been employed in the investigations.

Hedenström and Alerstam (1992) have recently investigated climbing performance of migrating birds and found an inverse correlation between body size and climb rate. The lowest mean climb rate, $0.32 \, \mathrm{m \, s^{-1}}$, was observed in the mute swan (*Cygnus olor*) and the highest, $1.63 \, \mathrm{m \, s^{-1}}$, in the dunlin, with maxima up to $2.14 \, \mathrm{m \, s^{-1}}$. These results indicate substantial species differences which may considerably affect migration strategies with respect to longer or shorter flights. In another approach, Gudmundsson *et al.* (1992) investigated flight speeds of about 7600 Arctic terns based on radar observations during northbound migration in the Antarctic. Terns increased their air speed when flying into head winds and decreased it with following winds. The terns' air speeds with no wind effect were significantly faster than the predicted maximum range speed.

*(b) Gliding and soaring*

Besides flapping flight, gliding as the second basic type of avian flight (Kerlinger, 1989) is of prime importance for migration in large birds like raptors, pelicans, cranes, albatrosses and gulls. The following reasons exist for the phenomena that many large birds utilize low-cost gliding flight or soaring (i.e. gliding in circles, where wings are held outstretched from the body in a fixed position without flapping). The power required for flight changes with weight according to the proportion $P = W^{1.17}$ (Pennycuick, 1975; Kerlinger, 1989), or mechanical power increases more rapidly with mass than does resting and feeding metabolism (which scale as $M^{0.75}$; Rayner, 1990). Thus, flight is disproportionately more costly for larger birds, and large fat deposits in order to compensate for these high costs would raise even more problems (in the extreme up to flightlessness). Therefore, the development of soaring-and-gliding performance was the best way to allow large species to participate in long-distance migration. The energetic costs of gliding, about 1.5–3 BMR, are only about 15–30% of that of powered flight. Sometimes, larger birds mix gliding with bouts of flapping flight. This 'partially powered glide', 'power glide', 'power-assisted glide', 'undulating' or 'intermittent' gliding consumes about 6 BMR (Kerlinger, 1989).

Many broad-winged migrants use upcurrents to minimize energy expenditure while on long-distance cross-country flight. They often follow traditional, narrow routes, taking advantage of topographical features which provide sustained orographic uplift and reliable thermal sources – 'thermals' streets'. This, of course, largely precludes the use of long sea crossings (Elkins, 1988b; Figure 2.49).

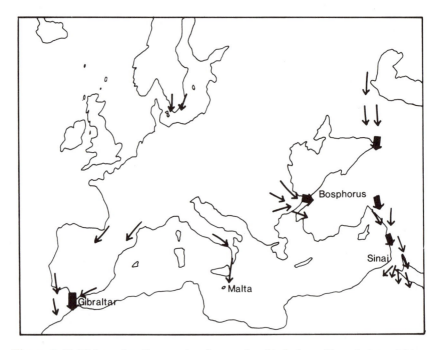

**Figure 2.49** Major migration routes for soaring birds from Eurasia into Africa. (Source: Elkins, 1988b.)

Within gliding there is considerable plasticity and adaptability that allows for various types of migratory behaviour and strategies. As Kerlinger (1989) and Elkins (1988b) have reported, most raptors rely on thermal soaring (thermalling flight) and gliding flight during migration, infrequently resorting to flapping flight or ridge gliding. Soaring flight can often be used throughout the day and may account for up to 90% of locomotion. Some species, like the European sparrowhawk (*Accipiter nisus*), rely on soaring only 15–30% of the time, do not always soar when thermalling conditions are good and often migrate early in the morning before these conditions are well developed. Other species, like harriers, have been recorded flapping across the Sahara desert. Lastly, some species, like the honey buzzard, perform longer sea crossings even at night [as frequently recorded over Malta, Elkins, 1988b; and also found in the hen harrier (*Circus cyaneus*), Russell, 1991]. In these species it remains to be shown in which way different types of locomotion are controlled in relation to migration stages and specific environmental situations.

It could be demonstrated in the field, but also in a wind-tunnel [with a freely gliding Harris' hawk (*Parabuteo unicinctus*); Tucker and Heine, 1990] that during gliding, birds, like raptors, can adjust their wings so that more or less surface area is exposed. Migrating hawks usually reduce wing span up to about 35%. By closing or opening the tail, wing loading can also be altered by about 30%. Due to such alterations wing loading and aspect ratio become 'functional'. Kerlinger (1989) concluded: 'By virtue of their ability to vary wing and tail planform by behavioural means, raptors exploit a wider range of air speeds and glide ratios than would be possible if wing span, wing area, and tail area were fixed and wing-tip configurations were constant'. Several researchers have found that soaring species have anatomical adaptations to allow them to 'lock' their wings in place, freeing them from using muscles for this. In albatrosses and others a tendon sheet is associated with the pectoralis, preventing the wing from being elevated above horizontal (Kerlinger, 1989). In comparison to sailplanes, gliding bird species are smaller and weigh less and thus are slower and have steeper glide polars. However, their turning radii are comparatively smaller, especially in small species (minimum in raptors about 6 m). Therefore, combined with a high degree of manoeuvrability, they are able to use weaker and even isolated updrafts for soaring.

Wanderers of the oceans, such as albatrosses and petrels, cover large distances without flapping their wings, in so-called 'dynamic soaring'. This gliding-on-waves results from the skilful use of wind currents over the waves, which are used alternately for climbing into the wind and gliding with it after a leeward turn. Dynamic soaring and leeward gliding might also play a role during migration along mountain ridges (Nachtigall, 1987).

With respect to a distance maximization theory, it was found that common cranes and raptors flew faster between thermals when lift was strong than when it was weak. It is likely that this behaviour is widespread among soaring migrants. In fact, mean air speeds of migrating hawks flying at high altitudes between thermals was close to that predicted from so-called MacCready tangents, i.e. tangents in a glide polar that determine the optimal interthermal speed for distance maximization (Kerlinger, 1989). As for powered flight strategies, much remains to be done to understand gliding strategies. So far, empirical determinations of the cost of gliding flight are not available for raptors. Further, it is unclear to what extent behavioural deviations from predictions may be related to the aerodynamics, energetics and perceptual (cognitive) abilities of migrants (Kerlinger, 1989). This author also suggested that 'a simple, long-term optimization strategy' may override part of local conditions.

### (c) Bounding flight

Many small passerines and other small species with low aspect ratio wings (section 1.6) show a characteristic 'bounding' flight in which the bird introduces pauses between bouts of flapping. At this time the wings are folded against the body for a short portion of the flight time ('ballistic gliding'; Oehme, 1990a). According to Rayner's (1990) studies, bounding or intermittent flight is a unique method of balancing the mechanical power output by the flapping wings and the power input by the flight muscles. This is especially relevant when the body mass of a migrant varies up to 100% during a long flight (section 2.1). Here, the bird can regulate the mean power output at the required level by introducing pauses whereas the muscles remain histologically homogenous and give a constant quantity of energy with each contraction. It is not clear whether bounding flight evolved as a response to migration; it does not appear to reduce the energy costs of migration, but it does permit greater fuel load, and hence greater range, than would be possible with steady flapping flight (Rayner, 1990). When bounding flight was modelled by means of computer simulation it could be shown that the intermittent flight improves the utilization of the bird's energy budget in comparison with sustained horizontal flapping flight (Oehme, 1991).

With respect to migration speed [distance covered per time unit, not flight speed (for figures see next section)], Alerstam (1991) made the following consideration. It is to be expected that large birds generally migrate more slowly than do smaller ones. This is because the metabolic scope for fuel deposition relative to BMR is about the same for birds of different sizes, while the power requirements in flight increase relative to BMR with increasing body mass. Migration speed as a fraction of the flight speed is given by the ratio of energy deposition rate to the sum of this rate and flight power. Hence, for an Arctic tern with an energy deposition rate corresponding to 2 BMR and flight power to about 7 BMR the speed of migration will amount to 2/9 of the flight speed. By comparison, a swan with flight power perhaps reaching 60 BMR will attain a maximum speed of migration only about 1/30 of its flight speed. Although optimal flight speeds are faster for larger than for smaller birds, they are not sufficiently so to offset the expected reduction in overall migration speed with increasing body size. This is another aspect of migratory movement that requires extensive tests in the field.

## 2.15 MINIMUM AND REAL STOPOVER PERIODS: THEIR CONTROL AND MIGRATORY EPISODES

It is known from banding and trapping data, sight observations and, more recently, from satellite-tracking studies, that most migratory flights, especially towards the winter quarters, are regularly made in stages or batches and thus may often last for months, up to about a half a year (e.g. marsh warbler; section 2.3). Few species actually cover the whole migratory distance during the autumn migratory period in one bout like some waders on transoceanic migration (Williams and Williams, 1990a,b; section 1.1). On the other hand, homeward migration in spring appears to more often be performed extremely rapidly and without interruption (Berthold, 1993). In the above-mentioned marsh warbler, for instance, the return migration takes only about six weeks (Pearson and Lack, 1992).

Both the relatively slow forward movements towards the winter quarters and the high proportion of retraps of migrants in banding stations indicate the regular occurrence of stopover periods during migration. However, we have little knowledge of the exact length of staging periods, their variance according to season, physiological state or immediate environmental conditions. The principal control mechanisms are simply not known. From what has been studied, only very rough patterns have begun to emerge.

Among the rare and 'conspicuous' species data has accumulated fairly well for how long small groups remain in certain staging areas during migration. Prominent examples are, in North America, the whooping crane, one of the superstars of avian conservation measures, *en route* from Canada to Texas (Doughty, 1989). In a number of wetlands it has been possible to determine staging periods of individually marked waterfowl and waders. For instance, in thousands of neckbanded greater snow geese (*Chen caerulescens atlantica*) it was possible to study the chronology of autumn migration in staging grounds in Canada

(Maisonneuve and Bédard, 1992). Individual birds tended to migrate annually at the same relative dates and staged for similar periods each year. For the geese observed on the staging grounds, mean length of stay varied between 15.5 and 19.1 days between years, and was not related to age, sex or status. Comparable intensive studies have been made with ruffs (*Philomachus pugnax*) in Germany (Harengerd *et al.*, 1972), and with other shorebirds elsewhere. In some of these studies it was found that juveniles, during their first autumn migration, stop at more places and stay longer at each than adults (e.g. dunlin; Rösner, 1990). Whilst stopover site fidelity is fairly common in larger birds, like cranes or geese with traditional migratory habits and strong social bonds (see above), or in birds with specific habitat preferences for staging like wetland species (e.g. Yesou, 1989), it is very rare in migrant passerines (Ellegren and Staav, 1990; Berthold *et al.*, 1991; Winker and Warner, 1991).

Surprisingly, among passerine migrants, tactics of the course of migration and interposed staging periods are best known from desert areas where analyses are not disturbed by local resident conspecifics or by individuals performing premigratory movements (Jenni, 1984). Especially intensive studies in the western and eastern Sahara (Bairlein, 1991b; Biebach *et al.*, 1986, 1991; section 2.16) have demonstrated that there are two or three strategies. Many nocturnal migrants with fat reserves sufficient to overcome the whole desert make a landfall and rest during the day and take-off for the next all-night flight the following evening. In this way, they cross the desert in a few nocturnal hops. Birds with fat reserves insufficient for desert crossing apparently search out oases in order to replenish fat deposits. There, they may rest for up to three weeks. In several species, like lesser whitethroats, willow warblers and yellow wagtails, median values of minimum stopover periods were 4, 5 and 5.5 days (Biebach *et al.*, 1986, Figure 2.50). Similar values have been obtained by Bairlein (1991b) and Lavée and Safriel (1989). A third strategy may be to overcome the desert in an extended non stop flight, but this has not been convincingly demonstrated (section 2.16).

There are several lines of evidence indicating that migrants, and particularly passerines, regularly interpose extended stopover periods when migrating through continental areas. Firstly, the movements forward to the winter quarters are fairly slow, as estimates from banding recoveries have shown. According to Alerstam and Lindström (1990), median migration speeds through Europe in passerines are only 75 km/day in early departing long-distance migrants (13 species), 53 km/day in later migrating regular migrants (19 species) and 27 km/day in partial migrants (11 species). For waders, values range between 60 and 200 km/day, with a median value of 79 km/day (13 species), which

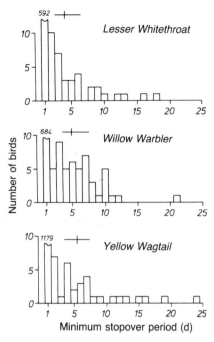

**Figure 2.50** Frequency distribution of the minimum stopover period of retraps and of single traps (minimum stopover period, 1 day) in the oasis Sadat Farm. Sample size for single traps are given above the respective columns. Medians ±95% confidence intervals for retraps only are given as vertical and horizontal lines. (Source: Biebach *et al.*, 1986.)

is higher than in most passerines. For the Arctic tern and the swallow values account for about 150 km/day. Hildén and Saurola (1982), and others (Berthold, 1993), obtained similar values. These rather short distances are certainly flown in a few hours, so that considerable resting time must remain. This is confirmed by the most comprehensive study of the subject published by Ellegren (1993) on speed of migration and migratory flight lengths of 62 passerine species ringed during autumn migration in Sweden and recovered within 1–3 days after ringing. The average migration speed was lowest among short-distance migrants (mean = 28 km/day) and highest among long-distance migrants (mean = 60 km/day). The maximum individual speed exceeded 100 km/ day in several species. For many species the migration speed increased along the route and late migrants also seemed to attain a higher speed than early conspecifics. In some cases adult birds migrated significantly faster than juveniles. The average flight length among nocturnal migrants (177 km) was significantly longer than that of diurnal migrants

(111 km). Most flights made by nocturnal migrants lasted for less than half the dark period. Such relatively short flights made during the first part of migration may relate to the slow initial migration speed found in many species.

Secondly, typical continental passerine passage migrants often have low fat deposits (Alerstam and Lindström, 1990). In southern Germany this would only allow for a theoretical mean flight distance of about 150 km to be flown in a few hours (11 species; Kaiser, 1992). Thirdly, regular and numerous retraps at banding stations clearly indicate that at least a proportion of passage migrants regularly interpose stopover periods. Retraps do not only consist of local birds and individuals performing premigratory movements. This has been documented in individuals which were found staging in resting areas halfway between restricted breeding areas and winter quarters (Berthold *et al.*, 1991).

Extremely little is known about the real duration of stopover periods and their control. It is clear, however, that staging periods established by sight observations, or capture and recapture of banded individuals, mainly represent minimum periods. Birds may often be present in an area before they are seen or captured, and they also may stay for some time after the last observation or recapture.

Minimum stopover periods have been established in many studies. It has been most intensively done in the MRI-program (section 1.3). In this standardized program in the 10-year period from 1974 to 1983 more than 200 000 first traps and over 30 000 retraps of 37 species have been investigated. From these retraps, detailed figures of the stopover periods and their alterations during the premigratory and the autumn migratory season, in relation to moult, intensity of migration and body mass increase, were obtained (Figure 2.51). Overall, an average minimum stopover period of about four days was found. This decreased slightly during the migratory season (Figure 2.52). Kaiser (1993a) tried to estimate the real staging periods on the basis of the same material. He used Jolly–Seber models to come to realistic estimates. According to his calculations, the real mean stopover period for all passerines was about 13 days. For different species, the mean stopover periods during autumn migration varied between about 10 and 15 days. The real resting community was about 5–12 times the figure of the actually

---

**Figure 2.51** Various data of blackcaps trapped during 1972–1983 in the MRI-Program in S Germany. EF, first traps; KG, body mass; FL, third primary length as wing measure; $EF_{WF}$, potential retraps (first traps that later occur as retraps); VD, stopover period; WF, retraps within the same season; KM and GM, body feather and wing feather moult, respectively. (Source: Berthold *et al.*, 1991.)

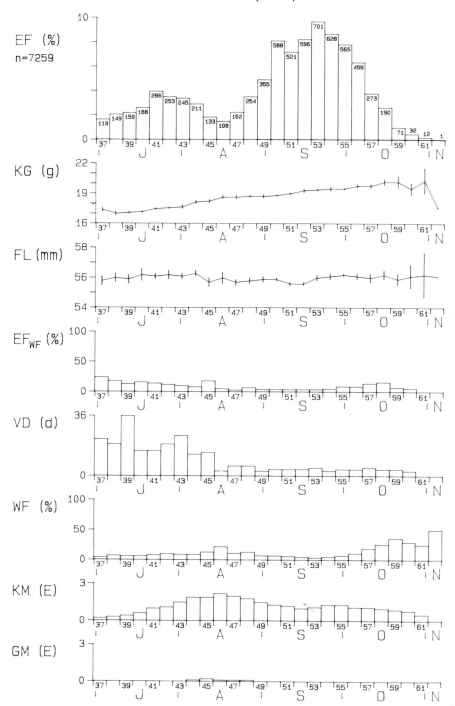

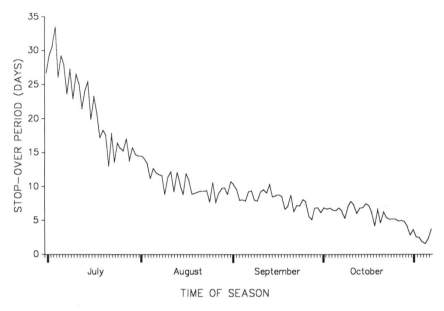

**Figure 2.52** Seasonal changes in the length of stopover periods in songbirds during the autumn migratory period in S Germany. Daily means of the period 1972–1992 based on more than 100 000 individuals of 37 species.

retrapped individuals depending on the assumed recruitment area. Similar calculations using a Jolly–Seber model have been made by Lavée *et al.* (1991) for a Sinai desert oasis. They found that more birds made a stopover and that they stayed longer: (1) in autumn than in spring; (2) in early autumn than later; (3) in late spring than earlier; (4) in species and seasons with large stop over populations; (5) in females than males (found in two sexually dimorphic species) in spring; and (6) in species encountering the oasis earlier on their autumnal route than those later.

There are many reports which more or less convincingly demonstrate that the length of the stopover is inversely related to the bird's fat deposits and that the staging period is controlled, at least to an extent, by the fat stores (Bairlein, 1985c; Biebach *et al.*, 1986; Moore and Kerlinger, 1987; Lavée and Safriel, 1989). Interesting details have been found in bluethroats at stopover sites in Sweden (Ellegreen, 1990b, 1991; Lindström *et al.*, 1990). Adults carried more fat and stopped over for a shorter period than juveniles. In contrast to juveniles, adults did not lose mass after arrival and started accumulating fat earlier than juveniles. Further, fat deposition rates were positively correlated with dominance status. The fact that dominance influences fat deposition rates may be important for the spatial and temporal pattern of migration in birds that compete for resources at stopover sites.

There is extreme variation with respect to body mass changes at stopover sites after arrival. These vary from initial losses in one-half of the species studied (Walsberg, 1990), to remaining constant to immediate increases (Berthold, 1993; Kaiser, 1993a). Alerstam and Lindström (1990) calculated mean values for 11 passerine species and found an average delay before fattening started of 1.5 days and a mean loss of mass (fat) during this period of 4.4% (0–13%). These differences are only partly understood. Castro *et al.* (1991) found in shorebirds that mass loss after capture is independent of bird species, but strongly influenced by temperature, especially above a threshold of about 30°C.

It appears that a number of factors influences the length of stopover periods, above all migratory state and moult phase, amount of fat reserves, age, dominance status, locality and season. In such a complex situation many more detailed studies are needed before clear rules can be determined. It is highly likely that endogenous components as part of endogenous time-programs for migration are also involved (section 2.3). This is indicated by the fairly constant species-specific lengths of stopover periods in many species over the years, which appear as preprogrammed time intervals (Berthold *et al.*, 1991). Safriel and Lavée (1988) came to a similar conclusion. In cross-desert migrants they found that trends of mass changes of individuals during stopover were usually inconsistent, and they proposed that stopping-over birds do not always resume their migration, only after their fat reserves have been replenished. They concluded that 'their decision to take-off, or the reappearance of the migration impulse, are also controlled by a time-program incorporated into their endogenous migration scheme, which constantly updates the time left for sampling and refuelling'. As many other researchers, they also found that in spring less time is allotted for the whole migration program. The time taken for return migration is in many species at least about one-third to one-half shorter, and thus faster, than outward migration (e.g. Berthold, 1993; and above). Safriel and Lavée (1988) stated that 'the time constraint overrides all other tactical considerations, such as the state of fat reserves, and the weather'. Studying semipalmated sandpipers, Dunn *et al.* (1988) found that fat content at capture was a poor predictor of length of stay during autumn migration, which suggests that other factors are more important in determining its length – possibly also endogenous factors. From the viewpoint of time-selected migration, stopover decisions should be governed by a combination of two major factors: the migrant's actual fat status and the prospects for rapid fat deposition at the potential stopover site. In accordance with the marginal value theorem, a migrant should, in order to maximize migration speed, depart from a stopover site when the marginal rate of gain in flight distance drops to the expected average speed of movement along the migration route (Alerstam and Lindström, 1990).

## 2.16 OVERCOMING MIGRATION BARRIERS

Migrants are regularly faced with six kinds of major ecological migration barriers which may be potentially dangerous: oceans, deserts, high mountain ranges, Arctic ice fields, tropical rainforests (for many species; section 2.21) and temporary adverse meteorological conditions. The geographical barriers can be regularly overcome by a number of specific strategies. Only rarely they do prevent the continuation of migration (Alerstam, 1990). Meteorological barriers are overcome either by specific adaptations (sections 2.11 and 3.7) or they can be avoided by an interruption of the migratory journey. Oceans, mountains, ice fields and rainforests are regularly transversed by extended non-stop flights. Deserts are overcome in the same way or, as has recently been elaborated, by an intermittent strategy. This phenomenon will be treated first.

### 2.16.1 Crossing deserts

Moreau (1961, 1972) proposed that Eurasian migrants wintering in tropical Africa cross the Sahara, the most inhospitable desert in the world, in an extended non-stop flight of about 40–60 hours; such a concept of non-stop transdesert migration was also applied to other regions (e.g. Schüz *et al.*, 1971). In the meantime, detailed studies in four desert areas, i.e. the Sinai Desert (initiated by Lavée and Safriel, 1973), central Asian deserts (Dolnik, 1985), the western Sahara (Bairlein *et al.*, 1983) and the eastern Sahara (Biebach, 1985), have clearly shown that many migrants regularly land in the deserts for resting or feeding. Those individuals can be by no means regarded as stragglers but have obviously developed an intermittent migratory strategy. In the most extensive studies carried out in the Sahara, the following rules have begun to emerge. Individuals of about 25 different migratory species, above all willow warblers, lesser whitethroats, red-backed shrikes (*Lanius collurio*), spotted flycatchers and yellow wagtails, were regularly found grounded; also, in central parts of the desert, in more or less large numbers in normal physical conditions and with fat reserves sufficient for a successful continuation of transdesert migration. In the open desert, and in smaller oases, trapped individuals have significantly higher fat reserves than in larger oases (Figure 2.53). Individuals of the first group, which represent the great majority, appear to rest only during the daytime. They seek shelter during the day and take-off for the next whole night flight in the evening. Observations and tests of

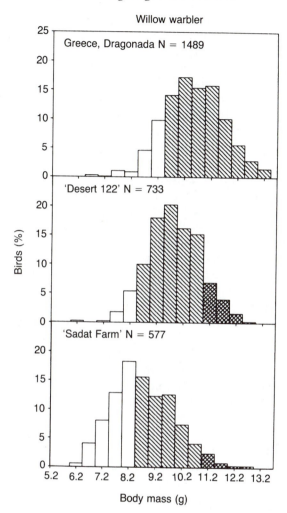

**Figure 2.53** Frequency distribution of body mass of willow warblers on autumn migration, before the desert crossing, on the isle of Dragonada in Greece and from two study sites in the Libyan desert. In still air, only birds represented by the cross-hatched mass classes are expected to manage the crossing without refueling, with a tail wind of 8 m/s both those of the dashed and cross-hatched mass classes are expected to manage the crossing without refueling. (Source: Biebach, 1990.)

caged individuals show that these fat birds regularly rest completely inactive during the day, ignoring food and water if available, and become active the following night. Individuals of the second group, a minority with only moderate or low fat reserves, obviously select larger oases for resting. Here, these individuals stay on the average for 2–4 days, maximally up to about three weeks (Figure 2.50). In these areas they are generally able to refuel, due to extensive foraging. The intermittent strategy in fat birds, which apparently have sufficient fuel for a non-stop transdesert migration, have most probably evolved in order to provide for the thermoregulatory demands of a daytime rest (section 2.10). The spotted flycatcher, however, feeds quite extensively on insects during its transdesert migration and most likely, therefore, only develops low premigratory and migratory fat reserves. At present, some intriguing questions are: (1) to what extent is the intermittent strategy, which so far is almost exclusively established in autumn, used during spring migration; (2) does exclusive non-stop transdesert migration, as first proposed, also exist; (3) and if so, what are the proportions of the two strategies among species, populations, and possibly individuals, and their causes and control mechanisms? For more details see the special reviews in Lövei (1989), Bairlein (1992), Biebach *et al.* (1991) and Lavée and Safriel (1989). As Herremans (1991) pointed out, it is likely that there is no single optimal trans-Saharan migratory strategy but a variety possibly including those yet to be discovered. Figure 2.54 schematically demonstrates various strategies. Recent radar studies in the eastern Sahara indicated that there is tremendous, thus far undiscovered, non-stop migration in a wide range of altitudes from about 200 m to 7 km (Biebach and Heine, pers. comm.). Further, recent flight-range estimates suggest that capacities in passerines, based on the observed fat deposits, were too low to cross the whole Sahara in still air. Birds appear to depend on tail winds for a successful crossing of this desert, independent of a non-stop or an intermittent migratory strategy. Weather conditions, at least in autumn, appear to allow them to rely on tail winds (Biebach, 1992; section 2.10).

### 2.16.2 Crossing other barriers

Arctic ice-fields and tropical rainforests are presumably mainly overcome by non-stop flights, but corresponding reports are scarce. Clearly, transoceanic migration in those species which are unable to rest on the water is exclusively performed by non-stop flights where extremes of up to 7500 km within 100 hours may occur (section 1.1). Non-stop transoceanic migrations were first described for the Gulf of Mexico at the beginning of the century (Cooke, 1905). Recently, strategies have been analysed in more detail by radar studies (Williams and Williams, 1990a,b). Transoceanic migrants show three different phases of flight behaviour. At take-off they are sensitive to local weather conditions and tend to depart in a period of favourable conditions. Once transoceanic flight has been initiated, migrants ascend to altitudes of

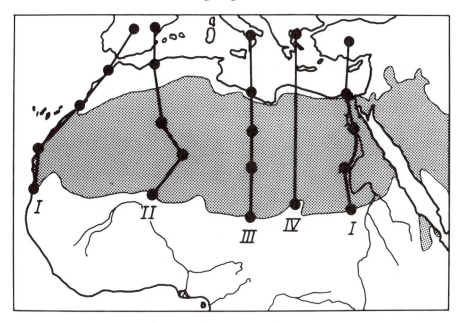

**Figure 2.54** Hypothetical strategies for crossing the Sahara. I and II, intermittent migratory strategies with stopover in vegetated areas where birds can feed (Nile, Atlantic coast, oasis); III, intermittent migratory strategy with stopover in the desert without feeding possibilities; IV, non-stop flight. (Source: Biebach, 1990.)

up to 6 km (section 2.12). There, they appear to be largely insensitive to immediate environmental conditions and landmarks (chapter 3). During the final phase of a transoceanic flight, when migrants descend considerably, they again become sensitive to local conditions which facilitate landfall, e.g. by locating a wintering area composed of islands like the chain of Hawaiian Islands. Transoceanic migration is, of course, performed much more easily by sea birds which can regularly rest on the water. Land birds may have a chance to fly away again after accidental watering; when exhausted, they often take a real chance to recover during an extended rest on boats or flotsam. Very rarely, they may continue their journey in a 'piggyback flight' on the back of a larger bird (Mead, 1983).

Many species, like raptors, are characterized as reluctant to cross water barriers. Among raptors, only species with high aspect ratios perform long-distance migration, some, such as some falcons, harriers or ospreys, make longer crossings, whereas others suffer from considerable mortality when attempting even short water crossings (Kerlinger, 1989). Dunn and Nol (1980), studying autumn warbler

migration across Lake Erie, found that juvenile birds were more prone
to return to land after starting to cross the lake than were adults.

Migratory birds also regularly cross the highest mountains, often at
extremely high altitudes (sections 1.1 and 2.11). The Himalayas, for
instance, are a barrier regularly passed by many species. We do not
know how these species cross the ranges, whether regularly in non-
stop flights or with periodic rests in high valleys or on passes. On Col
de Bretolet in the Swiss Alps, during studies over several decades
(Dorka, 1966; Bruderer and Jenni, 1990; Jenni, in litt.), about 75 species
have been found to rest regularly and about 130 species occasionally. A
variety of complex strategies used to cross or circumvent mountainous
barriers have recently been described for passerines in the Alps
(Bruderer and Jenni, 1990). A considerable number of migratory species
have apparently adapted strategies to avoid mountain ranges and
appear to have programs that guide them along the mountains (Figure
2.55). A direct crossing is a common strategy in larger species, as it is
in species with strong north–south migration routes, long-distance
migrants, in migrants travelling at relatively high altitudes, in
individuals in advanced stages of migration with high fat deposits and,
possibly, to some extent preferentially in adult birds (Jenni and Jenni-
Eiermann, 1987). Further, migrants crossing the Alps generally fly
faster than those migrating over adjacent lowlands (Liechti, 1992).
According to recent radar studies, the Pyrenees do not act as a barrier
for many European passerine long-distance migrants, at least not in
spring (Hilgerloh *et al.*, 1992).

## 2.17 IMMEDIATE EFFECTS OF WEATHER AND CLIMATE

Migration primarily evolved as a response to seasonal changes in food
resources (section 1.1). Food abundance does, however, vary with
climate and, hence, meteorological factors are expected to be still involved in the control, and possibly the immediate stimulation, of avian
migration. In addition, most migratory movements in birds are performed by flight and thus atmospheric phenomena from microscale
turbulences to global circulations can exert considerable modifying
effects on their timing, duration, route and success. The relationships

**Figure 2.55** (a) Distributions of track directions; and (b) headings in S Germany
and in the Swiss lowlands during four periods of autumn 1987. At N, the exact
location of the radar is indicated by a point between the distributions from the
four observation periods. In the other cases, the circle marks the radar site. The
letters P and F indicate the nearest village; N, R, A and S are larger towns in
the area. (Source: Bruderer and Jenni, 1990.)

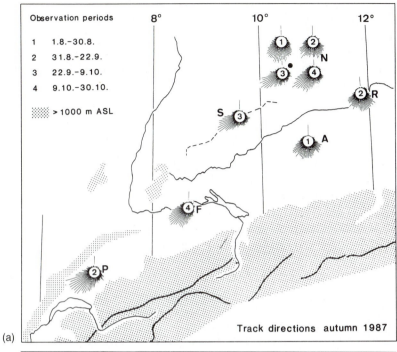

(a)

Observation periods

1  1.8.–30.8.
2  31.8.–22.9.
3  22.9.–9.10.
4  9.10.–30.10.

▓ > 1000 m ASL

Track directions  autumn 1987

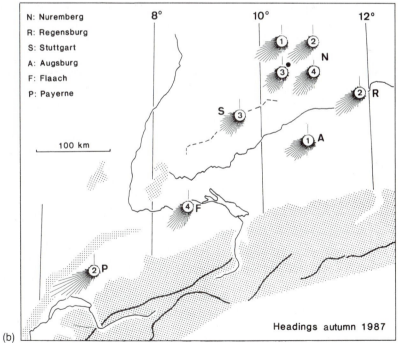

(b)

N: Nuremberg
R: Regensburg
S: Stuttgart
A: Augsburg
F: Flaach
P: Payerne

100 km

Headings  autumn 1987

between meteorological factors and bird migration are extremely complex. To treat them in detail would go far beyond the scope of this section. What will be considered here are the main rules of weather-related migration which have emerged from the extensive literature available. Fortunately, reviews by Elkins (1988b), Richardson (1990), Rabøl (1990), Walker and Venables (1990) and Gauthreaux (1991) provide excellent starting points.

### 2.17.1 Primary stimulatory effects

Primary stimulatory effects of meteorological factors are most prominent in so-called cold-weather movements. Elkins (1988b) and Ridgill and Fox (1990) summarized information on this type of migration for the British Isles. Plovers, thrushes and skylarks (*Alauda arvensis*) are known to exhibit some form of these movements in almost every spell of cold weather. Among the waders, lapwings (*Vanellus vanellus*) and golden plovers (*Pluvialis apricaria*) often show countrywide movements in response to adverse weather. Fleeing from winter (e.g. ice or snow) may lead lapwings from the British Isles down to Spain, where they are known as the 'Avefria' or the 'birds of the cold'. In exceptional circumstances, long-distance hard-weather movements can stimulate birds to carry out a complete transAtlantic crossing. For instance, lapwings were reported to have fled to North America in 1927 and again in 1966. A typical case for these movements involve the avoidance of ice, as in waterfowl, which leave their wintering areas in response to freezing temperatures. Some examples here are southern movements of whooper swans (*Cygnus cygnus*) from Iceland, bean geese from Sweden and teal (*Anas crecca*) which move from Central Europe to the Mediterranean (Nilsson and Pirkola, 1991; Ridgill and Fox, 1990). Snow avoidance is also a migratory cause, as in common buzzards (*Buteo buteo*) which go through altitudinal movements to lowlands. Many species, like wrens (*Troglodytes troglodytes*), or extremely resident species, like crested tits (*Parus cristatus*), do not react in a comparable way.

Hard-weather movements may then be initiated by ice, snow or low temperatures, and also to an extent by wind. Whether these meteorological factors or detoriating nutritional resources are the proximate factors for movement is still uncertain.

In the northern hemisphere return movements are stimulated by winds from the west and south, and thus birds generally move downwind when returning to their initial areas (Elkins, 1988b). Climatic factors may also have been involved in some recent changes in migratory behaviour. For instance, the migratory Canada goose, introduced in Europe, is still migrating in Sweden but has stopped migrating in England (Ringleben, 1956). The song thrush became resident when introduced to Australia and New Zealand (Wagner and Schildmacher, 1937). Still, the roles of nutritional or microevolutionary factors in these developments are unknown. In blackcaps, which have developed novel migratory habits, in Central Europe (section 4.1) nutritional factors appear to have been most important, although meteorological effects cannot be excluded. Finally, in extremely mild winters, like that of 1974/1975, unusual midwinter sighting of long-distance migrants like swallows, cuckoos, garden warblers and terns have been made (Elkins, 1988b). The significance of these sightings is unclear. Perhaps they are simply misdirected individuals which would have normally perished.

### 2.17.2 Modifying effects on actual migration and migratory disposition – general aspects

For a number of species the close correlation between the onset of migratory activity in captivity and that during actual migration indicates that endogenous factors essentially trigger departure and that environmental factors, including meteorological factors, are of minor importance (section 2.2). However, once on migration, weather can ground, delay or hasten migrants, or deflect them from their headings, or even kill them. Other species may already be strongly influenced in their timing of departure by specific meteorological conditions (Elkins, 1988b). Effects of short-term variations on migration were intensively studied in the 1960s and 1970s (e.g. Able, 1973; Muller, 1976; Nisbet and Drury, 1968). In comprehensive studies up to 15–30 weather variables were considered in multivariate analyses. As Richardson (1990) pointed out, despite the advances in analytical procedures, it is virtually impossible to determine from field data which of the inter-related variables, like pressure, temperature and humidity, birds respond to. Multivariate analysis normally cannot reliably separate causal from coincidental relationships. On the other hand, experimental work in the field has rarely been done, and experimentation on caged or released birds may have its own biases. Nevertheless, some general rules have emerged (Elkins, 1988b; Richardson, 1990): (1) there is a distinct tendency to avoid migrating in inclement weather (overcast, precipitation, poor visibility), and avoidance may result in arrested migration or migration congestion, normally for a day or a few days; (2) there are specific synoptic weather features that specifically promote migration in different migratory seasons; (3) migrants presumably react to locally-measurable variables (like wind, temperature or rain) rather than to synoptic features; and (4) wind and precipitation are likely to be the two weather factors which affect migration the most. As a whole, reactions of migrants may be understood as 'well-adapted responses to weather by birds, canalysing migration into windows of relatively favourable weather in a highly complex and rapidly varying weather environment' (Alerstam, 1981).

Meteorological factors actually do not only influence migratory behaviour, they may also affect migratory disposition, e.g. body mass development and migratory fattening. Significant negative correlations were found between daily measurements of variables, such as temperature, barometric pressure, precipitation and some others, and lipid mass. It would appear that temperature is the most significant factor determining lipid stores in migrants, but very little is known about proximate influences of environmental factors on fattening. Immediate responses of small passerines to low ambient temperatures appear mainly to concern non-migrants and wintering birds (Blem, 1990). For other examples see below.

### 2.17.3 Main aspects of autumn and spring migration

Intensive studies in Europe, North America, China and elsewhere have shown that mass autumn migration occurs with cool northerly tail winds in situations when low pressure systems are being transposed by anticyclones. A few radio-telemetry and 'release-aloft' results are consistent with this finding (Richardson, 1990). These rules apply for both regular and irruptive migrants (Elkins, 1988b). In sea birds, migratory movements are largely interwoven with feeding behaviour and relationships may be more complex (Elkins, 1988b). A number of species that have to make long transoceanic flights, for example from Greenland to the British Isles, make use of 'cyclonic approach' (Williamson, 1961). They follow the wind flow in polar air masses behind large Atlantic depressions to obtain the maximum benefit from tail winds (Figure 2.56). For swans Elkins (1988a,b) has calculated that birds which have departed from Iceland at the onset of very cold weather were able to fly the about 1200 km to Ireland in approximately 7 hours using an appropriate altitude of about 8000 m with its favourable wind flow.

A rather unique behaviour in this respect is the massive cyclonic weather movements of swifts. In order to avoid prolonged rain during the breeding season, these sensitive aerial plankton feeders fly in large flocks of up to 50 000 individuals against the wind on the flanks of approaching low-pressure areas. In this way, they might enter the warm sector and a less rainy area. Such adverse weather movements may lead individual birds over distances of up to 2000 km within a short period, and rapid return movement in fine weather may take place at altitudes exceeding 3000 m (Glutz and Bauer, 1980; Elkins, 1988b). It is open whether, for these movements, swifts are triggered

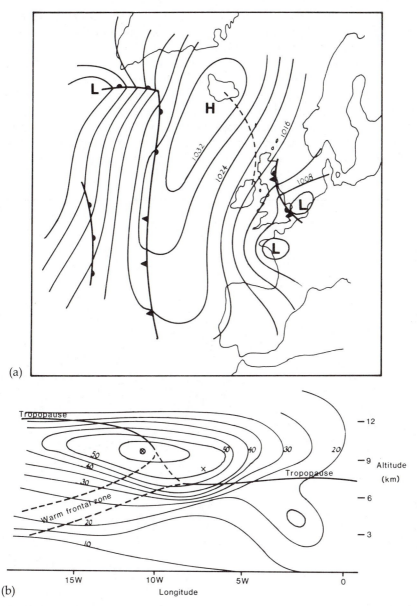

**Figure 2.56** (a) Weather situation between Greenland, Iceland and the British Isles, 9 December, 1967, 1200 hours. Broken line, probable track of southward migrating whooper swans. (b) West–east cross-section of wind speed across northerly jet stream, afternoon of 9 December, 1967. Solid lines with figures: wind speed in metres per second; crossed circle, core of jet stream; cross, position of migrating whooper swans located at 56.30 N 7 W. (Source: Elkins, 1988b.)

directly by weather-related factors (e.g. a drop in temperature, electromagnetic impulses, i.e. 'atmospherics') or indirectly by a reduction in food supplies (Berthold, 1993).

For the spring migratory period, intensive studies in several areas have shown that mass northward migration occurs with warm southerly tail winds as high pressure areas move away and lows approach (mild-weather migration; Richardson, 1990; Figure 2.57). However, due to the need for a rapid return in breeding areas, migrants often fly under poorer conditions and are hence less influenced by weather variables (Elkins, 1988b; Grazulevicius and Petraitis, 1990). Many species are

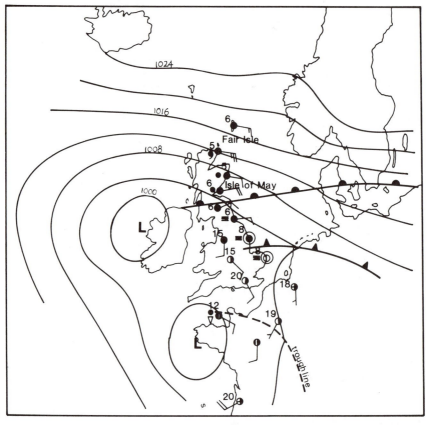

(a)

**Figure 2.57** (a) Spring migrant fall, 3 May 1969, 1200 hours. Large falls of Scandinavian-bound spring migrants in northern Scotland. (b) Typical situation delaying spring migration, 11 April 1973, 0600 hours. Blocking anticyclone west of Britain with cold northerlies prevailing over Britain and the North Sea. Frontal troughs running south in flow. Temperatures in °C. (Source: Elkins, 1988b.)

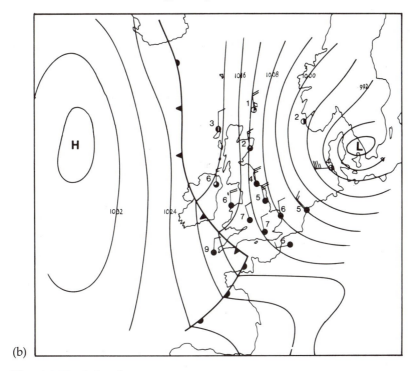

(b)

**Figure 2.57** *continued*

known to perform loop migration (section 1.1), taking shorter routes in spring to return faster, or flying other routes to take advantage of suitable seasonal wind regimes (e.g. red-backed shrike; Zink, 1973–1985).

In the past, arrival times of returning migrants plotted on maps, so-called isochronuous lines (isepipteses, isochrones, isophenes) were shown to match more or less specific isotherms (Schüz *et al.*, 1971).

When the weather in spring is unusually warm and sunny, as in mid-April 1980 in Central Europe, a migratory phenomenon called 'overshooting' may occur. Birds which generally do not breed above southern or Central Europe, like black terns (*Chlidonias niger*), bee-eaters (*Merops apiaster*) or hoopoes (*Upupa epops*), appear to be misled by the fine weather and overfly their breeding areas up to northern Europe. This may possibly be different from the prolongation of migration known in unmated males of several species (Schüz *et al.*, 1971). Over-shooting in quail (*Coturnix coturnix*) occasionally reach invasion pro-portions and many birds may then stay in northern areas to breed. In the case of extremely fine weather in early spring (as in early March

1977, indicated by falls of Saharan dust up to northern Ireland) rare summer visitors can, in addition, appear extremely early (Elkins, 1988b).

Hagan *et al.* (1991) investigated the relationship between latitude and the timing of spring migration of North American land birds. They confirmed the former observation (Morse, 1989) and found that tropical-wintering species show significantly less intra- and inter-year variability in the timing of migration. This suggests that the mechanism regulating their migration is primarily endogenous. On the other hand, temperate-wintering species showed higher interspecific correlations in timing, suggesting that external cues modulate spring migration in this group. Corresponding data have been presented by Kemlers *et al.* (1990) for late and early arriving migrants in Latvia. These results indicate that the migration regulation system employed may ultimately be deter-mined by wintering latitude and associated environmental cues. These findings are also in agreement with earlier concepts of more exogenously controlled short-distance 'weather migrants' in contrast to more endogenously controlled long-distance 'instinct migrants' (Berthold, 1975). From the first group, Emperor Frederick II reported as early as in the 13th century that the return of migrants like starlings, thrushes and finches depends on the time when it is possible to 'follow food supply and warmth'. For the second group, a striking example has been obtained by Mildenberger (Bairlein, 1995) for the garden warbler. During a 38 year period, first arrivals in the Rhine Valley were observed on May 1 with a standard deviation of only 5 days. For similar data obtained in the spotted redshank see section 1.1.

### 2.17.4 Individual factors

A very brief overview of the most relevant relationships between bird migration and individual weather factors will be given here following the reviews by Richardson (1990) and Elkins (1988b).

### (a) Wind direction and speed

Most types of birds are likely to migrate in peak numbers during calm periods or with following winds. Smaller numbers of individuals migrate with side or light head winds, and few birds move into strong head winds. Moreover, the number of individuals aloft is negatively correlated with opposing wind speeds (some examples are demon-strated in Figure 2.58). Wind selectivity can, however, vary widely among species. Why is uncertain. Some migrants may, for instance, adjust their altitude to select favourable winds (section 3.7). On the other hand, some experimental biases also confuse results. Radar studies, for instance, overestimate associations with wind, whereas

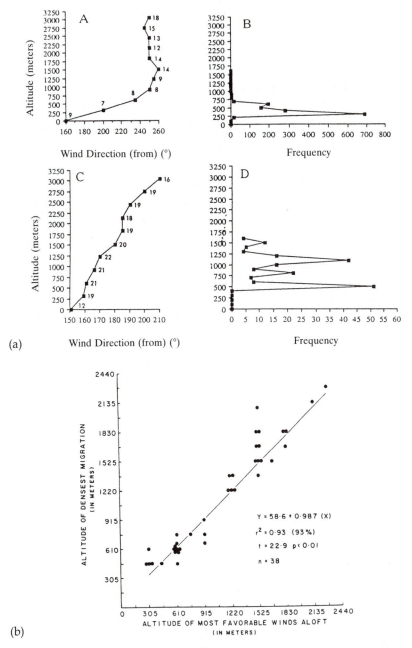

**Figure 2.58** (a) Winds aloft and the altitudinal distribution of trans-Gulf migrants arriving on the northern coast of the Gulf of Mexico in the spring of 1983. A, winds aloft for 7 May (speed in knots); B, altitudinal distribution of migrants aloft on 7 May; C, mean winds aloft for 6, 11, 12 and 13 May (speed in knots); D, altitudinal distribution of migrants aloft on 6, 11, 12 and 13 May. (b) Correlation between the altitude of densest bird migration at night and the altitude of most favourable winds aloft. (Source: Gauthreaux, 1991.)

counts underestimate them. Many diurnal studies show an apparent preference for head winds which is obviously for several reasons. Species that regularly fly into the wind, like swifts and swallows (preferably in autumn), may do so to facilitate feeding as aerial insectivores. During the day, flying very low in flat inland areas or along coasts as leading lines with relatively low head winds may be a strategy that cannot be followed in darkness by diurnal migrants. Soaring birds also often fly with light head or side winds if updraughts are present (Elkins, 1988b). Proper choices of wind conditions are often important because successful non-stop long-distance flights may require wind assistance (sections 2.10 and 2.16). Some pelagic species such as albatrosses may make use of planetary wind systems for migration as well as extended foraging flights. A recent satellite-tracking study in this group (Jouventin and Weimerskirch, 1990) has elucidated many details of the effects of specific weather situations. Arctic terns, with the longest distances to migrate, have been shown to be carried to the wintering area by southern ocean storms, where late birds miss them and remain in African waters (Elkins, 1988b). The relationships between wind and orientation behaviour which are important here are discussed in chapter 3. Finally, it should be noted that there is, so far, no systematic experimental investigation of the relationship between wind and migratory activity which has yielded conclusive results (Rabøl, 1990).

*(b) Temperature*

Some studies have shown clear correlations among peak autumn and spring migration, and low and high temperatures, respectively, or of the timing of migration and the earliness or lateness of the season as measured by average temperature (Richardson, 1990). However, it is not clear whether migrants respond directly to temperature or whether wind and energetical factors are simultaneously involved. Experimental studies of zugunruhe under various ambient temperatures are consistent with field data (Weise, 1956; Viehmann, 1982; Czeschlik, 1976; Berthold, 1975), but do not rule out multifactorial influences. More systematic investigations of temperature influences on zugunruhe under fully-controlled conditions are necessary (Richardson, 1990), especially since many of the earlier studies are not conclusive for various reasons (Berthold, 1975; see below).

*(c) Precipitation, cloud, fog and visibility*

Reactions of migrants to these factors are extremely variable depending on other factors like type of migration (autumn or homeward), species, migratory state and food availability. Normally, precipitation and cloud cover reduce numbers of birds aloft but rarely suppress migration totally. Zalakevicius (1990) reported 18 nights with 'zero migration' during the autumn migratory period of 1986 above continental Lithuania. For many species then, overcast conditions alone are sufficient to reduce migration intensity, while others may well migrate or depart under solid overcast. A few studies have distinguished between different cloud types and found strong correlations between migratory volume or behaviour and cloud type or the 'thickness' of the weather front (Gauthreaux, 1991; Zalakevicius, 1990). For instance, Gauthreaux (1991) noticed that strategies of migrants from landing to riding the leading edge of the front, or shifting direction (Figure 2.59; section 3.7), depend on the thickness of the front and the season. The relationship, here, to visibility and fog are variable, but a wealth of data suggests that migrants can react directly to visibility (Richardson, 1990). Rain or snow, depending on their intensity, are known to either reduce migration volume or interrupt it (e.g. Alonso *et al.*, 1990). Snow storms in late autumn, however, may also lead to forced departures. In light rain some migration often continues, especially in diurnal migrants. If precipitation stops migrants at the normal time of departure they may take-off later in more favourable conditions, or *en route* sometimes divert around a belt of rain. Schindler *et al.* (1981) experimentally studied the influence of rain and overcast (darkness) on the migratory restlessness in garden warblers. Both factors reduced migratory intensity considerably. For similar results see Gwinner *et al.* (1992a).

### (d) Barometric pressure and other factors

Both field and laboratory studies suggest direct reactions to pressure. The amount of autumn migration is often correlated with increase in pressure as well as overall weather trends. In spring, the amount of migration tends to increase when pressure decreases (Richardson, 1990). Mascher and Stolt (1961), Stolt (1969, 1977) and Viehmann (1982) found, in experimental studies, that a number of species react to pressure. However, further controlled experiments must be done to understand the behavioural patterns of such reactions with respect to migration. There is experimental evidence that birds can sense even extremely small changes in pressure (Kreithen and Keeton, 1974). It could be selectively advantageous for migrants to respond directly to pressure rather than to a synoptic weather situation (Richardson, 1990). Another factor that has been tested is relative humidity. None the less, there is no proof that migrants respond directly to it, nor that it could serve as a key factor affecting migration. Finally, atmospheric stability and effects of small-scale atmospheric features and updraughts have been

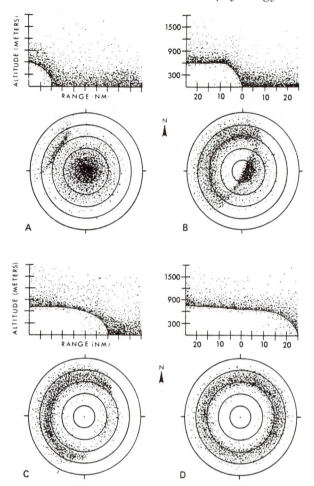

**Figure 2.59** Composite drawings based on radar displays and photographs showing the responses of migration birds during the passage of weak, shallow cold fronts in spring. A–D show the formation of the 'doughnut' display of bird echoes when the birds are flying higher than usual; nm = nautical miles. (Source: Gauthreaux, 1991.)

considered. Although data is scarce, there is little evidence that atmospheric stability affects migratory timing in non-soaring birds (Richardson, 1990). Updraught presence is, however, a key factor for soaring birds aloft (Kerlinger, 1989).

Richardson (1990) pointed out that in a list from 1978 where 11 topics deserving more study in order to understand weather-migration relationships were proposed considerable progress has only been made on four of them. Thus, despite the elaboration of many strong cor-relations, causal relationships are still rather suggestive. Therefore, not surprisingly, forecasting ability (Geil *et al.*, 1974; Richardson, 1990; Zalakevicius, 1990) is limited – comparable to that of the weather forecast. A large number of experimental studies have been done with some limited success in elucidating causal relationships (Farner, 1955; Berthold, 1975; Rabøl, 1990). Many of them should be repeated using more properly controlled conditions and larger numbers of experi-mental birds in better defined migratory states. In addition, experiments should be better designed to test individual meteorological factors.

### (e) Reversed movements

Reversed migration is a common behaviour which improves survival among early migrants in spring. It can be stimulated by falling temper-atures, snowfall or freezing conditions. In case of extremely changeable weather, migrants may finally reach their breeding area after a number of to-and-fro movements. Reversed movement is also found in several species in autumn. It may be initiated by rising temperatures and opposed winds, but may include dispersion movements (Elkins, 1988b), navigational insufficiency (chapter 3) or the development of novel migratory habits (chapter 4). Still, the phenomena of reversed move-ments in autumn are poorly understood.

### (f) Displacements of migrants

Exceptionally adverse weather, including intense storms, regularly lead to displacements of migrants and often to spectacular landfalls of disoriented or exhausted birds. Long-distance vagrancy, like transAtlantic vagrancy, is a phenomenon invariably correlated with weather which results in the sightings of 'extra limital' species or vagrants which are highlights for birders. Elkins (1988b) summarized some of these events. In Europe, such vagrants are known to arrive regularly from Asia (warblers, thrushes and buntings). Their occurrence is generally correlated with the distribution of high pressure areas which begin to form over Siberia in early autumn. In vagrants from North America (like waders, wildfowl and American robins) a con-nection has been made with wave depressions moving rapidly across the Atlantic or with the upper transAtlantic windflow. Disturbed weather patterns have also resulted in a tropical African bird reaching northwest Europe – the lesser gallinule (*Porphyrio alleni*), and the related corncrake (*Crex crex*) being blown off course about 20 times up to Greenland. Still, it should be kept in mind that part of these vagrancies may be attributed to ship assistance, population changes, spreading movements or long-distance dispersal, and range expansion and colonization processes [by founder populations, e.g. cattle egret (*Bubulcus ibis*) and others; Berthold, 1993], navigational errors (chapter 3) or newly evolving migratory habits (chapter 4).

Displacements and disorientation can lead to remarkable examples of falls of migrants (e.g. Sharrock and Sharrock, 1976). Heligoland was famous for millions of grounding migrants in former times when the old lighthouse had attracted huge masses of disoriented migrants under conditions of poor visibility (Schüz *et al.*, 1971). A spectacular example of a massive fall may be cited from Elkins (1988b). In September 1965 an estimated half a million birds landed along 40 km on the coast of East Anglia. A complex frontal situation with crosswinds and heavy rain caused this immense fall. Along a 4 km stretch, 15 000 redstarts, 8000 wheatears and 4000 pied flycatchers were observed. Walker and Venables (1990) cited another 'mega-fall' on the British Isles in October 1982.

### (g) Mass mortality

Mortality in migrants is generally not high due to exceptional risks *en route*. On the contrary, many migratory species are able to maintain their populations with relatively low reproductive efforts (section 1.1). However, extreme weather situations can cause mass accidents. Elkins (1988b) has listed such events like violent precipitation, hailstorms, thunderstorms, dense fog and low cloud, or hurricanes and other intense storms. In the extreme even coastal roosting flocks have been blown into the sea and drowned. During hurricanes, vessels have reported their ship decks being thickly covered with exhausted land birds. Mortality in such cases, where there was no chance of a landing, must be very high. Schüz *et al.* (1971) and Alerstam (1981) reported on mass mortality in white storks in sandstorms and of songbirds in poor visibility (section 3.7). Many birds may be killed while flying in fog and low cloud by collision with tall structures such as power lines. Nocturnal migrants may be especially sensitive to these effects, e.g. by collision with lighthouses or TV towers where often thousands of victims are found within short periods (Elmore and Palmer-Ball, 1991). Sudden cold spells immediately after the arrival of summer visitors, which do not allow for reverse movements, can also lead to serious mortality (Elkins, 1988b).

As already mentioned, meteorological factors can also influence body mass in migratory birds. Ormerod (1989) found that in swallows 84% of the variance in daily mean mass is explained by meteorological factors which are likely to affect the supply of aerial insects. In severe conditions, as in October 1974, when a succession of depressions moved over Central Europe, swallows can be caught by sudden cold spells to an extent that their migratory disposition collapses ('swallow disaster 1974'). In a study of delayed autumn migration of swifts from Finland in 1986, Kolunen and Peiponen (1991) found that part of the population was not able to accumulate fat deposits due to adverse weather. These birds did not depart. They started winter moult in October in Finland and perished in November when the frost started. Lehikoinen and Lindström (1988) showed that Chernobyl radioactive radiation was not the cause of this fatally delayed autumn migration, as had been suggested.

*(h) Long-term and global changes*

There are most likely several long-term and global changes in climate which obviously have already exerted considerable effects on avian migration. One is the so-called 'Sahel drought', i.e. a severe desertification in the Sahel zone savanna along the southern belt of the Sahara desert. It is most likely attributed to changes in the vertical atmospheric circulation brought about by anomalies in tropical ocean temperatures in combination with human mismanagement. This drought affects many Eurasian–African migrants through a lack of food, water and shelter (Folland *et al.*, 1986; Elkins, 1988b; Adejuwon *et al.*, 1990). The global increase in the mean temperature of about 0.6°C since 1900, primarily due to a build-up of carbon dioxide (known as 'global warming' or 'greenhouse' effect), has possibly already changed migratory behaviour in a number of species (Gatter, 1992). A further global rise of temperature could have rather dramatic effects on avian migration. Most likely, residents would generally benefit from a decrease in winter mortality, partial migrants would shift to sedentariness and, as a consequence, long-distance migrants would decline further (Berthold, 1991d; chapter 4).

## 2.18 EFFECTS OF POPULATION DENSITY AND SOCIAL RANK

As briefly outlined in section 2.4, Miller (1931) and Kalela (1954) developed the idea that the migratory drive in partially migratory species may be weakly developed. As a result, environmental factors could play an important role in stimulating their actual migration. Kalela (1954) suggested 'an evolutionary connection of true bird migration with some kinds of irregular migration or irruption', as could be the case in birds like the great tit (*Parus major*). He further stated that it is generally agreed that irruptions may be due to two factors, either a food shortage or high population density. With respect to the suggested connection between both types of migration he continued: 'The author believes that true migration originating as a consequence of autumnal territorial behaviour (or increased aggressiveness in general) is not uncommon in birds'. The basic idea is that due to conflicts in autumn, losers are forced to leave as migrants. Gauthreaux (1978, 1982) has extended these ideas to a comprehensive theory which he (1982) summarized: 'It is generally agreed that food availability is an important ultimate factor in the evolution of migratory strategies and that intraspecific competition is an important proximate factor that mediates the differential movements of birds in dispersal, irruptions, and migration. The outcome of intraspecific competitive contests can be predicted on the basis of asymmetries in fighting abilities of the contestants . . . , and the asymmetries are commonly those associated with age (experience) and sex, the same asymmetries that determine an individual's status in a dominance hierarchy . . . Therefore, it is not surprising that a number of ornithologists have suggested that the dominance status of the individuals in a population is important in determining which individuals leave and which individuals stay when resource levels decline'. This 'behavioural dominance hypothesis' predicted that the dominance status would also have immediate effects on how far migrants move and how wintering distribution is regulated with respect to sex and age classes. This idea, the so-called 'behavioural-constitutional hypothesis' is currently being intensively debated (Swingland, 1984; Lundberg, 1988; Schwabl and Silverin, 1990; and below).

Behavioural dominance does not appear to play an important role in cases of obligate partial migration for which genetic control mechanisms have been experimentally established, as in the blackcap or song sparrow. In these obligate partially migratory species only a few individuals shift from initial sedentariness to migratoriness, indicating fairly rigid genetic programs (Berthold, 1984a). Changes in the opposite direction are more common but may be explained by maturation processes (section 2.4). However, more weakly programmed individuals may also underly some conditional control, like European robins (see below).

What is the evidence for immediate behavioural-constitutional effects in other migrants? If such effects exist, they would be expected in facultatively partial migrants (like tits) and in species performing irregular irruptive movements (like crossbills or waxwings). In these cases, immediate environmental input in triggering migratory movements is generally believed to be essential, although the question still remains of whether it is more related to food, population density or social status.

Several authors have suggested that factors related to population dynamics and population density can be of prime importance. As the result of a detailed analysis of irruptions, Lack (1954) concluded that: 'the proximate stimulus is sometimes food shortage and sometimes a behaviour response to high numbers'. In the latter, he assumed that crowding factors of population pressure provided a proximate stimulus for departure, comparable to the (experimentally established) situation in migratory locusts. This view of density regulation was deduced from the observation that 'the really big movements sometimes start before the fruit crop on which the bird depends is ripe, the irrupting flocks sometimes pass through areas where their food is plentiful' and 'the birds seem abnormally restless and excited . . .'. Berndt and Henß (1967) studied great tits and found immigrations related to high reproductive output (Figure 2.60). They suggested 'for the inducing of such a density-dependent invasion a psychologically acting pressure factor, surely strengthened through territorial conflict and probably food shortage . . .'. The 'pressure factor' is thought to be mediated via a 'hypersensitization' leading to 'a neuroendocrine shifting from resident or local-migrant behaviour to a migratory drive . . .'.

In a recent analysis of movements in a partial and irruptive migrant, the blue tit in Sweden, Smith and Nilsson (1987) found that among juveniles more than 40% of the females and a significant proportion of the males migrated. At the same time, this was found in considerably fewer adult females and virtually no adult males. The authors concluded that the 'findings are consistent with the "dominance hypothesis" as an explanation of partial migration, i.e. that the individuals lowest in rank migrate'. This correlation alone, however, does not prove a cause-and-effect relationship between social status and migratory habit. In partially migratory Mediterranean blackcaps, for instance, first-year females, as the lowest ranking age and sex group, show the highest

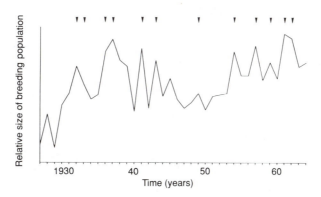

**Figure 2.60** Eruptions, indicated by arrows, of great tits in central northern Germany and their dependence on populations size. (Source: Berndt and Henß, 1967.)

proportion of migratory active individuals but the expression of the habit is based on a sex-linked inheritance (section 2.4).

In many partially migratory species intraspecific increase in migratoriness from adult males through females to first-year males and females is found. None the less, dominance status is not necessarily involved in its expression. This fact is supported by results on obligate partially migratory European blackbirds. According to studies of Schwabl and Silverin (1990), it is unlikely that in this species competition for autumn territories controls partial migration 'since aggressive interactions are rather infrequent during this time'. Further, it is unlikely, at least for juveniles, that possession of a territory acts as proximate factor in the control of partial migration. In several species late juveniles are apparently not able to establish territories in autumn and then migrate (Schwabl and Silverin, 1990). But also in these cases it remains unclear whether this failure immediately causes migratory movements or whether other factors are involved. Due to these uncertainties, Schwabl and Silverin (1990) concluded in their review that: 'Although these correlations are suggestive, experimental evidence that social dominance proximately controls partial migration is still missing'. van Balen and Hage (1989) studied the annual degree of movement in great tits and blue tits for the period 1957–1986 on the basis of ringing recoveries and by correlating them with environmental variables. For the great tit, the proportion of 'long'-distance recoveries decreased in the course of the study period, increased at high population densities and was not related to beech crop index or winter severity. The authors suggested that population density probably affects the degree of movement in a proximate way, while the beech crop influences movements indirectly, by its effect on population size through juvenile survival.

Adriaensen and Dhondt (1990) and Adriaensen *et al.* (1990) found the following in detailed studies of partially migratory European robins in Belgium. In a park and gardens most males were resident, in the woodland 70% of breeding males were migratory, as were almost all females in all habitats. Mean local survival of resident males was higher than local survival of migrant males, and mating success decreased with settling date from early settling residents to later settling migrants. With respect to this unbalanced reproductive success of the resident and migratory fractions 'the migrants are making the best of a bad job'. From these results the authors concluded: 'In this case partial migration is obviously not maintained by natural selection as a mixed evolutionary stable strategy [mixed EES, as proposed by Lundberg, 1988; but which possibly is inaccessible, Nowak, 1990], with the two strategies having equal pay offs at equilibrium'. Instead, they argue, partial migration could be a conditional strategy in which external, non-genetic factors strongly influence the migratory behaviour of individuals, and they conclude that socially dominant individuals are more likely to be resident. On the other hand, they state that 'this conclusion does not contradict the strong genetic basis of migratory restlessness in the species' (section 2.4). They further report: 'In our study, no changes were observed in the migratory behaviour of an individual during the course of its life'. Such changes, however, would be among the main expectations in case of a conditional strategy. In an earlier study, Adriaensen (1988) found in an analysis of ringing recoveries in Belgium that: 'Some recoveries suggest that robins may change their migratory behaviour from one year to another'. In my opinion, the most likely explanation for the control of partial migration in these robins is in accordance with the genetic data obtained for this and other species with similar relationships: largely genetic determination of migrants and residents with a strong bias towards increased migratoriness in females based on sex-linked inheritance (as in blackcaps of S France; section 2.4) in a population which, as a whole, is in fairly rapid transition towards a more and more resident population (as presently is commonly occurring, section 4.2).

Difficulties in demonstrating the immediate effects of population density and social status on migration behaviour are also seen in studies on species with pronounced differential migration like the dark-eyed junco. As will be treated in more detail in section 2.19, this species tends to segregate in the winter range according to sex and age. It provides a good test species for dominance factors because, in contrast to most species, young birds winter farther north than older birds. North to south, the most abundant classes are young males, adult males, young females and adult females (Nolan and Ketterson, 1990b). A number of studies have so far yielded the following results. Terrill (1987) experimentally tested the effects of restricted food and social dominance on nocturnal migratory activity (restlessness; Figure 2.61). As will be treated in section 2.20, food restriction tended to increase restlessness in both dominant and subordinate birds, however, subordinates showed significantly more migratory activity than dominants or solitary controls. Further, subordinate birds continued restlessness after dominants. Solitary controls ceased this activity for the remainder of the winter. The author suggested, from these results, that migratory behaviour in this species can be affected by social conditions. Presumably, this type of migratory regulation enables dark-eyed juncos to minimize migratory distance while enabling the prolongation when necessary. Rogers *et al.* (1989) have made the following tests. The behavioural dominance hypothesis of differential migration was examined by comparing interactions within two types of junco dyads. Dyads consisted of either birds of the same sex–age class, caught at

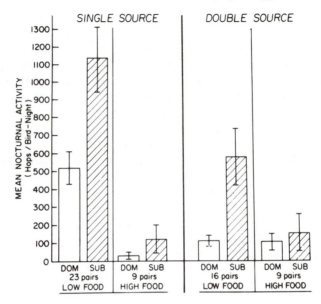

**Figure 2.61** Effects of social environment, food abundance and distribution on zugunruhe of dark-eyed juncos. Low food, 8 g of food per day of each pair; high food, 14 g. Single source, food was placed in a single source at the centre of each pair's cage; double source, food divided into two sources. SUB, subordinate members of pairs; DOM, dominant partners. (Source: Terrill, 1990.)

different latitudes in the winter range or of young and adult birds caught at different latitudes. The results were inconsistent with the predictions of the dominance model. The authors concluded that they: 'indicate, at the very least, that in migratory *J.h. hyemalis*, dominance does not play as an important a role in determining latitude of winter residence as has been suggested'. They also stated that the process determining wintering sites may be complex and that, therefore, the results obtained do not completely rule out a role of age-sex-specific dominance relationships which may be an important factor. In another analysis, Nolan and Ketterson (1990a) compared first-capture dates to determine whether young juncos establish winter residency at earlier dates than adults and whether prior residence effects make them dominant before adult birds pass through their wintering area. Residents were, however, not caught earlier than transients, and adults were caught earlier, not later, than young among transients. Thus, settlement in the winter range does not proceed from north to south, and 'dominance established through prior residence cannot account for the concentration of 1st-year males in the northern part of the winter range'. Relative effects of both prior residence and age in relation to dominance hierarchies on the wintering grounds have been tested by Cristol *et al.* (1990) in aviary experiments with dark-eyed juncos. When given prior residence, young birds clearly dominated adults. If these results apply to wild populations, they indicate that young juncos might gain in dominance status during winter if they timed their autumn migration so as to arrive on the wintering grounds before adults. That prior residence confers an advantage on both adult and first-winter birds was also confirmed by Holberton *et al.* (1990). However, they conclude that in nature adults are more likely to experience this advantage as a consequence of earlier arrival at stopover sites, of site fidelity to the wintering grounds and of a relative higher dominance success due to age-related darker plumage. Cristol and Evers (1992) found, in extended experiments, no evidence that social dominance is a mechanism responsible for variance in migration distances within classes of juncos. Wiley (1990) finally reported on a 'coat-tail' effect in dark-eyed juncos: an influence of dominant individuals on the relationships of their familiar subordinates. If this effect applies in the field, it would create an advantage for subordinates associating with familiar dominants. It is amazing how many factors may be involved in the control of migration and wintering strategies of this species. However, many of these obviously conflicting and contradictory findings may now be seen from a new point of view, since Holberton (1993) has detected that differential migration, also in this species, is based on endogenous factors as outlined in section 2.3.

A few other recent attempts to test the behavioural dominance hypothesis also exist. Senar *et al.* (1990) investigated whether the observed aggression in feeding flocks of siskins (*Carduelis spinus*) was related to residents attempting to exclude transients from feeding patches or whether it occurred between transients to permit access to food. Observations at bird feeders showed that residents were dominant over transients and that transients interacted with other transients more often than expected. However, whether dominance among transients reduced further movements in this group remains open. Belthoff and Gauthreaux (1991) tested the behavioural dominance hypothesis and other hypotheses in the house finch (*Carpodacus mexicanus*). This species is especially suitable because females dominate males despite their smaller size. Examination of records of banded individuals throughout the eastern United States, following the introduction of the population in 1940, produced data which suggested the evolution of partial migration (chapter 4) with a trend in recent years for females to migrate farther than males. The authors concluded therefore that only the body-size hypothesis (section 2.20) correctly predicts the pattern observed among wintering finches. Prescott (1991) analysed banding data of evening grosbeaks (*Coccothraustes vespertinus*) to demonstrate differential migration in males and females. No conclusions were drawn whether social dominance behaviour, body-size differences or other factors proximately caused the more southern wintering range in females. Choudhury and Black (1991) studied aggression and dominance behaviour in pochards (*Aythya ferina*). Wintering flocks of this species, like other diving ducks, often show a great disparity in sex ratio with males predominating in the north. The explanation of the behavioural dominance and 'dispersal hypothesis' is that males dominate and exclude females from limited food sources, forcing them to migrate further south. Under manipulated feeding conditions, male attacks were directed more at females than at other males. Male dominance appeared to influence the timing of female foraging activity rather than foraging location. The authors suggested that, in the long run, feeding second may be energetically too expensive for females and hence cause them to migrate further. Lindström *et al.* (1990) offered bluethroats food *ad libitum* in the field. Prior to migration, lean birds won significantly more interactions than fat birds, probably owing to higher motivation. Dominance between individuals also shifted and was not correlated with size. During migration, however, dominance between individuals was constant, positively correlated with size, and fat deposition rates were positively correlated with dominance status (section 2.7.1). Sandell and Smith (1991) tested dominance, prior occupancy and winter residency in the great tit in aviary experiments. Dominance proved to be affected by prior residency and thus, the authors concluded, selection likely operates in favour of winter residency in this species. Holmgren and Lundberg (1990) investigated two competition mechanisms [dominance correlated with traits (competitive weights) and dominance due to prior occupancy] with respect to the evolution of migration patterns in dominance-structured populations. In the case of dominance correlated with traits, mainly leap-frog migration pattern developed. In the case of prior occupancy, chain migration patterns were most likely to develop.

To sum up this discussion, Schwabl and Silverin (1990) concluded that, despite suggestive correlations, experimental evidence that social dominance proximately controls (partial) migration is still missing. This also appears to be largely valid for differential migration. In many cases, like irruptions, it is extremely likely that social factors are proximately involved in triggering movements. However, fully convincing evidence is still lacking. Cornwallis and Townsend (1968) have suggested that for invasions of great tits the imbalance between food supply and population level is decisive rather than the absolute level of the two factors. If this is true it will be a difficult task to sort out all behavioural and energetic components in future studies, which should also include observations of social interactions during migration.

## 2.19 INFLUENCES OF THE RESOURCE AVAILABILITY

It is generally accepted that most bird movements are ultimately caused by seasonal changes in resource availability, mainly by changes in food availability (Cornwallis and Townsend, 1968; Baker, 1978, 'initiation factor model'; Gauthreaux, 1982; Berthold, 1993). Even part of moult migration can possibly be explained in this way ('food quality hypothesis', Owen and Ogilvie, 1979). Terms like 'contingency movements', 'deficiency movements' or 'follow-up movements' (section 1.1), related to changing food situations, often imply that nutritional conditions may proximately control migratory movements. This, however, has only been clearly demonstrated in a few cases. The rather complex situation will be treated here. First, evidence from field studies can be considered and secondly those from experimental studies.

A number of cases of movements within the wintering area are possibly immediately related to food conditions. They will be treated in the following section. A peculiar recent change in migratory behaviour, which is possibly proximately controlled by food availability, refers to eiders (*Somateria mollissima*). This species has shown an exponential population increase over the last few decades, especially in the Baltic area. The population increase appears to be due to eutrophication. Good breeding success and extreme densities can lead to novel mass exodus of juveniles into Central and southern Europe which is thought to be controlled by food shortage as a deficiency movement (Helbig and Franz, 1990).

In cases of hard-weather movements, or escape migration, when birds are immediately leaving an area due to a cold spell or snow cover, including food shortage, it is practically impossible to judge from the field situation whether food shortage itself, low temperatures or other factors were the proximate factor for departure. A reasonable judgement appears somewhat easier in the case of irruptions which normally start less precipitately. Lack (1954) tried to precisely analyse the possible proximate role of the food in irruptions. He stated that two main theories have been put forward to explain irruptions: on one hand, they occur following an unusually successful breeding season and are due to overpopulation (section 2.18); on the other hand, they may be correlated with failure of the fruit crop or other basic food of the species in its normal haunts and not with high bird numbers. He went on to state that: 'Neither view provides a full explanation, and there is some truth in both'. The best evidence that migration in irruptive species may be linked with their food supply is, according to Lack (1954), that of Reinikainen (1937). He travelled over the same (120 km) distance in Finland on skis each Sunday in March for 11 years. He recorded the number of crossbills (*Loxia curvirostra*) and estimated at the same time the size of the cone crops of spruce and Scots pine. The number of breeding crossbills was strongly correlated with the size of the spruce cone crop (Figure 2.62). Crop failure resulted in emigration. In other cases correlations were not so clear. This was precisely demonstrated by Newton (1975) in his review on the movements of crossbills. He also found that irruptions occurred in years of widespread crop failure. In addition, two lines of evidence have suggested that irruptions cannot be attributed to crop failure alone and that population changes are probably also involved. First, not all poor crops result in irruptions; and secondly, in some years emigrations begin before the new crop is ripe. In Sweden, where movements have been studied in detail, irruptions occurred in years of varying crop sizes, except the largest, but less than half the poor crops have resulted in irruptions. Hence, Newton (1975) concluded that, while food has a strong influence, it cannot be the only factor involved, and high bird numbers may also be needed. Further, while most movements are in July and August, quite a number begin in late June or July and some even in May. This suggests again that mass emigration is not simply linked with the failure of individual crops. Unfortunately, Newton (1975)

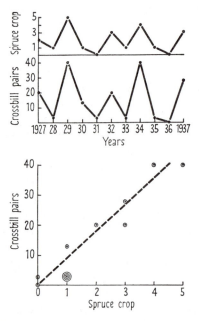

**Figure 2.62** Relation between population density of crossbills and spruce cone crop. Crossbills in number of pairs per 120 km; spruce crop classified in six categories. (Source: Lack, 1954, based on data from Reinikainen, 1937.)

found little information about crowding as a secondary stimulating factor. However, he concluded: 'The hypothesis that best fits these various observations is that high numbers are necessary for irruptive movements, but that the size of the spruce crop modifies this. Once the population is high, emigration probably occurs in response to the first inadequate crop . . . The problem remains, however, as to whether crowding alone will stimulate emigration, irrespective of food. The earliness of some irruptions suggest that it might, but none of these early movements have preceded good crops (which is needed to test this view), and it is possible that the birds can assess a developing crop before it is ready'.

The possible close relationship between food supply and population level lead to the hypothesis of Cornwallis and Townsend (1968) (already cited in the last section) that: 'It is neither the absolute level of food supply nor the absolute level of population that is decisive, but the balance (or, rather, imbalance) between the two'. This hypothesis, derived mainly from results on great tits, suggests that high population levels are not related to emigration with plentiful food supply and that crop failure does not lead to emigration at low population levels. According to this hypothesis, even a stable population may turn to a critically imbalanced high one when food supply decreases.

So far, for none of the irruptive species like crossbills, tits, waxwings or others, has the exact proximate role of food supply for movements been demonstrated unambiguously. Thus, from the field studies it remains open whether food shortage, population level or an imbalance between both factors acts as the most important trigger for departure. In some special cases, food does appear to play a crucial role. Fieldfares (*Turdus pilaris*), for instance, are often capable of staying during a period of harsh winter conditions in rowanberry or apple trees and to maintain themselves on remaining berries or rotten fruits. The last individuals stick to their place and often winter without any competition. Their residency or departure appears to depend solely on the remaining crop. In most other cases additional experimental evidence is needed to draw any conclusions.

Experiments have shown that altering food availability can clearly effect migratory behaviour in captive birds in several ways: the pattern of migratory restlessness can be changed during the migratory period and facultative migratory restlessness can be induced after normal migratory periods. It has been known for a long time that animals, including birds, when deprived of adequate food, exhibit hyperactivity. This can turn into 'pathological' hyperactivity and lead to death (McMillan *et al.*, 1970). A similar behaviour to this 'hunger restlessness' (section 2.1) in case of water deprivation was not found in several migratory bird species tested (Berthold, 1988b). 'Thirst restlessness' may, however, well occur in lean birds in which metabolic water production due to fat metabolism (section 2.10) is insufficient.

Merkel (1966) was one of the first to systematically investigate how temporary food deprivation increases activity in migratory birds. When white-crowned sparrows were fed only in the afternoon, locomotor activity increased strongly during the morning hours of food deprivation ('restlessness set in'). In some individuals restlessness also appeared during the night, apparently as a result of starvation. A systematic study of food deprivation in garden warblers showed that food removal at any 2 hour period during daytime resulted in a sudden rise of locomotor activity up to about threefold of the normal level (Figure 2.4). This simple experimental stimulation of 'hunger restlessness' may, according to Merkel (1966) 'represent a model of how primitive migration flights were caused'. Increasing locomotor activity in case of food deprivation must, however, not necessarily be linked with migratory behaviour because resident species also behave correspondingly (Aschoff, 1987). Yet experiments with pied flycatchers suggest such a relationship (Thalau and Wiltschko, 1987). In this species, food deprivation produces a threefold increase in diurnal locomotor activity, as in the garden warbler, and possibly also an increase in nocturnal migratory restlessness. Tests of orientation behaviour during the first part of the autumn migratory season showed that both experimental birds and controls were significantly oriented towards southwest in the early morning (which represents their normal migratory directions). Later during the day, only the experimentals continued to be active in a southwesterly direction. Thus, food deprivation obviously induced an increase in directed day activity. This could mean that birds of this species, normally migrating at night, can continue migration into the day when they land on a stopover site with an inadequate food supply. Similar results were obtained in food deprived blackcaps (Terrill and Berthold, unpublished). The occurrence of restlessness as a result of (enforced or voluntary) fasts was recently also reported for some waders and penguins (Piersma and Poot, 1993).

In series of experiments with spotted flycatchers and garden warblers Biebach (1985) and Gwinner *et al.* (1985, 1988) were able to demonstrate a number of details of how food deprivation and food supply may influence migratory activity. In both species, food deprivation was extended over several days until body mass reached premigratory levels with only a small fat reserve (Figure 2.63). This was followed by a period of refeeding in which the birds were given only enough food to allow a daily mass gain of about 0.5 g. In the nights following food deprivation there were slight, but significant, increases in migratory restlessness. During the refeeding phase, when body mass increased, migratory restlessness decreased sharply. This reduction was more

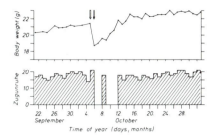

**Figure 2.63** Body weight (curve) and intensity of zugunruhe (columns) of an individual garden warbler. The bird received no food on October 5 (open arrow) with the effect that within 24 hours its body weight dropped by about 3 g. Starting October 6 (solid arrow) the bird received controlled amounts of food so that its body weight increased by about 0.5 g per day over the following 9 days. Zugunruhe was slightly increased in the night following food deprivation but was drastically reduced in most of the nights during the first part of the refeeding phase. The slight decrease in body weight from October 8 to 9 was accompanied by a burst of zugunruhe in the night between these two days. (Source: Gwinner *et al.*, 1985.)

pronounced during the first half of the refeeding phase. Intense restlessness usually resumed before the birds had attained final body mass. The authors suggest that the increased migratory restlessness during food deprivation may mimic the situation of a bird which spends the day at a resting place with poor feeding conditions, e.g. in a desert. Under such circumstances a bird would do best to migrate further on the following night. The reduced migratory restlessness during fat deposition may reflect the situation of a bird in a staging area where it can effectively refuel. In such a situation a migrant would do best to stay there until it has regained sufficient energy reserves for a longer stage, possibly including overcoming of inhospitable areas. These experimental data are in good accordance with field data on body mass and resting behaviour of birds investigated in the Sahara desert (Biebach, 1985; section 2.16). Similar results as in garden warblers were also obtained in dark-eyed juncos (Terrill, 1987). Highly restricted food tended to increase migratory restlessness, especially in subordinate individuals (section 2.18). Yong and Moore (1993) tested several American thrushes immediately after spring passage across the Gulf of Mexico. Lean migrants displayed less night activity than migrants with undepleted fat stores but were more active during the day (presumptive feeding behaviour). As lean individuals replenished fat stores, nocturnal restlessness resumed whereas day activity decreased. These results also support the idea that a migrant's energetic status modifies the programmed course of (homeward) migration.

In another experiment (Gwinner *et al.*, 1988), garden warblers were food-deprived at intervals from the autumn migratory period through winter to the winter moult period (Figure 2.64). As in the experiments mentioned above, food deprivation consistently caused an increase in

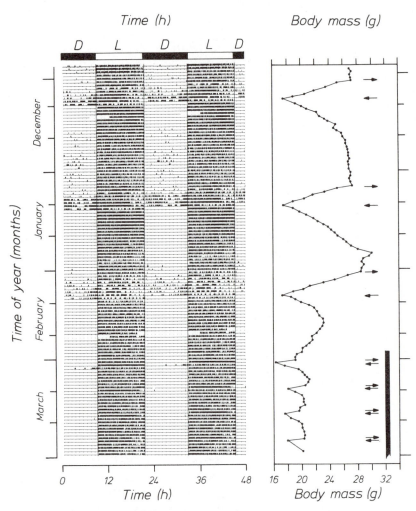

**Figure 2.64** Diurnal activity (dense recordings during light phase, L) and nocturnal restlessness (fewer recordings during dark phase, D) of garden warblers; also depicted are changes in body mass (above) and winter moult (solid bar). Reduced food supply and body mass reductions (arrows) trigger nocturnal restlessness until winter moult sets in. In order to give a clearer picture of the activity at night, the whole pattern is duplicated. (Source: Gwinner *et al.*, 1988.)

duration and intensity of nocturnal migratory restlessness. After the end of the normal migratory season, and the termination of the endogenously programmed migratory restlessness, food deprivation still, however, led to a re-induction of nocturnal restlessness. This continued until the winter moult had started, when facultative migratory activity could no longer be induced. These results again suggest that nocturnal migratory activity is enhanced in migrants staging at unprofitable stopover areas. They further suggest that nocturnal migratory activity may be reactivated, and facultative migration or resettling movements may be initiated, in birds that have settled for the winter in an area with deteriorating food resources. Such a situation was described for yellow-rumped warblers which moved at night to more southerly wintering areas when food supply decreased at the initial site (section 2.20).

To conclude, according to experimental evidence, food shortage can increase programmed migratory activity and induce facultative migratory activity, and presumably facultative migration, after normal migratory periods. Food richness can reduce migratory activity and favour energy restoring for further migratory flights in fat depleted individuals. In order to understand the situation in the wild, the supplemental role of intra- and interspecific competition for food resources in these relationships has to be analysed in future studies. The critical energy levels which make birds sensitive to change programmed migratory behaviour have to be established, and the extent of alterations in relation to endogenous programs remains to be shown.

## 2.20 CHOICE OF GOAL AREAS AS A RESULT OF PROGRAMMED AND FACULTATIVE MOVEMENTS

Goal areas for migrants are first of all wintering grounds and breeding areas, and to a lesser extent special staging areas (or intermediate goals; Schüz *et al.*, 1971). Reproduction normally restricts migrant's movements after return to a specific breeding site, most often within a population-specific breeding area. In contrast to this, wintering grounds choice reflects a far more complex situation (Terrill, 1990).

Migrants' wintering strategies range from a high degree of site fidelity and tenacity to largely mobile site choice. Ketterson and Nolan (1990) have proposed use of the term 'site attachment' for the processes leading to formation of a bird's preference for a location and defined 'site fidelity' as the act of a migrant in returning to a location occupied earlier. Schmidt-Koenig (1975) used the term 'navigational imprinting' for breeding site fidelity.

The traditional idea that migrants cover the distance between the breeding area and the wintering grounds as quickly as possible to stay there in full winter residency has changed dramatically. The discovery of large-scale midwinter migration in Africa, America, Mediterranean areas and elsewhere 'has given a new dimension' to the phenomenon of bird migration (Curry-Lindahl, 1981). The broad behavioural spectrum from winter residency to almost continual movement appears to reflect an environmental gradient in resource availability, ranging from stable to predictable to unpredictable (Terrill, 1990).

### 2.20.1 Winter-site fidelity and tenacity

Intensified bird ringing in recent decades has produced increasing evidence that not only individuals of large species like cranes or swans migrate annually to, and remain in, the same wintering sites but also that many small birds behave similarly. Analyses of ringing recoveries (Zink, 1973–1985; Curry-Lindahl, 1981) have revealed hundreds of cases of winter-site fidelity or winter recurrence. Many seasonally breeding passerines establish territories and also sing during the non-reproductive phase. In some species, males and females defend individual winter territories and both sexes sing, in other species pairs defend territories (Kriner and Schwabl, 1991). Overwinter site attachment may be so strong that all sex and age groups occupy territories throughout the winter, with most individuals remaining within <50 m, as has been demonstrated in American warblers (Holmes *et al.*, 1989). Or, as another example, in white-throated sparrows, it was found that wintering ranges of most individuals extended for 100 m or less, birds tended to return to the same location and to have ranges of similar sizes from year to year, that dominant birds tended to have small ranges and that range size decreased with age (Piper and Wiley, 1990). At present, in most migrants, the proportions of individuals that use fixed wintering areas and those which move around cannot be quantified. This is only possible in a few species like rare cranes where practically the whole wintering population has been monitored over a number of years. For many other species, however, tendencies are well established. According to Debussche and Isenmann (1984) and others, there are clear species differences obviously related to nutritional ecology. Some species like European robins are fairly territorial during large parts of the wintering period, others like the blackcap are more vagrant and may move over wide areas depending on the availability of certain fruits (Berthold *et al.*, 1990b). In wintering dunlins (Ruiz *et al.*, 1989) resident and mobile groups within the same area in California were found. In contrast to findings in passerines at staging areas during autumn migration (section 2.15), more mobile birds showed higher body mass and fat deposits. Similar results were obtained in wintering wood thrushes (*Hylocichla mustelina*) in Mexico where telemetry studies demonstrated the existence of both sedentary individuals and wanderers (Rappole *et al.*, 1989; Winker *et al.*, 1990). The latter had higher fat reserves but seemed to incur greater mortality due to predation. In white-throated sparrows, however, high levels of subcutaneous fat were found in site-faithful birds which were predominately dominant individuals (Piper, 1990).

Ketterson and Nolan (1990) have discussed basic questions of goal determination, goal finding, site attachment and site fidelity in migrants. They conclude, mainly referring to experiments of Sniegowski *et al.* (1988), where indigo buntings were caught on their breeding territories and later released there again that 'at least some individuals recognize the site and this recognition overrides migration'. They further suggested that there is little evidence to justify the general conclusion that natal site attachment is an imprinting-like process and that the data seem to point just as consistently to a process of gradual learning. Similarly, with respect to the attachment to the winter site, they found evidence for a sensitive period – the most important criterion for imprinting – 'slim at best'. Nolan and Ketterson (1991) tested whether winter-site attachment in young birds occurred during a sensitive phase as the result of an imprinting-like process (comparable to breeding-site attachment; Löhrl, 1959). Experiments in which trapped dark-eyed juncos were released at the capture or confinement site after having been caged did not produce unequivocal results. At the same time, they did not rule out the existence of a brief sensitive phase for site attachment. Also, the question whether recognition of the winter site in this species may result in a suppression of both migratory fattening and activity is still open (Nolan and Ketterson, 1990).

The lowest winter-site fidelity has been found in species performing wide-angle migration. In species like the redwing (*Turdus iliacus*) or brambling, with late breeding seasons, wide-angle migration, and annual and seasonal winter nomadism, this type of movement may reduce competition pressure during wintering. Individuals of these species may winter as far west as Britain in one year and in the Middle East or Georgia in the next, as ringing recoveries have shown (Alerstam, 1991a).

## 2.20.2 Winter movements

Intensive ringing, trapping and censusing have shown that many species formerly believed to start wintering at fixed places in late autumn actually have a far more protracted 'autumn' migration. Many species of Old World warblers, for example, are regularly found on heavy southward nocturnal migration through East Africa in

December and January, and to some extent even in February (Pearson and Lack, 1992; Pearson, pers. comm.; Figure 2.65). The marsh warbler offers a rather striking example in which the 'autumn' migratory period can last 6–7 months and overwintering only 2–3 months. Spring departure of these birds begins in March (for details see Terrill, 1990). During the wintering period, however, marsh warblers defend territories for longer periods than during the breeding season as this period includes moult (Kelsey, 1989). There is experimental evidence (Berthold and Leisler, 1980; section 2.3) that, in species with such regularly extended autumn migratory movements, endogenous migratory programs may regulate the expression of these activities. In other cases, however, extended migration within the wintering grounds is clearly facultative and thus exogenously controlled. In many species, the immediate association between midwinter migration, unusually cold weather, drought or changes in food availability, supports this view of facultative-route prolongation or abbreviation (Terrill, 1990). In several cases, the facultative nature of movements has clearly been demonstrated.

Probably the first experimental evidence for variable choice of goal areas comes from Perdeck's (1964) displacement experiments with

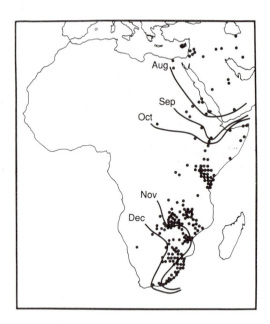

**Figure 2.65** The spatiotemporal course of migration of European marsh warblers to South African winter quarters. (Source: Dowsett-Lemaire and Dowsett, 1987.)

starlings. Juveniles of this species, which do not show high degrees of winter-site fidelity, terminate their autumn migration if environmental conditions at a particular site, reached at the appropriate time and resulting from the displacement procedure, are favourable for survival. If conditions are unfavourable, they prolong migration. Very clear results were obtained on the yellow-rumped warbler (Terrill, 1990). Numbers of these warblers wintering in the southwestern United States vary greatly between years. Decreases are typical for winters when unusually cold conditions occur. Further, the magnitude of change in numbers of these warblers correlated with the availability of insects and corresponded to increases at more southerly wintering sites in Mexico. The clearly facultative winter movements between the two wintering areas involve nocturnal migratory behaviour, as was demonstrated by orientation tests and studies of nocturnal tower kills.

There is further evidence for facultative winter movements from experiments with caged individuals. Several stressful conditions can reactivate, or extend, nocturnal migratory activity in winter during a restricted period. Dark-eyed juncos, especially subordinates, showed increased or induced nocturnal activity during periods of food restriction in December and January. After late January, migratory activity could not be reactivated (Terrill, 1990). Similar results have been found in garden warblers (Gwinner *et al.*, 1988) and blackcaps (Terrill and Berthold, unpublished). Food deprivation during the migratory period may increase nocturnal migratory activity, and after the migratory period may reactivate it. But after mid- to late winter, with the onset of winter moult, food deprivation no longer stimulated migratory activity. This is approximately the time of year in which winter-site fixation has been shown to occur in the species tested.

From these results, Terrill (1990) suggests that migratory programs in the experimentally-tested species change from: (1) endogenously stimulated migratory behaviour during the regular autumn migratory period; to (2) facultatively stimulated migratory behaviour in midwinter; and finally (3) to non-migratory behaviour in late winter. Winter movements may thus be performed either on the basis of final parts of endogenous programs or facultatively as an immediate response to environmental conditions (Figure 2.66).

In many species, individuals of different sex and age classes share the same wintering areas and also occupy the same habitats (Holmes *et al.*, 1989). In many other species, however, differential migration leads to considerable latitudinal variation in sex and age classes on wintering grounds. There are five hypotheses that attempt to explain this variation (Belthoff and Gauthreaux, 1991; and below). The behavioural dominance hypothesis predicts that intraspecific competition forces subordinate members of sex or age classes to migrate farther; it has been

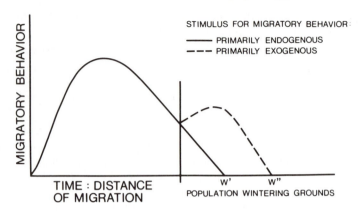

**Figure 2.66** Model of a change from a period in which the stimulus for migratory behaviour is fundamentally endogenous to a period in which the stimulus for further migration is facultative. If environmental conditions are favourable, and relatively stable, no facultative migration occurs and birds terminate their autumn migration when the endogenous program is terminated. (Source: Terrill, 1990.)

treated in detail in section 2.18. The arrival time/sexual selection hypothesis suggests that intrasexual selection drives individuals of the sex that establishes breeding territories to winter closer to the breeding area, in order to arrive earlier on the breeding grounds. The body size/physiological hypothesis predicts that individuals of the larger sex winter closer to the breeding range in higher latitudes because of their greater tolerance of low temperatures and food shortage. A prior residence effect hypothesis proposes that, as in dark-eyed juncos, residence of early departing individuals might make this set of birds dominant over those passing through later (Nolan and Ketterson, 1990). These hypotheses are presently being intensively tested and debated (Terrill, 1987; Rogers *et al.*, 1989; Nolan and Ketterson, 1990; Belthoff and Gauthreaux, 1991; Choudhury and Black, 1991; Prescott, 1991). An example supporting the sexual selection hypothesis has been published by Arnold (1991) concerning the American kestrel (*Falco sparverius*), whereas in sparrowhawks the body size hypothesis appears to be valid (Payevsky, 1990). For some species, like house finch or evening grosbeak, evidence more or less supports the body size hypothesis (Belthoff and Gauthreaux, 1991; Prescott, 1991), for others, to a varying degree, the behavioural dominance hypothesis (section 2.18). Senar *et al.* (1992) have studied patterns of residence, site fidelity and nomadic tendencies of siskins at wintering sites in Britain and Spain. Residents were consistently heavier, and this may also fit the dominance hypothesis. It has, however, to be kept in mind that differential migration may well have an endogenous basis and that differences in dominance, body size and related ecophysiological characteristics may act as ultimate, rather than proximate, factors. Terrill and Berthold (1989) have shown that in German blackcaps, where females tend to migrate farther than males, hand-raised females held in constant conditions showed significantly more, and significantly longer, autumnal migratory activity (section 2.3; Figure 2.67). In partially migratory blackcaps in the Mediterranean area sex-linked inheritance of migratory behaviour has been found to cause a high proportion of migratory first-year females (section 2.4). Holberton (1993) obtained similar data in dark-eyed juncos, suggesting that sex-biased differential migration is also heritable in this species. Thus, an 'endogenous program' hypothesis may also explain intraspecific differences in the choice of goal areas in some species. However, even in species in which endogenous components could, in theory, essentially be responsible for differential migration, they do not preclude the possibility that exogenous factors might also play a proximate role. In order to better understand such details the very complex matter of choosing goal areas needs much more investigation. Such a very complex situation recently emerged from a band-recovery analysis in wood ducks (*Aix sponsa*) by Hepp and Hines (1991). Winter distributions of male and female ducks did not differ, but adults migrated shorter

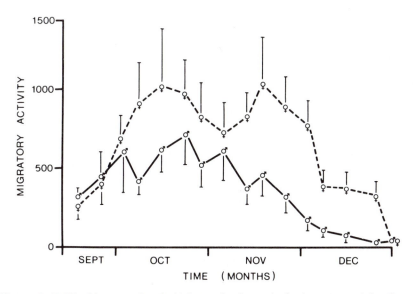

**Figure 2.67** Weekly mean levels (±1 standard error) of migratory activity (hops per bird per night) shown by male (solid lines) and female (dashed lines) blackcaps from SW Germany. (Source: Terrill and Berthold, 1989.)

distances than young. Further, variation in winter distribution was associated with early-autumn habitat suitability. Adults were recovered further south when precipitation was below average, and young birds occurred to the south when rainfall in spring–summer was less than normal and when average November temperatures were lower than normal.

Root (1988) has calculated the metabolic rate at the northern boundary of the distribution (NBMR) for a number of passerines known to have range boundaries associated with a particular average minimum January temperature isotherm. He found that NBMR is 2.45 BMR and suggested that the northern winter-range boundaries of several species are associated with the physiological costs of maintaining thermal homeostasis (Root, 1989). Repasky (1991) analysed additional recent data of winter-bird distribution. He found that species of all sizes occur at the lowest temperatures, indicating that a ceiling on metabolism does not generally constrain the distributions of birds wintering in North America. Thus, these metabolic aspects also require more analyses.

As mentioned above, many migrants return from wintering areas to breeding grounds close to their hatching sites (breeding-site attachment). Ketterson *et al.* (1991) obtained data from dark-eyed juncos which suggest that male site attachment forms before or during autumn, and that familiarity or a prior residence effect confers an advantage in territorial competition the following spring. Premigratory preselection of potential breeding territories has also been discussed earlier (e.g. Löhrl, 1959). It remains an interesting task to study to what extent post-juvenile dispersal (section 1.1) might be directed to preselect prospective nesting sites.

## 2.21 HABITAT PREFERENCES, NICHE SEGREGATION AND COMPETITION DURING MIGRATION AND WINTERING

Migrants which leave their parental breeding areas as juveniles during the premigratory period of dispersal (section 1.1) are increasingly confronted with new habitats or unfamiliar habitat composition the further they move. Still, the high demands during migration to replenish depleted energy reserves, to rest and to avoid predation, along with the resource competition, require quick and effective habitat selection. Appropriate habitat choice may be equally important during wintering, above all during the widespread period of a complete moult. These selection processes during migration and wintering can be controlled both ultimately and proximately. The most important features would be floristic composition of habitats and their structural characteristics (namely, physiognomy), food availability and, last but not least, the degree of niche occupancy by competitors (MacNally, 1990; Leisler, 1992). Although there is need for further rigorous investigation (Leisler, 1992), the major strategies of habitat selection and resource use by migrants and the main separation mechanisms have begun to emerge, especially for the periods of autumn migration and wintering.

### 2.21.1 Autumn migratory period

It is actually almost platitudinous to maintain that many migratory birds tend to rest during the autumn migratory period in their 'typical' habitats, i.e. in habitats of a type which are normally used during the breeding period. Thus, for instance, staging areas of waders during migration are mainly found in wetlands or flooded agricultural habitats (Hands *et al.*, 1991), harriers hunt during their journey in open country and water birds rest on lakes, rivers, etc. On the other hand, many exceptions are also almost a rule when, for instance, reed warblers are fairly regularly found in shrubbery and forests, or wood warblers in reed beds etc. Morse (1989) reported that American warblers living in spruce forests during the summer can be found foraging in lush grass during autumn migration – 'a situation for which mice might seem better prepared'. But all these cases are rather extreme. The crucial questions are: (1) do species that share fairly similar, but nevertheless different, habitats during the breeding period also select (slightly) different habitats during migration? (2) if so, are they able to select species-specific habitats even when only fairly unfamiliar habitats are available? and (3) have migrants been able to develop mechanisms of habitat selection and partitioning which *a priori* help to avoid competition, or does habitat segregation result mainly from direct competition?

These questions have been intensively studied in the MRI-program (section 1.3). Over several decades, the distribution of more than 250 000 individuals of about 40 different migratory songbird species have been recorded in up to eight different habitats at three Central European bird trapping stations throughout the entire autumn migratory period from end of June to beginning of November (Berthold *et al.*, 1976; Bairlein, 1981; Streif, 1991). The following main results were obtained. All species investigated are characterized by species-specific distributions of capture over the different habitats. These distributions are the expression of distinct species-specific habitat preferences. Inter- and intrageneric differences in these preferences, as well as in the vertical distributions, are similar in their basic properties at the three stations, although the stations are more than 500 km apart. Furthermore, the observed patterns of habitat choice remain extremely constant from year to year (Figure 2.68). Finally, similarity in habitat preferences is pronounced in both a species and individual level. By comparing

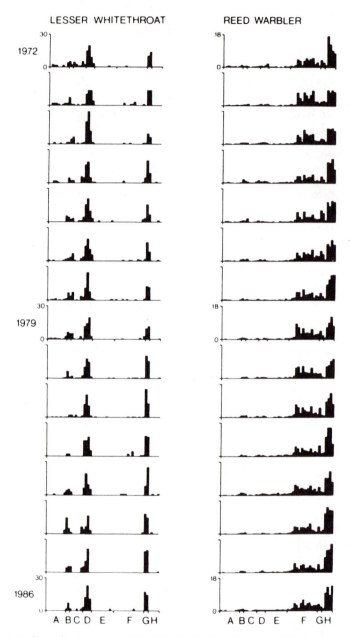

**Figure 2.68** Distributions of trapped individuals of lesser whitethroat and reed warbler along a 500 m line of nets during the summer and autumn migratory period from June to November in 15 successive years. Data are from the Mettnau station, S Germany, of the 'Mettnau–Reit–Illmitz Program'. A–H along the abscissa are habitats: A, bush, dense; B, bush, light; C, wood; D, 'savannah' (bushes, small trees, grassland); E, wetland, F, reed, dry, scattered bushes; G, natural dam with bushes; H, reed, on shore. (Source: Berthold, 1988a.)

first captures and subsequent recaptures the patterns are almost identical (Figure 2.69). These distinct preferences are presumably not found by trial and error, because the distribution patterns in early morning, after arrival in the staging area, were essentially as distinct as those later in the day. Since mainly nocturnal migrants have been investigated, which either arrive in the early morning or at least seek a suitable habitat at that time, one must conclude that early-morning arrivals are quick and precise in their habitat choice. It can be excluded that this decision results from interspecific competition or from attraction from conspecifics, e.g. by calls. Rather, habitat decision-making is controlled by some type of search image ('habitat conception', 'gestaltwahrnehmung') when approaching the area in the early morning or after having landed during the night. Similarly, it is highly likely that during transdesert flights those migrants which prefer to stage for a while in oases, in order to replenish fat deposits, may often have a chance to find suitable places via appropriate search images over distances of up to 100 km or more. Even tiny islands may be correspondingly detected during transoceanic migration.

Detailed analyses at one of the stations (Mettnau peninsular, S

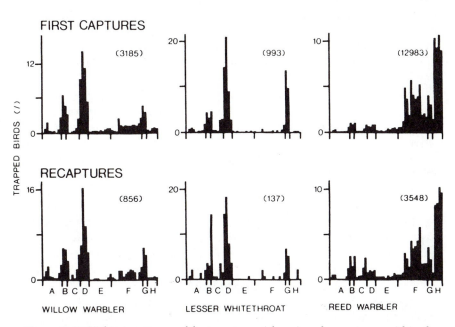

**Figure 2.69** Habitat patterns of first captures (above) and recaptures within the same season (below) of willow warblers, lesser whitethroats and reed warblers. Total trapping figures are in parentheses. Habitat abbreviations are as in Figure 2.68. (Source: Bairlein, 1981.)

Germany) with three large trapping sites in one resting area of about 100 ha has provided a data set to examine habitat utilization on a microhabitat scale. Over a small distance of a few kilometres, the numerical distribution of staging individuals of different species has been shown to depend on the amount of suitable habitats available (Figure 2.70). In small patches of a given habitat the density of individuals of the species which prefer it can be significantly higher than in larger areas (Streif, 1991). This high density, however, can be lowered by diurnal movements of staging individuals within preferred habitat zones (Bastian, 1992). Finally, species whose preferred habitats are not represented in the areas of the trapping stations of the MRI-program are either not trapped at all or are numerically underrepresented, like crossbills, due to the almost complete lack of coniferous trees, or wagtails and pipits, due to the lack of larger meadows. Corresponding results for a number of species investigated were also found by Spina *et al.* (1985) in Italy at a station that has been run according to the rules of the MRI-program.

With respect to the importance of different habitats as staging areas for migratory songbirds the MRI-program yielded the following results. Individual numbers per trapping unit (mist-net) were highest in the shrub habitat, while species numbers were higher in reed beds. The diversity index was highest in the former and lowest in the latter habitats (Bairlein, 1981). Similar results were also obtained during spring migration (see below). Similar results, as found in the MRI-program, have been reported for leaf and reed warblers by Ormerod (1990b) and Pambour (1990).

Up until now we have not known how the species-specific habitat choices are controlled. However, the above-mentioned hypothesized search image ('habitat conception') appears to be based, to a considerable extent, on preprogrammed mechanisms. Comparisons of the habitat and vertical distributions of juveniles and adults (within the migratory as well as the premigratory period) demonstrated that knowledge of the species-specific habitat is either largely innate or fixed early by imprinting processes or learning, and is only little influenced by subsequent learning (Bairlein, 1981). This is in agreement with experimental results obtained from at least a few species which, in part, have also been studied in the stations of the MRI-program (Ley, 1985; van Patten and Price, 1990). The key factors for the search image appear to be morphological characteristics of the birds and structural properties of habitats. 'Habitat conception' thus means a lock-and-key mechanism where habitat selection requires structural prerequisites. Habitat choices may then be due to physical constraints of species and individuals. Detailed ecomorphological studies within the MRI-program have compared 36 morphological characteristics with habitat structures.

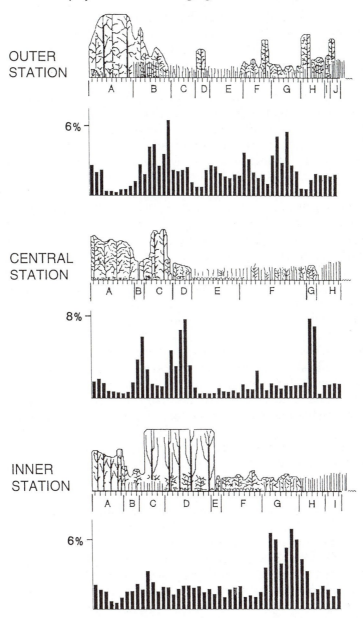

**Figure 2.70** Distributions of trapped individuals (trapping figures) during the summer and autumn migratory period from June to November 1988 and 1989. Data are from three neighbouring stations on the Mettnau peninsula, S Germany, of the "Mettnau–Reit–Illmitz Program". A–H: habitat abbreviations as in Figure 2.68; I and J, two additional wetland habitats. (Source: Streif, 1991.)

The following relationships have been found. The height of capture is related to morphological characteristics relevant for locomotion. There is a less pronounced relationship between the type of habitat and the wing shape and no relationship between beak structure and habitat (Bairlein, 1981; Bairlein *et al.*, 1986; Figure 2.71). According to various degrees of ecomorphological relationships, it is clear that the staging community of songbirds in staging areas comprises a wide range, from specialists to generalists, with quite different niche breadths. As a result of these ecomorphological relationships, distinct habitats are characterized by ecological groups of heterogeneric species rather than by systematic groups (Bairlein, 1981).

In long-lived species it may well be that initial habitat choice, by programmed 'habitat conception', may be modified in later years by additional habitat learning (Catterall and Kikkawa, 1989). Artificial induction of landfall in nocturnally migrating passerines, by playing tape lures overnight (which is extremely effective), raises the question whether an assessment of suitable habitats may be possible on the basis of song recognition (Herremans, 1990b). This most interesting finding needs detailed analyses, although song during the night appears not to play any role in resting areas like those investigated in the MRI-program.

Habitat selection based on structural properties may permit choices of staging areas in quite unfamiliar habitats. Maize fields appear to be a typical example for this, in which many staging migrants of various species can be found. They may, however, soon be disbanded by reed warblers or preferred by species like chiffchaffs (Degen and Jenni, 1990).

Habitat selection during the autumn migratory period shows some interesting adaptive seasonal and diurnal shifts in a number of species, particularly late migrating insectivorous and omnivorous forms. With respect to seasonal shifts, there are two main categories. One represents omnivorous species like blackcap, garden warbler and European black-bird, which with the advancing season increasingly move to habitats offering ripened berries. Another one concerns species like European robin, chiffchaff and wren, which move to reed beds during the later migratory season. These shifts most likely occur in order to utilize nutrient-rich habitats at a time when many reed warblers are leaving. Another typical alteration observed in a number of species is that with the advancing season individuals are increasingly observed lower in the vegetation, most likely due to reasons of food supply, microclimate and shelter; for more details see Bairlein (1981), Degen and Jenni (1990) and Streif (1991). In many species, diurnal changes in the niche breadth have been found. Some may represent an increasing concentration on mostly preferred habitats, others may be related to diurnal feeding

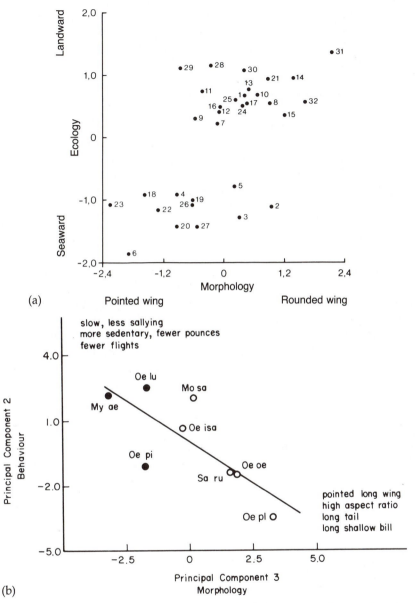

**Figure 2.71** (a) Second canonical correlation between morphology and ecology in 32 songbird species. The morphological axis means more rounded wings. The ecological axis means net position (increasing distance from the lake shore). (b) Relationship between morphology and behaviour of eight turdid chats ($r = -0.77$). Axes are principal component values of 15 behavioural and 25 morphological traits, respectively. Filled circles, African residents; open circles, Palaearctic migrants. (Source: Leisler, 1992.)

patterns (Bairlein, 1981). Alerstam and Lindström (1990) reported a habitat shift in bramblings from summer rape fields to beech forest. This surprising shift, with a reduced possibility for energy intake, appears to be due to frequent attacks from raptors and thus may be understood as a mechanism to minimize mortality.

Clear-cut species-specific habitat selection, including those of distinct vertical vegetation layers (Figure 2.72), act as a mechanism to prevent competition among potential competitors during the course of migration. Programmed habitat conceptions enable passage migrants to stage fairly smoothly by habitat partitioning based on coexistence. Competition during staging in potential competitors is also reduced by differences in: (1) main migratory seasons (e.g. Berthold *et al.*, 1991); (2) diurnal activity patterns (Brensing, 1989; section 2.2); and (3) food choice (see below), all of which can function as isolating mechanisms. Nevertheless, if suitable habitats are small, active competition may occur and possibly even separate age classes of the same species, as indicated in the marsh warbler in a resting area in NE Germany (Mädlow, 1992). Local concentrations of migrating, fat-depleted passerine migrants in North America were found to be in increased competition for food at a time when energy demand is high (Moore and Yong, 1991). Individuals of other species that strongly depend on patchy resources may even defend them during migration. This was, for example, reported for migrant rufuous hummingbirds which can defend nectar resources (Heinemann, 1992). For other examples see section 2.13.

### 2.21.2 Wintering period

Habitat partitioning and competition during wintering has been extensively studied in various areas, and in a number of Eurasian migrants

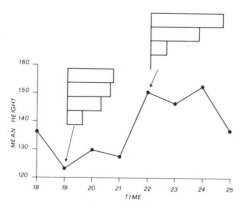

**Figure 2.72** Variation of mean height in blackcaps for 5-day periods in spring, with vertical distribution in periods 19 and 22. (Source: Spina *et al.*, 1985.)

in relation to Afrotropical residents (reviewed by Leisler, 1992, and Pearson and Lack, 1992), in the Neotropics (Rappole *et al.*, 1983; Greenberg, 1986; Rappole, 1988; Terborgh, 1989; Finch, 1991; Kelsey, 1992) and elsewhere (e.g. Malaysia; Wells, 1990). The following main principles have been described. As during migration, terns, waders and other open-country species prefer quite similar habitats to those in the breeding areas (Hayes *et al.*, 1990). In the majority of migrants breeding in woodland the situation is quite variable. Generally speaking, migrants are characterized by a more opportunistic use of sporadically available resources in space and time, above all by use of only seasonally available niches. This means in detail that in Africa, the majority of migrants occur in dry bush, dry woodland and other savanna ('open') habitats. About 70% of the individuals and about 40% of the species remain in the tropics north of the equator, and, exceptionally, many of them in the Sahel zone (Moreau's paradox, 'paradoxical' or 'green' Sahel; Fry, 1992; Morel and Morel, 1992). This holds true for all species which eat seeds and for most fruit eaters, and essentially it is only the pure insectivores which migrate further south. For many species this means a considerable shift in habitat use in comparison to the breeding area. For instance, in Britain and Ireland the highest migrant ratios were recorded in mature upland woods (Fuller and Crick, 1992). Also, more than half of all nearctic land bird migrants winter north of the equator (Kelsey, 1992), but forests play a more important role in the Neotropics and elsewhere. Tropical rainforests, whether lowland or montane, are used only insignificantly in Africa by about 4% of the wintering migrants, e.g. by wood warblers (*Phylloscopus sibilatrix*), and golden orioles in the lowlands and by blackcaps and chiffchaffs in the mountains. Utilization of mangroves is similarly restricted (Altenburg and van Spanje, 1989). This is presumably caused by the fact that these largely stable habitat niches are fully occupied by resident species. In the Neotropics and Malaysia, however, a much higher (10–20-fold) percentage of the wintering migrants occupy rainforests and are able to coexist with huge numbers of resident species (Lynch, 1989; Kelsey, 1992; Mönkkönen *et al.*, 1992), although wintering in 'open' habitats is also not uncommon, even in intratropical migrants in the Neotropics (Levey and Stiles, 1992). With respect to the preference of 'open' habitats, in many migrants it is not surprising that in forests modified by humans, where many resident species have been expelled, a substantial proportion of migrants is found. Morel and Morel (1992) have emphasized that Palaearctic migrants can take advantage of the new-savanna/forest mosaics and agricultural developments which have replaced the rainforests in W Africa. Hence, habitat loss for native species may result in habitat increase for immigrants. The situation may be different in central and south America where migrants are more forest-related (Hutto, 1989;

Terborgh, 1989), and the effect of disturbance will clearly vary with respect to its intensity (Kelsey, 1992).

Wintering groups in open dry habitats are not significantly fixed to distinct habitats, rather specific habitat use depends largely on the habitats predominating in the reception area. Also, migrants and residents did not differ in their degree of habitat specialization in the Bahamas and Greater Antilles (Wunderle and Waide, 1993). Within these open habitats, migrants can form up to 25–50% of the individuals present (Kelsey, 1992), and they are generally more eurytopic and exploit more open parts than ecologically-equivalent tropical species. In citrus plantations on the Yucatan peninsula, Mills and Rogers (1992) found 50–81% (mean 61%) overwintering migrants among mist-netted birds. In a detailed comparative study, including 14 shrike species and some other habitat occupants in southern African savannas, however, Bruderer and Bruderer (1993) found that wintering red-backed shrikes are not restricted to dryer and more open habitat parts by interspecific segregation, as observed in other groups. They are probably able to establish their temporary wintering territories in a variety of habitats on the basis of different prey size.

Adaptive food selection and foraging strategies appear to be generally important for migrants to allow them to act as an integral part of the tropical avifaunas rather than as 'intruders' (Kelsey, 1992). Discussed are that migrants feed more on conspicuous prey and possibly are more generalized in their diet than residents (i.e. polyphagy versus monophagy and opportunism versus stereotypy, which may be established by calculating 'population dietary heterogeneity'; Sherry, 1990). In some ecological guilds it could be shown that migrants use a broader array of foraging tactics, their foraging speed and capture rates are higher, they use their wings more often and are more aerial, whereas residents are more commonly ground feeders (Figure 2.71). One hallmark of success of migrants is their ability to switch food resources opportunistically (Leisler, 1992). This is above all conspicuous in those species which are insectivorous during breeding but become omnivorous during wintering. This is mainly a characteristic of a number of larger species. It is further strongly expressed in many migrants which, by performing step migration, use various seasonally available food resources in succession. In a few species investigated in detail a positive correlation between the amount of food available and the actual settling reaction of migrants was found; in other cases vegetation structure was found to outweigh the importance of food supply as a proximate factor. The black redstart, however, has a diverse repertoire of prey-capture techniques in its breeding habitat, while in its winter quarters it becomes more specialized as a 'ground cleaner' (Zamora, 1992), and thus appears to differ from many other migrants.

The complex of actual competition and of possible niche shifts of residents induced by migrants has been only superficially investigated. However, the available evidence suggests that, in general, actual competition does not seem to be severe, and there is only weak evidence that niche shifts of residents are induced by the presence of migrants. Although 10–28% of the Palaearctic migrants in Africa are considered to be potential competitors for residents or other wintering species, overt interspecific interactions between Afrotropical and Palaearctic species seem to be infrequent. To what extent territoriality and age- and sex-related separation of migrants, frequently described for the Nearctic–Neotropic system, results from actual competition remains to be studied. In a group of chat-like thrushes with considerable overlap in habitat requirements and diet in Africa, interspecific interactions were found to be exceptionally frequent and important for niche sharing. In this case, African residents showed a clear 'home advantage' in interspecific disputes (Leisler, 1992). Blake and Loiselle (1992) obtained evidence that some Neotropical migrants on Costa Rica may rival permanent residents in the diversity of fruits they consume. In many species, however, intraspecific competition [like in grey plovers (*Pluvialis squatarola*), leading to displacement over several 100 km, Evans and Davidson, 1990; or in Pacific golden plovers, Johnson *et al.*, 1989] appears to be much more critical for migrants than interspecific rivalry. For more details see Leisler (1992), Greenberg (1986), Ornat and Greenberg (1990) and Rappole (1988). Future studies on problems of habitat utilization and competition during wintering should surely concentrate more on morphological and behavioural characteristics, prerequisites and constraints, on genetically-based differences in neophobia and related phenomena (Leisler, 1992). Rappole (1988) has presented a directive example. He has measured micro- and macrogeographic distribution by sex in forest-related migrants and proposed, for the non-breeding season, that male-like plumage in females represents a type of andromimesis wherein the female mimics male plumage to compete more effectively for limited winter-food resources, and partial non-overlapping use of microhabitats by males and females on the wintering grounds results from slight structural differences between the sexes.

### 2.21.3 Spring migratory period

The few studies carried through (e.g. Spina *et al.*, 1985; Fasola and Fraticelli, 1990; Moore *et al.*, 1990; Dinse, 1991) suggest quite similar results to those obtained during autumn migration. American warblers exploited sites resembling those of their breeding grounds during spring migration, more than would be predicted by chance. Most species that were habitat generalists or habitat specialists during migration had

similar patterns in their breeding grounds. In some species niche choice –
i.e. the selection of certain areas within a habitat – might be of greater
importance than the habitat alone (Parnell, 1969). Overlap in habitat
utilization between species was higher during migration than during
the breeding season (Morse, 1989). In Italy it was found that blackcaps
in spring use lower vegetation layers than in autumn and change
progressively to higher layers with increasing foliage or food supply
(Figure 2.72; Spina *et al.*, 1985). In another study (Fasola and Fraticelli,
1990), intraspecific competition affected the niche of two species out of
nine in that their niche breadths increased significantly with increasing
density of conspecifics. However, no interspecific competitive effects
were detected. Migrants appear to exploit foraging resources opportunistically and do not use mechanisms of segregation from residents.
Laursen (1978) investigated interspecific relationships between some
insectivorous passerines with respect to their diet. The species showed
great flexibility with regard to prey (mainly arthropods) selection, which
was affected by interspecific competition. Segregation into different
microhabitats and/or different types of foraging behaviour reduced
competition.

## 2.22 INTERACTIONS WITH OTHER ANNUAL EVENTS

A migratory bird often has to fit a sequence of up to five main seasonal
processes into its annual schedule. As a regular migrant, two migratory
periods, a reproductive period and often two moult periods (pre-
and post-nuptial) are a basic part of these schedules. In long-distance
migrants a high degree of precision among the individual processes is
necessary for temporal and energetic reasons. In these birds migratory
periods can be very long and breeding seasons often are extremely
short, and precisely predetermined. In the Arctic, the restricted
length of stay for the short breeding season is mandatory. In this case
migratory periods have to be adapted strongly to the breeding season,
and the moult has to be adapted to both of them or, in case of a winter
moult, at least to migration. This hierarchy of adaptations and import-
ances appears to be widespread, at least among long-distance migrants.
In migrants breeding in more favourable areas where they only have
to leave during a short winter period, for example, short-distance
migrants, interrelations among annual processes may generally be different, or at least less pronounced. However, these interrelations have
not been sufficiently investigated. In the following, the three main
interactions of migration with other annual events will be treated, i.e.
with juvenile development, moult in adult individuals and the reproductive cycle.

## 2.22.1 Juvenile development and migration

Long before a migrant takes off for its first departure, adaptive processes in the course of the juvenile (or post-juvenile) moult may have started. Comparative studies of migratory and non-migratory species have demonstrated a number of adaptations to early departure. As a rule, juvenile moult in migrants starts earlier, and/or proceeds more rapidly, and, as a consequence, is of shorter duration compared to resident or less typical migratory forms. As a result, migrants have more time to prepare for migration. This general picture can easily be obtained from moult guides based on moult record cards (Kasparek, 1981; Ginn and Melville, 1983). It is not only the length of the juvenile moult that is adapted to different migratory habits, but also the temporal pattern of its intensity. In *Sylvia* warblers (Berthold *et al.*, 1970) and stonechats (Gwinner and Neusser, 1985) the detailed study of the moult of many individual plumage parts has shown that typically migratory forms reach the maximum of their moult intensity at the beginning of the moult period, i.e. their 'moult intensity curve' is strongly skewed to the right. In contrast, less typical migrants and non-migrants intensify their moult period slowly towards the middle of the moulting period, so that their intensity curve looks more like a normal distribution. The median value of the moult pattern is not reached before the middle of the moulting period, whereas in more typical forms it lies well before the middle of this period (Figure 2.73). The concentration of moulting efforts to the beginning of the moult period in typical migrants is thought to be adaptive in that it allows for preparations for migration while moult still proceeds during final phases with low intensities. In typical migrants, an accelerated course in the development of the juvenile plumage in nestlings, during the growth of so-called second and third body feather sets; is also found. These actually cover the remaining unfeathered body parts (e.g. Berthold *et al.*, 1970). Finally, most processes of juvenile development can be more 'compressed' in typical rather than in less typical migrants (Figure 2.74). A general acceleration of juvenile development in typical migrants can be interpreted as adaptive, as is the generally accelerated course of the juvenile moult. It comprises many interesting details. In the garden warbler, for instance, an early, long-distance migrant, feather growth was found in about twice as many plumage tracts on the second and third day after hatching as in the blackcap, a late middle-distance migrant (Berthold *et al.*, 1970). Hence, an adaptive acceleration of development for migration appears to start in the egg. As briefly mentioned in section 2.4, at least part of this adaptation in juvenile development among typical migrants is under genetic control. This was demonstrated in cross-breeding experiments with blackcaps

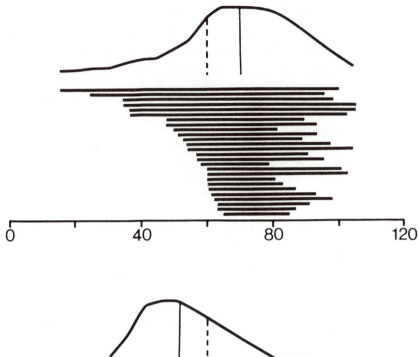

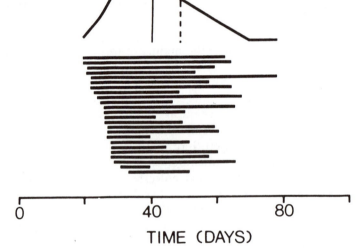

TIME (DAYS)

**Figure 2.73** Juvenile moult in hand-raised German blackcaps (above) and garden warblers (below) of about the same age. Bars, course of the moult in 26 individual plumage parts; curve, cumulation of contemporaneously moulting plumage parts; broken line, middle of the time course of the moult; solid line, median of the cumulation. (Source: Berthold, 1988a.)

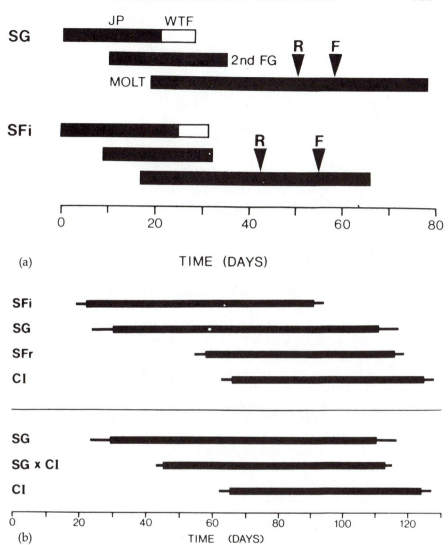

**Figure 2.74** (a) Juvenile development of two groups of hand-raised garden warblers from S Germany (SG) and S Finland (SFi). JP, juvenile plumage; WTF, wing and tail feathers; 2nd FG, second set of body feathers; R, onset of migratory activity; F, onset of migratory fattening. (Source: Berthold, 1988a.) (b) Time course of juvenile moult in hand-raised blackcaps of four populations (above) and of hybrids and their parental stocks (below). SFi, S Finland; SG, S Germany; SFr, S France; CI, Canary Islands, Africa. Broad bands show the duration of the moult; narrower bands show the additional standard error of the duration.

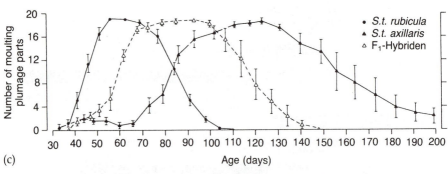

(c)

**Figure 2.74** (c) Temporal course of juvenile moult of European stonechats (filled circles), of African stonechats (filled triangles) and of their hybrids (open triangles). Mean values and standard errors. (Source: Gwinner and Neusser, 1985.)

for the course of the juvenile moult and with stonechats for its intensity pattern (Figure 2.74). Moreover, in garden warblers from Finland, which stay shorter periods in their breeding grounds than conspecifics in Germany, the processes of juvenile development were found to be more rapid, or to have a greater initial intensity (Figure 2.74). When hand-raised individuals were kept in photoperiodic conditions simulating those normally experienced by birds of the other populations, differences in the juvenile development as small as a very few days remained discernible. Thus, these differences may also reflect genetically determined processes; see also Berthold and Querner (1992).

Accelerated juvenile development, as described here for different migrants, is also regularly found in individuals of the same species and area which hatched late in the season compared to those which hatched earlier. These differences are, however, in contrast to those reported earlier, predominantly under photoperiodic control (section 2.5), although their adaptive significance is quite similar. However, despite these accelerating processes, there is considerable overlap in most migratory passerines between moult of body feathers and migration. This holds true for both short- and long-distance migrants, and even in the latter moult may last until about the middle of the migratory period (Herremans, 1990a; Berthold *et al.*, 1991).

### 2.22.2 Moult and migration in adult individuals

As a rule, wing-feather moult and normally tail-feather moult and migration are mutually exclusive (e.g. Stresemann and Stresemann, 1966; Stresemann, 1967; Payne, 1972; Mead and Watmough, 1976). According to Stresemann and Stresemann (1966), migratory birds avoid travelling with 'gappy wings' in order to provide optimum flight capacity. In fact, when gaps in the wings of raptors or storks are found during migration, their irregular occurrence indicates accidental feather loss rather than regular moult. There are very few exceptions to this 'rule of complete wings' in migrants. As an example, tertials are sometimes also renewed during incomplete post-nuptial moults (of body feathers only) and may still be growing during initial phases of migration. The same can occur in the juvenile moult, above all in less typical migrants or individuals hatched early in the season. In such cases, one or a few of the inner secondaries may occasionally also be replaced, and their regrowth may also overlap with migration. Numerous examples are given in modern moult guides as, for example, in Ginn and Melville (1983). In subarctic breeders, the overlap of secondaries moult and initial migratory movements can be more pronounced. Ryzhanovskii (1987) reported that in subarctic passerines investigated in northern Russia, 14–65% of birds (warblers, wagtails, buntings and others) were still moulting some inner secondaries at departure. Only a few species were able to complete wing moult before the onset of migration. These exceptions, none the less, do not effect the case for the most important flight feathers, the primaries.

There are a number of specific adaptations which guarantee the general rule not to migrate with gaps in the wing. According to Stresemann and Stresemann (1966), six strategies have evolved. (1) Moult before autumn migration (post-nuptial moult, normally on the breeding grounds, sometimes in areas towards the winter quarters reached by a specific intermediate migration, or in other areas reached by specific moult migration; section 1.1). (2) Moult in the winter quarters. (3) Moult twice by type 1 and 2 in succession (in a few species only, see below). (4) Moult during parts of the breeding cycle. (5) Staggered moult (beginning before the onset of autumn migration and ending before homeward migration. Feathers of up to three successive moult cycles can be found in a wing; this is a characteristic of holarctic terns). (6) Suspended moult ('interrupted moult', 'arrested moult', 'split-moult' or 'moult with moulting pause', which means a moult with temporary interruption of feather replacement). These different strategies have evolved both among different species and different populations. One classical example is the ringed plover (*Charadrius hiaticula*), where the west and Central European populations start their wing moult shortly after breeding, the Arctic populations, however, do so only after arrival in the African winter quarters (Stresemann and Stresemann, 1966). Many intraspecific differences have evolved in interrupted moult strategies (see below).

In some species and populations the time between breeding and departure does not allow for a complete post-nuptial moult, in others the situation during wintering may hinder a complete winter moult. This difficulty has been overcome by adapting one of the specific moulting strategies. Variations can also be found in the relative timing

of moult and migration. Some species, like raptors or terns, start moult
during the breeding cycle. Such an adaptive overlap between breeding
cycle and moult is also found in small birds. In subarctic breeders, loss
of primaries starts at the latest moment, immediately when feeding of
fledgelings ends. The earliest extreme is when birds are still incubating.
This moult, however, often proceeds so rapidly that individuals may
temporarily lose their ability to fly (Ryzhanovskii, 1987). The crag
martin (*Ptyonoprogne rupestris*), which winters in Mediterranean countries where food is seasonally scarce and temperatures are low, does
not carry out a complete moult then. Wing feathers are replaced while
rearing the young. Firecrests (*Regulus ignicapillus*) and goldcrests start
their complete moult with respect to their extremely long breeding
season regularly within the second brood from June/July on. Wing
moult is then largely terminated when the offspring of the second
brood fledges (Thaler-Kottek, 1990).

If the time span between breeding and departure only suffices for
part of the wing moult, interrupted moult may be an appropriate
strategy. In case of interrupted moult, wing moult starts on the breeding
grounds, is then arrested before departure at various stages (Nikolaus
and Pearson, 1991) and continued in the winter quarters. It may
again be interrupted there and then not be continued before the following post-nuptial moulting period. Mead and Watmough (1976) and
Swann and Baillie (1979) have put together long lists of species using
this moulting strategy. According to later records (Kasparek, 1981;
Riddiford, 1990) this strategy appears to be extremely widespread.
Interrupted moult in migrants is characterized by considerable variation. With respect to interspecific variation, Mead and Watmough
(1976) established three categories: (1) species in which most individuals
migrate with suspended primary moult; (2) species usually completing
moult before migration, but a small proportion migrates with some old
feathers; and (3) species usually moulting after migration, with a limited
moult in a few individuals. Intraspecific geographical variation also
exists. It is expressed, for instance, in a higher incidence of suspended
moult among trans-Saharan migrants or in birds with SE European
passage than in SW European individuals. These differences are ascribed to the fact that Eastern European birds have less time to moult
before migration because of their comparably later and shorter breeding
season (Swann and Baillie, 1979). A typical example here is the whitethroat (*Sylvia communis*), whose incidences of suspended wing moult
were 70% in SE and 5–40% in SW Europe. Finally, there is obviously
individual variation indicating that there are adaptive pressures currently affecting suspended moult (Mead and Watmough, 1976).

In order to justify the highly variable suspended split-moult systems,
Norman (1991) has proposed to cover the different interrupted moult

strategies in the categories: (1) seasonal split-moult (covering all suspended moult systems); (2) abridged moult; and (3) arrested moult. It remains to be seen whether this new terminology will be accepted.

So far, there have been very few experimental studies on suspended moult. Stresemann (1967) stated that the moult interruption had been noticed in captive hobbies (*Falco subbuteo*) and young quails, contemporaneous with the suspension of moult in wild birds, but no details of these observations have been published. In a special study carried out with Orphean warblers it was found that suspended moult can be adapted on a facultative basis by the conditions experienced. Adults, caught during breeding, all suspended their moult. The same individuals completed their moult before the migratory period in a second experimental year when they were prevented from breeding and moult started weeks earlier (Berthold and Querner, 1982c). Experimental results obtained by Lofts *et al.* (1967) from turtle doves (*Streptopelia turtur*) are similar. Although the endogenous control of suspended moult presently cannot be excluded, a high degree of facultative adaptability in this strategy appears highly likely.

A widespread phenomenon is sex differentiation in post-nuptial moult related to breeding and migration. In many species, females, engaged in second or replacement broods, start moult later than males because their onset of moult is more synchronized with the offspring development. Often, however, females can make up for their initial delay by a more rapid course during the moult (Ginn and Melville, 1983; Ryzhanovskii, 1987; Norman, 1990). It is possible that photoperiod may play a role in the acceleration of the moult. Here, a short-day 'calendar' effect could be postulated (section 2.5), but this remains to be demonstrated. In other cases, females may be forced to suspend their moult before departure, whereas males may be able to complete it. Also, in the wintering areas the time available for moulting may be restricted due to a limited food abundance between arrival and the beginning of the dry season. As an adaptation, a number of Palaeartic passerines wintering in west Africa have developed extremely rapid moult periods (Bensch *et al.*, 1990).

The occurrence of biannual primary moult in passerines has been found in seven species comprising six separate families, and long-distance migrants as well as residents (Prys-Jones, 1991). In the willow warbler it was studied comprehensively in Europe and Africa (Underhill *et al.*, 1992). A key factor appears to be the rate at which the primaries are worn away, regardless of whether migration, structurally weak composition or other factors are basically responsible.

Many bird species visit specific areas for moulting either in special moult migrations (Jehl, 1990) or interrupt their migration towards the winter quarters after an initial 'early-summer' or intermediate migration (Schüz *et al.*, 1971; section 1.1). Moult migration is most common in waterfowl, but has also been described for birds of about 15 other families, except passerines. It is widespread in simultaneously wing-moulting birds which, during periods of flightlessness, need safe and food-rich areas. During such intense moult, the flight muscles underwent disuse atrophy (e.g. George *et al.*, 1987), which appears to be best explained by the 'use–disuse' hypothesis, although the possibility that mobilized breast muscle proteins are used for feather synthesis ('nutritional stress' hypothesis) cannot presently be excluded (Piersma, 1988). This moult-induced muscle degradation requires special efforts to readapt the flight muscles after the moult period to migratory performance (sections 1.6 and 2.7). Interruption of autumn migration for moult has also been reported for many bird species including passerines (Schüz *et al.*, 1971; Ellegren and Staav, 1990). Young (1991) gives another complicated example for the lazuli bunting (*Passerina amoena*). A limited 'prebasic moult' (post-nuptial moult; for terminology see Humphrey and Parkes, 1959) in hatching-year birds is followed by a nearly complete presupplemental moult at stopover places during autumn migration. During winter, only a limited prealternate moult occurs, leading to some sex differentiating head coloration, and the definitive prebasic moult takes place from the next autumn to October.

### 2.22.3 Reproductive cycles and migration

In several categories of migrants the fitting of a successful reproductive cycle between two subsequent migratory periods needs specific adaptations. In intercontinental long-distance migrants, that may be *en route* for more than seven or eight months a year (section 1.1), the time spent on the breeding grounds may be too short to allow for both complete gonadal development from quiescence to fertility and successful breeding. This holds true especially for migrants that breed in extremely high latitudes with very short seasons, or in species like the rose-coloured starling (*Pastor roseus*) or the Arctic warbler (*Phylloscopus borealis*) which remain in the winter quarters and *en route* for up to 10–10.5 months per year (Ticehurst, 1922; Stresemann, 1934). Similar difficulties may arise for nomadic migrants moving in areas with highly unpredictable environmental conditions where breeding performance is required immediately conditions become favourable.

Most, if not all, categories of migrants initiate their gonadal development before the homeward migration on the wintering grounds, presumably often based on endogenous circannual rhythms. This has been well established for long-distance migrants in Asia (Ticehurst, 1922) and Africa (Rowan and Batrawi, 1939; Marshall, 1952a; Marshall and Williams, 1959; Lofts, 1962) and for many short- and middle-distance migrants from America, Eurasia, Africa and Australia (Rowan, 1925; Farner, 1966; Bullough, 1942; Immelmann, 1963). It appears to be a rule of thumb that gonadal development at arrival on the breeding grounds is more advanced in late migrants which arrive shortly before breeding. The rose-coloured starling develops its testes to almost full-breeding size before departure (Ticehurst, 1922), whereas many short-distance migrants arrive with less developed gonads on their breeding grounds (Marshall, 1952a). Thus, premigratory gonadal development seems to be adaptive in relation to temporal constraints on breeding. An updated compilation of the relevant data here would be very helpful in understanding its control and adaptive significance.

As a second rule, it appears that premigratory gonadal development is more pronounced in males than in females. Development of oocytes in females seems to require to a higher degree of stimuli related to the breeding area (Farner, 1966; King *et al.*, 1966). A third rule is that gonadal development appears to be either arrested or at least delayed during actual migration so that the two may be largely mutually exclusive. This may be the reason why, in species that leave the wintering grounds with fairly developed testes still free, sperms are often only produced on the breeding grounds. In a number of species, gonadal development occurs quickly after arrival (Berthold, 1969). In the rose-coloured starling the suspension of spermatogenesis during migration appears to be effected by an excessive release of TSH and thyroid hormones (Naik and George, 1964), and the accelerated gonadal development around arrival has been shown to be paralleled by substantial increases in the pituitary gonadotropins (King *et al.*, 1966). Schwabl *et al.* (1984a,b) found suggestive evidence in the European blackbird that migratory disposition influences the recrudescence of the reproductive system and behaviour by selectively suppressing only the secretion of LH in males but of both gonadotropins (LH and FSH) in females.

Gwinner (1990) was able to demonstrate experimentally that garden warblers held in winter, under photoperiods of 20°S, showed a clear tendency to moult more quickly, initiate migratory activity and testicular growth earlier than in conspecifics held under an equatorial photoperiod (Figure 2.75). This acceleration by long photoperiods (section 2.5) is thought to be biologically significant for individuals which migrate to more southerly wintering areas. In order to return to their breeding grounds at the same time as conspecifics wintering farther north, it appears necessary for them to speed up moult and to begin northward migration earlier. Gonadal growth also starts earlier, either because it is associated with the onset of homeward migration (section 2.6) or in order to ensure a certain degree of reproductive maturity after migration. In the zebra finch (*Taeniopygia guttata*) it was

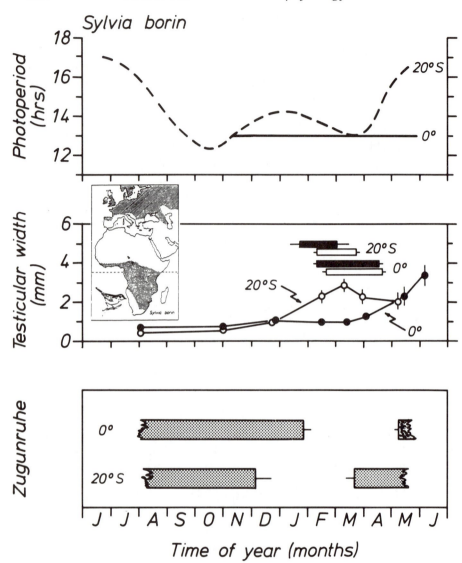

**Figure 2.75** Changes in testicular width (middle; closed and open circles), occurrence of moult (middle; black bars, wing feathers; white bars, body feathers) and migratory restlessness (below; dotted bars) in two groups of garden warblers exposed to the photoperiodic simulations of wintering grounds at 0° and 20° south, respectively, shown in the upper diagram. Horizontal lines at the bars and vertical lines at the curve points, standard errors of the mean. (Source: Gwinner, 1990.)

found that birds in central northwest Australia maintain a certain degree of gonadal activity and regularly show courtship displays during nomadic movements. This 'hypersexualism' is thought to strengthen pair-bond, to synchronize breeding preparedness of pairs and to minimize the time necessary to obtain full breeding performance. It can be seen as an adaptation to sudden rainfall, which triggers a new breeding season (Farner and Serventy, 1960; Immelmann, 1963). Similar adaptations appear to be widespread among species which perform irruptive movements in areas with highly unpredictable resources that immediately stimulate breeding (Immelmann, 1971). In crossbills, the ecological equivalents in temperate zones, an extremely long annual period of gonadal development of eight months has been described. Since testes begin development during December and breeding is concentrated from February on (Berthold and Gwinner, 1978) it is likely that many birds migrate with well-developed gonads. Other adaptations in species which arrive late in breeding areas and have only short periods for breeding are pair formation in the winter quarters or *en route*, nest building and egg laying immediately after arrival, a shortened incubation period and accelerated embryonic development [e.g. in Ross's goose (*Chen rossii*), Stresemann, 1934], or a shortened nestling period (in passerines; e.g. von Haartman, 1969). In some medium-sized, northern-nesting geese, and other species, nutrient and (especially) protein reserves accumulated during premigratory feeding appear to be crucial for the determination of clutch size because no additional assimilation is possible between arrival and egg laying (e.g. Blem, 1990). In other species, like greater white-fronted geese (*Anser albifrons frontalis*), food acquired on the nesting grounds can increase endogenous reserves necessary for reproduction (Budeau *et al.*, 1991). In American coots (*Fulica americana*) the question of whether birds arrive on the breeding grounds with enough endogenous fat stores for breeding or whether there may be far greater flexibility in the use of exogenous resources toward reproduction than has previously been noted is still discussed (Hill, 1989). But non-breeding by Arctic birds, due to late arrival or arrival in conditions too poor for nesting, is common (Johnson and Herter, 1990). From Tennessee warblers (*Vermivora peregrina*) insemination during spring migration has been reported (Quay, 1989), as detected by cloacal lavages. It was found in up to 25% of the females, although it is unknown to what extent this insemination contributes to fertilization in the nesting areas.

There are, as expected, also clear relationships between the breeding cycle and autumn migration. In the bluethroat, Ellegren (1990a) found a clear correlation between the timing of breeding in Swedish Lapland and the timing of autumn migration in S Sweden. Ellegren suggested that bluethroats start their autumn migration as soon as possible after breeding and moult because it appears advantageous to reach suitable stopover sites as early as possible. This view is supported by the fact that Arctic breeders, like Siberian white-fronted geese (*Anser albifrons*), which are adapted to large-scale breeding failure of regular occurrence, due to adverse environmental conditions (e.g. Elkins, 1988b), may depart extremely early (Marshall, 1952b). There are numerous reports which also demonstrate that extremely late broods are usually raised to fledging, even when this results in a delayed onset of migration (Salomonsen, 1967). In the swift, late termination of breeding in northern populations regularly causes a considerably later departure of whole populations compared to more southern populations (Stresemann, 1934). In Arctic breeders like geese successful breeding can be prevented both by delayed egg laying or an early onset of winter weather (Bauer and Glutz, 1968). There are also observations in passerines, above all swallows, indicating that extremely late broods can be deserted by the parents before the nestlings are fledged (Stresemann, 1934). In such difficult situations the urge to migrate prevails over useless or dangerous continuation of reproductive efforts so that at least the parental birds may have a chance to get to safety. It is well known that late-born offspring suffer from increased mortality due to inadequate preparation time for migration (e.g. Morton, 1992).

# 3

# Orientation mechanisms

This chapter contains a concentrated compilation of our present knowledge of orientation mechanisms. It is necessary to understand the subsequent synopsis of the control of migration and to make use of numerous cross-references in the previous chapter. Such a compilation can be relatively short, since a recent review on the subject is available which covers virtually all aspects of the field (Berthold, 1991b); the following compilation is based mainly on chapters of this review.

## 3.1 TYPES OF ORIENTATION, ROUTES AND ACCURACY, TRADITIONAL AND PREPROGRAMMED DIRECTIONS

### 3.1.1 Compass orientation and homing to a familiar site

Inexperienced individuals of many species that normally migrate alone are able to fly fixed courses ('normal directions'; section 1.5) in order to reach species-specific wintering areas. This directional or compass orientation is based on preprogrammed directions (section 2.4) and a number of environmental reference cues involving compass systems. Finding goal areas in these cases is most likely due to endogenous time-programs (vector-navigation hypothesis, section 3.5). Birds migrating in flocks with experienced conspecifics and relying on social bonds may also possess innate information on how to fly a compass course (e.g. white stork, section 2.4).

After the first migration, most migrants return more or less exactly to familiar breeding grounds or wintering areas (section 1.1; Figure 3.1), which appear to be fixed by imprinting-like processes (Wiltschko, 1990b; but see also section 2.20). This goal-oriented or goal-directed homing (or true navigation, navigation in its narrower sense) may simply be some kind of route reversal. However, whether route reversal or path integration plays a role in homing in migratory birds is unknown (e.g. Hilgerloh and Bingman, 1992). Since homing often is

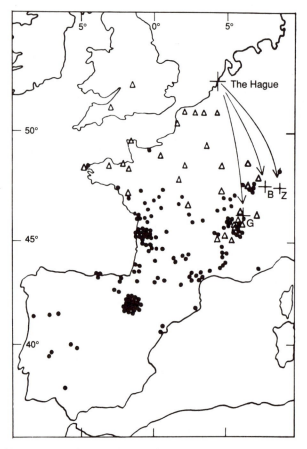

**Figure 3.1** Recoveries of European starlings coming mainly from northeastern Europe, which were caught in The Netherlands and transported to Switzerland, where they were released in Basle (B), Geneva (G) and Zurich (Z). The juveniles (dots) continued their migration parallel to their original migratory direction and reached new, atypical resting quarters on the Iberian peninsula; the adults (triangles) moved towards their former familiar winter quarters in northern parts of western Europe. (Source: Perdeck, 1958.)

performed directly from unfamiliar areas, or following different routes (Evans and Townshend, 1988), navigational mechanisms other than route reversal have to be expected. 'Gradient maps' (or 'grid maps') are thought to be used which are based on at least two gradients of a physical substrate which extend over a sufficiently large area (Wallraff, 1991a). Compass orientation (flying innate courses) in naïve migrants and true navigation based on gradient maps in experienced birds (i.e. goal-finding from various places) may, then, be the two basic orientation mechanisms of avian migration. Preprogrammed compass orientation could also be used in experienced birds. This would be important to cover main parts of long routes. In line with this, experiments with blackcaps have shown that genetically encoded migratory directions persist at least through the period of the second autumn migration and perhaps through life (Helbig, 1992).

### 3.1.2 Choice of routes and accuracy

Very little is known about what determines the exact migratory route, i.e. how directed and straight migration routes can be chosen, and how eventually a minimum of environmental disturbance by wind drift and other factors may be achieved. This kind of information is important in determining the degree of accuracy in orientation. Initial satellite-tracking studies (Nowak and Berthold, 1991; Berthold *et al.*, 1992b) demonstrated surprisingly straight routes and directedness over long distances (Figure 1.5) which are close to the formerly assumed 'ideal migratory direction' (Schüz *et al.*, 1971). Experimental tests showed, for example, that the sun compass of the domestic pigeon has only an accuracy of $\pm 3.4 - \pm 5.1°$. However, computer simulations suggest that this accuracy would be sufficient for homing success (Schmidt-Koenig *et al.*, 1991).

Theoretically, migrants have two basic possibilities to choose their routes. They can either follow great circles (orthodromes), i.e. the shortest routes between two points on the earth's surface, or the rhumbline routes (loxodromes). The latter represent longer distances but can be flown with a constant compass course, whereas great circle navigation requires continuously changing courses (Figure 3.2). The time-compensated sun compass furnishes migrants with a possibility of orienting along great circle routes, but whether birds make use of this possibility remains to be shown (Alerstam and Pettersson, 1991). Piersma and Jukema (1990) assumed that it is likely that bar-tailed godwits flying from Mauritania to The Netherlands try to follow a 'fixed' track that is reasonably close to the great circle route. Field observations along the flyway of knots destined for Siberia indicate that they are flying along a rhumbline route (constant compass direction) rather than the shorter great circle route between the Wadden Sea and Taymyr (Gudmundsson *et al.*, 1991).

Among social migrants, one can assume that juveniles normally migrate guided by experienced conspecifics. This is clear in species like cranes, where inexperienced birds regularly occur in family groups at specific resting areas (section 2.13). It is also documented in cases of fostering, which demonstrates parental influence on the migration route (Alerstam, 1991). Since such species may also well possess

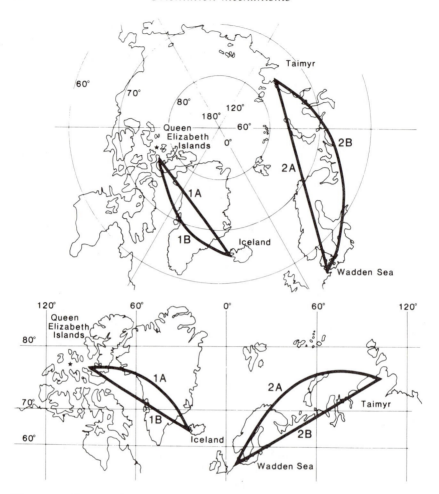

**Figure 3.2** Great circle and rhumbline routes between points of departure and destination for migration flights by certain high-arctic shorebirds and brent geese, drawn on an azimuthal stereographic map projection (above) and on a Mercator map projection (below). Between Iceland (65°N, 24°W) and the Queen Elizabeth Islands (77°N, 90°W) , great circle (1A) distance and courses are 2535 km and 328°/265° (initial/final course). Rhumbline (1B) distance and course are 2665 km and 300°. Between the Wadden Sea (54°30′N, 8°45′E) and the Taimyr Peninsula (76°N, 100°E) , great circle (2A) distance and courses are 4234 km and 23°/110° (initial/final course). Rhumbline (2B) distance and course are 4634 km and 59°. (Source: Alerstam, 1991a.)

preprogrammed migratory directions, the question remains to what extent inexperienced individuals may also contribute to the overall migratory direction? (For cooperative orientation see section 2.13). Recent results of satellite-tracking suggest that some juvenile white storks migrate largely independently from experienced conspecifics (Berthold *et al.*, unpublished). Analyses of ringing recoveries have demonstrated that inexperienced migrants often show a wider orientation scatter than adults. At present, it is not clear whether the orientation capability is poorer in naïve migrants or whether it may be fundamentally different in adult birds, due to *en route* experience or winter site knowledge (Alerstam, 1991). In experiments, orientation of young birds seems to be more disturbed by natural or artificial light sources than that of adults (Baldaccini and Bezzi, 1989).

## 3.2 LANDMARKS AND PILOTING

Theoretically, home-finding within a familiar area could be based on familiar, i.e. learned, landmarks. This type of navigation, with the aid of a 'topographical map' ('mosaic map', 'familiar area map', 'home-range map'), is called pilotage or piloting (Wallraff, 1991a; Terrill, 1991, formerly sight or parallactic orientation; Schüz *et al.*, 1971). Postfledging dispersal may play an important role in becoming acquainted with an array of local landmarks important for subsequent homing (Morton *et al.*, 1991). Landmarks may also be found in unfamiliar areas among visual or acoustic reference cues, like stars, call notes of other animals or even ocean breakers (Wiltschko and Wiltschko, 1991a). To what extent landmarks are really used by migrants is unknown. Even in domestic pigeons this rather simple question is still under debate (Streng and Wallraff, 1992). Some recent experimental results are in favour of their use in the vicinity of pigeon lofts (Bingman, 1988; Braithwaite *et al.*, 1990) but others are not (Rüttiger and Schmidt-Koenig, 1991). Actually, the finding that even North American corvids employ sun compass orientation rather than landmarks for local orientation to memorize cache sites of food (Jahnel *et al.*, 1990; Balda and Wiltschko, 1991) raises doubts as to whether landmarks may be of great importance for migrants. Temporary use of landmarks occurs, however, regularly in leading-line behaviour, preferably in conditions in which drift is highly likely. This method is mainly used by diurnal migrants which therefore tend to be less prone to drift than nocturnal migrants (Elkins, 1988b; section 3.7).

## 3.3 COMPASS MECHANISMS

The capability of using a biological compass for orientation has evolved in a large variety of systematic groups of animals, including birds. Biological compasses have been defined as mechanisms which provide animals with constant directions and a non-changing directional reference system. Their existence was (presumably) first proposed by Legg (1780a,b). For migratory birds, three compass systems have been demonstrated – a sun compass, a magnetic compass and a star compass. There may also be other environmental cues that could help to establish a compass direction (Terrill, 1991; Wiltschko and Wiltschko, 1991b; see below).

### 3.3.1 Sun compass

The sun is used as a reference system for compass orientation in a wide variety of animals and was first demonstrated in birds (Figure 3.3). The sun compass system requires time compensation, which has been demonstrated in birds by clock-shift experiments. This obligatory 'sense of time' is provided by internal circadian clocks (section 2.2). The crucial reference quantity of the sun appears to be the sun azimuth ('sun azimuth compass') and not sun altitude. So far the use of a sun compass has only been demonstrated more or less convincingly in less than 10 bird species. Domestic pigeons use the sun compass during initial orientation in homing experiments as well as under the solar conditions of equinox. The mechanism is not simple, as details of the sun compass have to be learned and the compass has to be updated to keep track of the seasonal variation of the sun's course. The update may possibly lag behind reality by a few weeks. In domestic pigeons, the sun compass has been established as an alternative compass which can replace the magnetic compass on sunny days. For our considerations it has to be emphasized: although the sun compass was originally studied (and discovered) in migratory starlings (Figure 3.3), very little is known about its use in migrants and during migration (Schmidt-Koenig, 1991; Terrill, 1991; Wiltschko, 1990a,b, 1991). Beside the starling, it has also been demonstrated in the meadow pipit (Wiltschko, 1990a).

### 3.3.2 Magnetic compass

The earth is a huge magnet and, thus, it is not surprising that animals have evolved capabilities to use the geomagnetic field as a means of distinguishing directions (Figure 3.4). The properties of the earth's magnetic field are theoretically almost ideal as directional reference within large latitudinal ranges. The magnetic field can provide a direction, namely magnetic north (or south). It does not require sophisticated mechanisms of time compensation and does not change with the season, time of day or weather (but see below). In effect, it is fairly constant over the entire globe in space and time. Thus, the magnetic compass represents the simplest, as well as theoretically the most reliable, compass. This may be the reason why it evolved as a phylogenetically primitive primary compass from the most simple

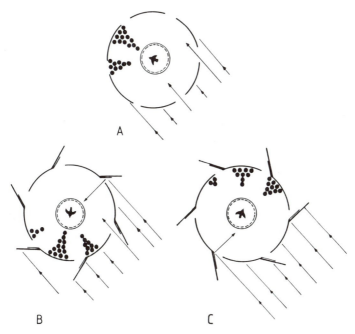

A

B                    C

**Figure 3.3** Gustav Kramer's (1951) mirror experiment with an European starling showing migratory restlessness in a circular arena with six windows. Each dot inside the arena symbolizes one average direction observed in a 10 sec interval. A, control condition; B, mirrors attached clockwise; C, attached counter-clockwise; arrows indicate the direction of incoming sunlight. (Source: Schmidt-Koenig *et al*., 1991.)

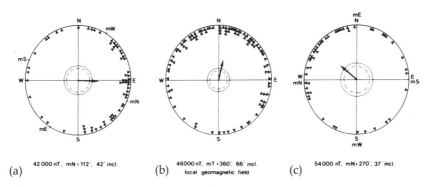

(a)     42 000 nT,  mN = 112°,  42° incl.

(b)     46000 nT,  mT = 360°,  66° incl.
        local geomagnetic field

(c)     54 000 nT,  mN = 270°,  37° incl.

**Figure 3.4** The magnetic compass of European robins. Orientation (a) in a magnetic field with magnetic north turned to ESE; (b) in the local geomagnetic field; and (c) in a magnetic field with north turned to W. Each symbol at the periphery of the circle gives the mean of one test bird in one night. The arrows represent the mean vectors and the two inner circles mark the 1 and 5% (dotted) significance border of the Rayleigh test. (Source: Wiltschko and Wiltschko, 1991a.)

prokaryotes on up through the vertebrates (Wiltschko and Wiltschko, 1991a; Terrill, 1991; Wallraff, 1991a).

The magnetic compass in birds does not use either the polarity of the magnetic field nor the intensity gradient from the magnetic poles to the equator for orientation. Instead, it relies on the axial course of the field lines and their inclination in space ('inclination compass'; Figure 3.5). Hence, birds do not distinguish between 'north' and 'south', but rather between 'poleward' and 'equatorward'. The compass operates within a narrow range of intensities which, however, can vary adaptively over a period of days (Wiltschko, 1972; Wiltschko and Wiltschko, 1991a). None the less it has been proposed that magnetic orientation is limited by sensory constraints (Wallraff, 1991a). For this reason it is mostly used in connection with or parallel to, other, mainly visual, cues (sections 3.6 and 3.7).

Transequatorial migrants at the magnetic equator experience field lines that run horizontally, so that the information from the magnetic inclination compass becomes ambiguous. After crossing the equator migrants should reverse their reaction with respect to the magnetic field lines. Experiments with garden warblers, by Wiltschko and Wiltschko (1992), indicated that this is the case. Beason (1992) has tested bobolinks accordingly, which breed in temperate North America

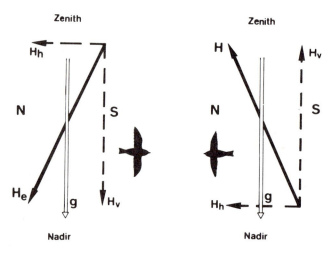

**Figure 3.5** The magnetic compass of birds is based on the inclination of the field lines: section through the geomagnetic field north of the magnetic equator – the polarity points downwards – and south of the magnetic equator – the polarity points upwards. N and S, north and south; g, vector of gravitation; $H_e$ and H, vector of the geomagnetic field; $H_h$ and $H_v$, horizontal and vertical component of the magnetic field. (Source: Wiltschko and Wiltschko, 1991a.)

and winter in temperate South America. Birds were tested in a planetarium under fixed star patterns in a series of magnetic fields, incremented each night from the natural field in the northern hemisphere through an artificial horizontal field to an artificial southern hemisphere magnetic field. The birds maintained a constant heading and did not reverse direction after the simulated crossing of the magnetic equator, as previous experiments indicated. In addition, Beason (1992) suggested that the ability to maintain a constant heading may be based on the use of visual cues such as stars.

The capability to orient using a magnetic compass has been demonstrated in at least 11 nocturnal migrants and to some extent in three diurnal migrants (Wiltschko, 1990a,b). A magnetic compass has been found in all avian species so far examined. It appears, therefore, to be a widespread phenomenon among birds.

### 3.3.3 Star compass

Many birds migrate at night. Hence, it is not surprising that stellar orientation is a common avian phenomenon (Figure 3.6). Beside birds, there is some evidence for stellar orientation in a moth (Sotthibandhu and Baker, 1979). The avian star compass is not based upon an innate knowledge of stellar configurations, as was initially believed. Rather, birds apparently derive directions from the star patterns in a similar way as we are able to find north from the constellation of Ursa Major. These crucial configurations are learned using the axis of stellar rotation as the polar direction (Emlen, 1967a,b; Wiltschko and Wiltschko, 1991a–c; Terrill, 1991). The use of a star compass has so far been shown in five bird species (Wiltschko, 1990a,b).

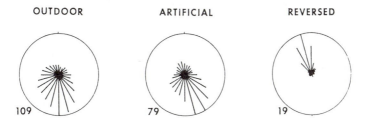

**Figure 3.6** Orientation behaviour of three indigo buntings during outward migration. Left, under natural night sky with view of the stars; centre, in the planetarium under an artificial sky imitating a natural starry sky; right, in the planetarium under an artificial sky, the north–south axis of which was shifted by 180°. (Source: Emlen, 1967a,b.)

### 3.3.4 Other possible cues

'Sunset cues', like the position of the setting sun – the sunset point, horizon glow or pattern of polarized light – appear to be highly informative for nocturnal migrants (Sandberg, 1991; Terrill, 1991). At present it remains open whether they may be part of the sun compass or whether they represent an independent system indicating the geographic direction 'west'. It is also unclear whether the light polarization pattern (most likely the *e* vector) can be used only during sunset or at other times too. Theoretically, it could provide axial information for nocturnal migrants and, for diurnal migrants, a reference for determining the sun position under cloudy skies ('polarized light compass' or 'polarization compass'; Able, 1991; Schmidt-Koenig, 1991; Terrill, 1991). Phillips and Moore (1992) exposed migratory warblers to a rotated polarized light pattern at sunset. The birds oriented at a constant angle to the axis of polarization. When polarized light cues were eliminated, this shifted orientation was maintained relative to the setting sun. The authors concluded that polarized light patterns appear to provide a calibration reference for the sun compass in nocturnal migrants, and may also play a role in calibrating other compass systems. In this context, it is interesting to note that flocks of waders departing for long-distance flights very often leave before sunset and whilst there is a visible sun (Evans and Davidson, 1990; Piersma *et al.*, 1990). In fact, Able and Able (1993) reported that, in young Savannah sparrows (*Passerculus sandwichensis*), visual access to natural skylight polarization patterns is necessary for calibration of magnetic orientation during daylight. Coemans and Vos (1992) concluded from recent studies that homing pigeons are not able to detect the polarization state of light.

In all cases where in-orientation test covers of glass, plexiglass, PVC or polarizers are used it has to be realized that these materials, appearing colorless to the human eye, may act as a colour filter to the bird's eye and this may influence orientation behaviour in various ways (Helbig, 1991b).

It is well known that moonlight can increase nocturnal restlessness in caged nocturnal migrants during its full phases. Thus, nocturnal migrants may fly further during full-moon phases due to improved visibility (Gwinner, 1967b; Martin, 1991). However, there is no evidence that the moon itself can be used for orientation. This contrasts with insects and other animals where a moon compass has been suggested (Baker, 1987). On the contrary, moonlight may lead to the disturbance of orientation due to phototaxis, especially in young birds (Baldaccini and Bezzi, 1989; see also 'nonsense orientation' in ducks, Schüz *et al.*, 1971).

Theoretically, infrasound can provide information for orientation.

Infrasound is produced by various sources like mountains and oceans. Theoretically, birds can obtain directional information from such frequences below 20 Hz (which are inaudible to humans) by perceiving Doppler shifts (Quine and Kreithen, 1981). At least, domestic pigeons and possibly some other bird species are capable of discriminating such sound frequency shifts (Beason and Semm, 1991). Recent experiments with domestic pigeons suggest that infra-sound may be involved in the system of orientation cues of that species (Schöps, 1991). Whether it is used in migrants remains open. For infrasound perception see the next section.

Wind has been discussed intensively as a directional cue. Since birds can sense wind direction before take-off, and possibly when flying (section 3.4), wind might be useful in maintaining or selecting a heading. There are many reports of upwind or downwind orientation which have been summarized by Richardson (1991). The complex issue to what extent migrants utilize tail winds to conserve energy, or tolerate wind drift for the same reason, or even use winds to orient *per se*, has not been resolved clearly (Terrill, 1991; Richardson, 1991), but will be treated in more detail in section 3.7. Odours as chemical cues to orient are mainly discussed with respect to olfactory navigation and rarely as a basis of an olfactory compass for olfactory orientation (Waldvogel, 1989); see section 3.5 for further details.

## 3.4 SENSORY BASIS

### 3.4.1 Vision

Vision in birds is well developed, including colour vision, which allows for a broad use of visual cues like celestial characters or other landmarks for orientation (Beason and Semm, 1991). However, according to Martin (1990), nocturnally migrating birds are unlikely to have any more visual information available to them than a human being would have under the same circumstances. However, these visual capacities at night probably suffice to detect gross details of the topography and/or star patterns. Thus, visual cues of some kind may play an important role in determining and sustaining the bird's migratory orientation, although they cannot function in cases of severe cloud cover or fog. So far, there is no convincing evidence of specific nocturnal adaptations in the visual system of temporarily nocturnal birds like sea birds or migrants (Brooke and Prince, 1990). Where flight in migrants occurs at night it nearly always takes place in open habitats (with the exception of certain nocturnal species; Martin, 1990).

Birds can also detect ultraviolet light which may be related to their polarized light detection (Beason and Semm, 1991). A number of bird species process four or five visual pigments with sensitivities in different regions of the light spectrum, whereas humans process only three pigments (Bowmaker, 1988). Since vision seems to play a secondary role in nocturnally foraging birds, and audition, olfaction and mechanoreception are the principal senses (Martin, 1991), a similar situation may be the case in nocturnal migration. With respect to sensitivity to infrared radiation, the available evidence suggests that birds are no more sensitive to it than humans (Martin, 1990).

Homing behaviour of hippocampus- and parahippocampus-lesioned domestic pigeons suggest that these regions of the brain are involved in the birds recognizing landmarks in the vicinity of the loft (Bingman and Mench, 1990).

### 3.4.2 Magnetoreception

Most interesting, of course, is the problem of how magnetic information is used for orientation. Beason and Semm (1991) have summarized the available evidence. So far, neither the structure nor the mechanism of the magnetoreceptor are known, and, unfortunately, no specific magnetic sensory organ could be detected. Experimental data support the following hypotheses for magnetic field transduction. Four candidate transducer substances have been proposed: magnetite, melanin, biological radicals and photopigments. In the domestic pigeon, the most intensively studied species, the consistent presence of magnetic material, like magnetite, is still doubtful. It might well be that the photoreceptors are the magnetoreceptors. This view is derived from the finding that pigeons need the presence of light and an intact retina to use the magnetic compass. It is also supported by electrophysiological studies of the optic system. However, following severing of the ophthalmic nerve, magnetic responses in both the visual system and the ophthalmic nerve have been found to persist. This implies that two different magnetic systems, with different receptors, are present in birds' central nervous systems. It is possible that the mechanism of perceiving the magnetic information for orientation is different from, and perhaps even independent of, that eventually used as the magnetic part of a map (section 3.5; Wiltschko and Wiltschko, 1991c).

With respect to magnetoreception by photoreceptors in the retina, a bi-radical hypothesis was proposed – biochemical bi-radical reactions of excited macromolecules. The assumption of magnetic resonance phenomena in the retina led to the question of whether magnetoreception is light dependent? Recent experiments have produced evidence consistent with such a mechanism: red light disrupts magnetic orientation of migratory birds, and there are similar findings for amphibians (Wiltschko *et al.*, 1993). The pineal gland and melatonin are apparently not involved in actual magnetic orientation. However, it seems that both are somehow involved in the biological timing of the mechanism used for the magnetic orientation system. Investigations on the pied flycatcher have indicated this type of connection (Beason and Semm, 1991). Thus, almost all information on the mechanisms underlying magnetosensory receptors and magnetic field transduction in birds is speculative and needs much more investigation. Edmonds (1992) demonstrated that a biological compass, with characteristics to use the magnetic field for orientation and true navigation, can naturally be formed if a biologically synthesized magnetite crystal becomes attached to a torque detector, such as a hair cell. Edwards *et al.* (1992) measured values for natural remanent magnetization and isothermally-induced remanent magnetization in the head and neck of eight bird species. Most of the magnetic materials appeared to be grains of magnetite. No differences were found between migratory and non-migratory species with respect to the amount of remnant magnetism. Finally, there is some indication that whole bird bodies may, to some extent, act as dipoles (see Berthold, 1993).

### 3.4.3 Sense of smell

With respect to the hypothesis of olfactory navigation (section 3.5), the capability of birds to use chemical cues to orient is of great interest. Unfortunately, study of the structure of the avian olfactory system and its functioning has been very limited. However, it is known that a considerable number of bird species, like kiwis, vultures or some procellariiforms, rely on an osmotactic mechanism to find food sources or breeding holes using various scent cues. Laboratory studies in birds have elucidated sensitivities ranging from 0.1 to 40 ppm. This indicates a moderate olfactory capacity compared to other animals. With respect to receptors, relatively large olfactory bulbs have been found only in nocturnal and crepuscular species. Projections from the olfactory bulb of domestic pigeons suggest that olfaction is an important element in their behaviour. In addition, lesions in the ventral telencephalon resulted in the impairment of pigeon homing from unfamiliar sites (Beason and Semm, 1991; Papi, 1991). At present, some most intriguing questions are still open: whether there are long-lived or short-lived olfactory cues in the environment that produce fairly stable gradient fields over larger areas which can be used for olfactory navigation by migrants over hundreds or even thousands of kilometres (Wallraff, 1991a,b), how sensitive birds might be to environmentally-relevant odours and how many different odours birds can perceive and distinguish (Beason and Semm, 1991; Papi, 1991).

### 3.4.4 Hearing

Hearing ability is generally very well developed in birds, allowing nocturnal migrants to perceive the flight calls of their conspecifics, as well as sounds and noises from their surroundings that could be used for orientation. With respect to infrasound and navigation, the following information about perception and transduction is known. Domestic pigeons, at least, can perceive infrasound, i.e. sound frequencies of below 20 Hz – probably via the ear. The method by which vibrational information, detected by the ear, is transduced to the nervous system involves hair cells within the cochlea. Electrophysiological recordings from afferent fibres have revealed that individual hair cells are tuned to individual infrasound frequencies (Beason and Semm, 1991). Echo location or active sonar (an acronym for SOund NAvigation Ranging), used by cave dwelling species (Martin, 1990) does obviously not play a role during migration. But it has been repeatedly discussed (e.g. Alerstam, 1990) that calls at night may not serve as contact between birds that migrate in flocks but for echo sounding. Guided by the echo's time laps and character the birds could possibly determine their altitude and the nature of the ground below.

### 3.4.5 Wind perception

The possibility that birds may use wind as a directional cue raises the question whether they can sense the wind while aloft. When flying in or above clouds, or at night over a landscape lacking discernible features, pilots of planes without modern instruments cannot sense lateral drift and thus wind direction. Birds, however, most likely can do so according to several lines of evidence summarized by Richardson (1991). The 'turbulence hypothesis' suggests that birds sense wind direction and speed by using anisotropic patterns of air turbulence, which are related to wind direction. Tsvelykh (1990) suggested interactions between wing-beats and the air flow around the head. The 'acoustic landmarks hypothesis' states that: localizable noise sources (on the ground, or flight calls of other migrants) could be used in detecting wind drift without visual cues. Brown and Fedde (1993) studied air-flow sensors in the avian wing. According to their studies, mechanoreceptors on or near feather follicles in the wings of birds may provide information about air-flow over the wing. They conclude that birds have the necessary sensor-feather mechanisms in the wing to detect an imminent stall, the location of the separation point of the air-flow from the wing's surface, and to measure air speed by detecting the frequency of vibration of the secondary-flight feathers.

## 3.5 NAVIGATION HYPOTHESES, MAP-AND-COMPASS CONCEPT

There are about seven different hypotheses on how birds can perform navigation (in its narrower sense, i.e. goal-finding). All but one of these concern goal-oriented returns to familiar sites (often from distant areas). One concerns finding an unfamiliar goal from a familiar site, i.e. reaching species-specific winter quarters by naïve first-time migrants. This is a 'vector-navigation' or 'time-and-direction program' hypothesis. It will be treated first in this section. Among the others, olfactory navigation is currently being most intensively discussed and magnetic navigation or path integration to a lesser extent. Kramer's map-and-compass concept as a hypothetical navigation model is still of great heuristic value.

### 3.5.1 Vector-navigation

Many naïve, first-time migrants, like young cuckoos, are capable of finding their species-specific winter quarters without any guidance. How they do that is not exactly known, but quite convincingly elaborated. Among a number of conceptual approaches, the vector-navigation hypothesis is the best supported. Results showing inherited patterns of migratory activity and genetically preprogrammed migratory directions (sections 2.3 and 2.4) have given use to the idea that migrants are equipped with innate time-and-direction programs for migration (spatio-temporal programs or a time-and-direction vector leading from the breeding grounds to the wintering areas). These programs, of a bearing-and-distance basis, may often bring migrants more or less 'automatically' to their unknown winter quarters without essential environmental disturbance ('automatic' control of migration). In other cases, for example, where exceptionally prolonged stopover periods are necessary, they may be balanced by a temporary faster flight due to higher body weight or by intermittently higher programmed migratory activity (section 2.19). Since winter quarters are normally a fairly large area quite a bit of environmental disturbance can be tolerated by the endogenous programs. It is also possible that part of the environmental interference has also been taken into account by these programs through selection processes. For instance, the long duration of the migratory period in naïve first-time migrants, their relatively short migratory episodes (which allow avoidance of adverse weather situations) and mechanisms of at least partial compensation of wind disturbance (e.g. Schindler *et al.*, 1981; sections 2.14, 2.15 and 3.7) may be adaptive in this respect, and may help to allow a relatively undisturbed course of endogenous migration programs. Liechti (1992) has discussed, in some detail, his integration model of how naïve first-time migrants should behave with respect to varying wind influences in order to optimize goal-finding based on vector-navigation. Berthold (1991c) has outlined in detail how endogenous migration programs may well fit into environmental conditions so that migrants should, as a rule, not be essentially embeded in following them. Presently, there are no real objections to the vector-navigation hypothesis. The evidence to date supports the view that a large number of migrants are essentially brought to their wintering areas by vector-navigation systems (Berthold, 1991c; Sokolov, 1990). Return migration is certainly, at least in its final part, another story (see below).

A few authors, above all Rabøl (1993), reported some observations that naïve first-time migrants also showed compensatory behaviour after displacements, which cannot be explained by a (simple) vector-navigation hypothesis as proposed above. Rabøl proposed a 'cross-axis system' and a 'goal-area navigation hypothesis' which are based on 'some sort of – possibly stellar-based – coordinate navigation'. So far, however, there is no convincing evidence for such navigation systems in naïve migrants.

### 3.5.2 Olfactory navigation

A long series of investigations, initiated by experiments by Papi in 1971, have led increasingly to the idea that olfaction is a possible sensory input for homing in domestic pigeons from distant, unfamiliar areas (Papi, 1991; Wallraff, 1991a). Wallraff (1991a) stated that 'totally unexpected empirical findings (respective smelling) forced man's mind (although not yet every man's mind!) to accept this idea'. The concept of olfactory navigation has been deduced almost exclusively from domestic pigeon studies. The basis was experiments with pigeons made anosmic by a variety of methods, or detour, site simulation or deflector loft experiments (in which wind is deviated; Papi, 1991; Figure 3.7). Contradictory results have also been gathered (Schmidt-Koenig, 1991), so that a general agreement has not thus far been obtained (Wallraff, 1991b). Fortunately, the present controversial discussion is not of great importance for our considerations here, since it is not certain whether olfaction plays any role in migration. In starlings, for

**Figure 3.7** The effect of anosmia on initial orientation of domestic pigeons from different lofts is shown by these pairs of vanishing diagrams. In each diagram, the bearings recorded in four to nine test releases are pooled. Control birds on the left, anosmic birds on the right. The smaller circles give the direction of the release sites with respect to home. (Source: Papi, 1991.)

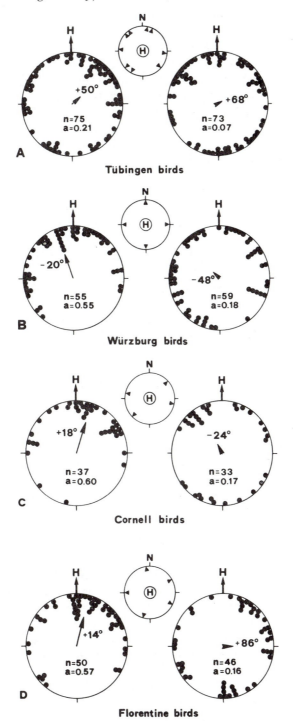

A     **Tübingen birds**

B     **Würzburg birds**

C     **Cornell birds**

D     **Florentine birds**

instance, subjected to bilateral cutting of the olfactory nerves, homing success was half that of controls. Papi (1990) suggested, therefore, that there may be an 'auxiliary mechanism' which helps to fix the homeward direction. Such a suggested auxiliary mechanism could easily be an important component in a redundant system. According to Wallraff (1991a): 'The now most problematic part of olfactory navigation is not the biological machinery but the physical environment of which it makes use'. Atmospheric odours coming with winds over thousands of kilometres, or fairly stable 'odour landscapes', which could be used as a navigational map for long-distance migrants require atmospheric structures that have not yet been shown to exist. Thus, for migrants, Wallraff's (1991a) statement appears to be appropriate: 'At the present stage, any conceptual approach to the mechanism underlying olfactory navigation is necessarily speculative; its heuristic value cannot yet be assessed'. Schmidt-Koenig (1991) summarized the current controversy on olfactory navigation in general as follows: (1) the necessary sensory capacities have not been discovered; (2) the suggested trace compounds (odorous substances) for orientation are not known; (3) it has not been shown whether the avian olfactory system only transduces olfactory information; (4) according to present knowledge, atmospheric odours are not available in an adequate distribution and concentration; (5) the use of gradients is not proven; and (6) the use of landmarks is also not clear. Similarly, Waldvogel (1989) concluded in a comprehensive review that 'the bulk of the evidence is currently against an essential role for odors in avian navigation'.

### 3.5.3 Magnetic navigation

Theoretically, two components of the earth's magnetic field could allow for magnetic navigation (Wiltschko and Wiltschko, 1991c). This is possibly the case in a migratory salamander, the eastern red-spotted newt (*Notophthalmus viridescens*), which possesses two magnetoreception pathways (one for the vertical component, one for polarity; Phillips, 1986). In birds, evidence for magnetic navigation is weak at best. So far, only the use of one component of the field – the axial course of the field lines as a compass – has been demonstrated, although two different magnetic systems with different receptors may be present (section 3.4). In a recent review on magnetic maps in domestic pigeons, Walcott (1991) concluded that it is hard to believe that pigeons use such a map. Anyway, the possibility of magnetic navigation ('Viguier's hypothesis'; Berthold, 1993) as one component in a redundant system should not be uncritically dismissed. In 1942 Yeagley proposed that domestic pigeons could navigate by the use of a grid-work of components of the earth's magnetic field and the Coriolis effect (the effect of the Coriolis force on the flight path of the earth's rotation, which changes with latitude). Yeagley's hypothesis of bi-coordinate navigation, based on geophysical grids, was generally rejected because the physiological requirements of birds are beyond known capacities (Wiltschko, 1990a; Berthold, 1993). In a review of animal navigation (Long, 1991), however, the discussion flared up again. It was proposed that domestic pigeons might indeed sense the earth's rotation, and James Gould commented 'we probably should go out and do Yeagley's experiments again'.

Kiepenheuer (1984) postulated the existence of a magnetocline compass which, as a 'navigatory algorithm', might, in theory, explain how a large number of species can reach their resting areas in arched migration, shifting their bearings increasingly south during migration from the northern hemisphere to more southerly latitudes. According to this theory, the birds are able to maintain a subjective inclination angle, which inevitably leads to directional changes if the inclination angle of the geomagnetic field changes. As the inclination angles decrease towards the south, birds move continuously further southwards, in a similar fashion to a hiker who is endeavouring to keep the angle of gradient constant when climbing a slope which is levelling out. With the help of a time-program this compass might enable birds to reach resting areas by vector-navigation. How the implied 'magnetic sensor' is characterized and how it might operate is completely unknown at present. This interesting hypothesis will have to undergo many more tests, as it is probably insufficient to explain some of the migration routes encountered (Alerstam, 1991a).

### 3.5.4 Other navigation hypotheses

These include the sun, star patterns, gravity and moments of inertia. The sun navigation hypothesis of Matthews and Pennycuick (Matthews-Pennycuick hypothesis, e.g. Berthold, 1993) proposed that birds, in addition to the sun azimuth, also use sun altitude, the sun's arc and their annual variation so that they can identify latitudes. However, there is no indication that birds pay attention to sun altitude as an navigational cue (Schmidt-Koenig, 1991). Sauer suggested that migratory birds use inborn patterns of star configurations to recognize specific goal-areas for wintering (e.g. Gwinner, 1971). There is, however, no evidence from quite a number of critical tests and, most likely, the use of star navigation can be neglected (Gwinner, 1971).

Theoretically, birds should be able to return to any familiar site by route reversal and path integration, as many animals do, especially arthropods and mammals. Displacement experiments with domestic pigeons, where all useful input channels (above all in the vestibular apparatus in the middle ear) were interrupted or disturbed during the outward journey, did not result in the failure to orient homewards. Thus, domestic pigeons are able to home without a path integration system. However, these experiments do not exclude that birds may use types of route reversal and path integration for homing in a redundant navigation system (Wallraff, 1991a; Wiltschko and Wiltschko, 1991c). For more detailed discussions of 'angle sense', geodetic and kinaesthetic orientation see Merkel (1978) and Creutz (1987). Ganzhorn and Burkhardt (1991) found some evidence that, in domestic pigeon homing, some kind of route reversal could be involved together with some olfactory component. Recent experiments by Tögel and Wiltschko (1992) suggest that information on the route of the outward journey is used together with local map information in the navigational process. Finally, gradients of gravity (from a 'gravitational landscape') have been thought to serve as navigational cues (Alerstam, 1990). Several tests have demonstrated some influence of gravity and its anomalies in homing in domestic pigeons. Still, any substantial function, if it exists, and the perception of gravitational acceleration, remain to be demonstrated (Wiltschko, 1990a). It may well be that other factors, not yet recognized, may also be of importance. Kreithen (in Long, 1991) stated: 'If we are going to understand animal navigation, we must discover a new sensory channel. Existing ones are not sufficient to explain the behaviour'.

Both theoretical considerations and empirical findings have led to the 'map-and-compass concept', first proposed by Kramer (1953) and on which detailed navigation hypotheses still are based. The idea is that birds use gradient maps (rather than mosaic maps, Wallraff, 1991a and above; Figure 3.8) and compass directions for homing on the basis of bi- or multi-coordinate navigation. From experiments with domestic pigeons it is clear that homing is performed neither by direct sensory contact with the goal nor random search. Domestic pigeons normally fly to their loft directly from the beginning, but they can also return to their lofts if their initial direction is shifted by experimental manipulation. The most reasonable interpretation for this complex behaviour is that the birds combine an initial compass direction and a gradient map for homing (Wallraff, 1991a; Figure 3.9).

## 3.6 ORIENTATION AS A COMPLEX REDUNDANT SYSTEM

Specialists in orientation research have agreed in recent reviews that the navigational portion of bird migration comprises a system integrating a number of different mechanisms, which themselves include different sub-systems. The system is effectively based on a number of redundant mechanisms (Wallraff, 1991a; Able, 1991; Wiltschko and Wiltschko, 1991a; Figures 3.10 and 3.11). The primary framework is

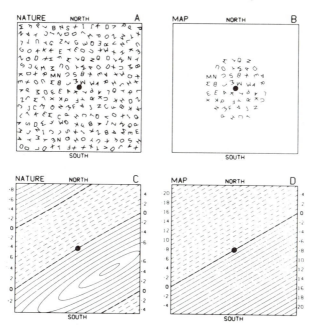

**Figure 3.8** A, mosaic of landmarks, symbolized by letters, surrounding a bird's home site (central dot); B, the corresponding mosaic map (topographical map), which is limited in extent by the bird's range of experience; C, iso-lines (arbitrary units) of a gradient field – the line running through the bird's home is designated as 0, higher values indicated by solid and lower values by broken lines; D, the bird's corresponding gradient map as established by extrapolation of home-site conditions. Notice that in C and D, for the sake of simplicity, gradients of only one variable are shown. For complete site localization, at least two gradient fields are required intersecting at sufficiently large angles. The sections shown are thought to be of different sizes; sides of the square in A, B, may be, at most, a few hundred kilometres, those of C and D, 1000 km or more. (Source: Wallraff, 1991a.)

most likely a time-and-direction program guiding the birds over very long distances to their winter quarters, and possibly also roughly back to the breeding grounds. For more detailed homing, e.g. to pinpoint the last year's breeding site, further mechanisms like olfactory navigation or visual landmark recognition are certainly expected to come into action (Wallraff, 1991a). So-called distance effects in displacement experiments (Schüz *et al.*, 1971) may also indicate interactions of various navigation mechanisms.

The star and the magnetic compasses appear to represent two largely independent mechanisms which seem to enable naïve first-time migrants to find their direction of migration. Under normal circumstances,

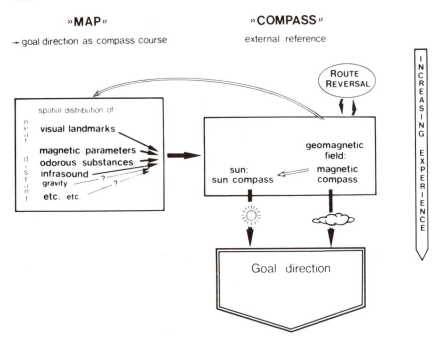

**Figure 3.9** Model of the navigation system of birds: The modern "map and compass" concept views bird navigation as a multi-factorial system. Open arrows indicate learning processes with long-term effects that lead to the establishment and updating of the respective systems; solid arrows indicate the actual use of information in determining the homeward course and the flying direction. (Source: Wiltschko, 1991.)

however, both are most likely integral parts of one complex system. The magnetic compass, as the most simple primary compass, appears to be present without any juvenile learning processes. The star compass may develop largely independent from any magnetic influence during ontogenetic processes. For this reason, celestial rotation and stars appear to play an important role during ontogeny and in establishing the direction of first-time migration (Able, 1991; Wiltschko and Wiltschko, 1991a–c). In this way, a primary direction for the onset of migration appears to be provided (Weindler and Wiltschko, 1991). The star compass, which seems to have developed largely independent from magnetic influence, is apparently later calibrated by the magnetic compass during migration, at least in a number of nocturnal migrants (Able, 1991; Figure 3.12). During the progress of migration, when new stars have to be calibrated, the magnetic field appears to increase in importance (Wiltschko and Wiltschko, 1991a). For the garden warbler it could be shown that the migratory direction during the first autumn

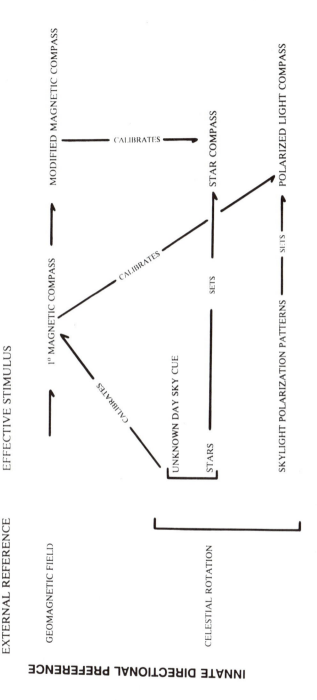

**Figure 3.10** Diagrammatic representation of interactions during the ontogeny of orientation mechanisms in migratory birds. The relationships represent a composite of results from studies on a variety of species, and must be regarded as tentative. Most are based on data from one or a very few species. (Source: Able, 1991.)

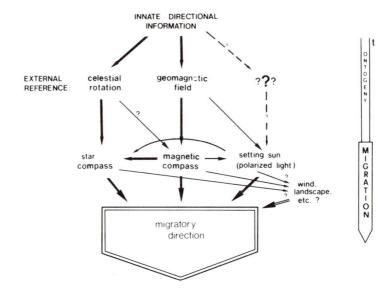

**Figure 3.11** Interactions of stars, magnetic field and sunset factors in migratory orientation. (Source: Wiltschko and Wiltschk, 1991a.)

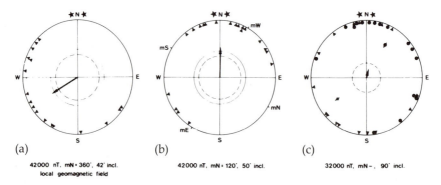

**Figure 3.12** During migration, magnetic information controls the star compass: orientation of wild-caught garden warblers tested under the clear natural sky (a) in the local geomagnetic field, (b) in an experimental field with magnetic north of 120° ESE and (c) without meaningful magnetic information. (Source: Wiltschko and Wiltschko, 1991a.)

migration is determined by the star compass and that of the return migration by the magnetic compass (Daum-Benz *et al.*, 1988). The magnetic compass is further involved in the learning processes establishing the sun compass. Thus, the most important role of the magnetic compass seems to be its function as a basic directional reference in an integrated orientation system (Wiltschko and Wiltschko, 1991b).

Factors associated with sunset, which are perhaps of importance for nocturnal migrants, appear to gain directional significance during ontogeny, presumably in combination with magnetic cues, and may provide a secondary system of orientation (Wiltschko and Wiltschko, 1991a; Berthold, 1993).

Whether true navigation to familiar goals (as regularly shown by birds transplanted in displacement experiments) is based on one mechanism, like olfactory navigation, or on several mechanisms is unknown. Displacement experiments have shown that, in homing, true navigation mechanisms (whatever they might be) leading to direct homing from various directions are clearly dominant over any innate directions. Yet possible interactions between different basic mechanisms in the normal homing of migrants from their winter quarters to the breeding areas may exist which remain to be elucidated (for more details see Berthold, 1991b).

## 3.7 ORIENTATION MECHANISMS AND ENVIRONMENTAL INFLUENCES

Normally, orientation mechanisms allow for undisturbed migratory journeys and homing up to the distance of half the earth's circumference. But there are many cases of misorientation and disorientation indicating temporary restriction of orientation ability, as has been summarized recently by Alerstam (1991a). These may cause substantial losses of migrants. Wind and wind drift are very important here. Their effects on orientation during migration have been reviewed by Richardson (1991) and will be treated in detail below, as will be other relevant factors.

Young birds often show a comparatively wider scatter of migratory directions than adults. In the coastal United States 1–10% of juvenile passerine migrants perish, possibly because of an offshore misorientation due to this scatter. It is unclear whether the wide scatter of orientation in juveniles reflects greater phenotypic variation (due to the high percentage of juveniles during the autumn migratory period), greater genetic variation (before reduction caused by selection processes), more environmental disturbance or less environmental sensitivity. In domestic pigeons, increasing local experience clearly reduces the scatter of orientation (Figure 3.13). Commonly registered reverse misorientation (180° misorientation, flying in the opposite direction) in free-living birds, as well as in experimental conspecifics, may be due to misuse of compass or navigation cues. It can be extremely common and may, for instance, account for up to 40% in raptors observed

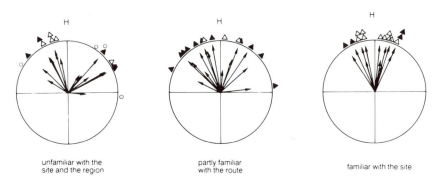

unfamiliar with the
site and the region

partly familiar
with the route

familiar with the site

**Figure 3.13** Effect, in pigeons, of increasing local experience on orientation at distant sites more than 100 km from the loft. (Source: Wiltschko, 1991.)

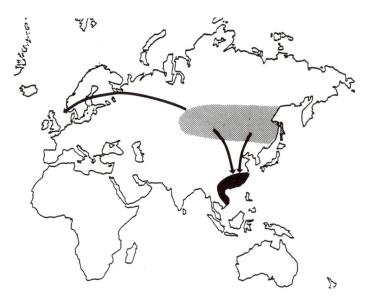

**Figure 3.14** Orientation and misorientation in Pallas's leaf-warbler (*Phylloscopus proregulus*). Some of the birds do not migrate in the normal southerly or southeasterly directions in autumn to reach the typical winter quarters (black), but in the opposite direction, mainly to northwestern Europe. (Source: Mead, 1983.)

during spring migration in Wisconsin (Mueller and Berger, 1992). In the special case of the so-called mirror-image misorientation (Figure 3.14), the birds appear to take the correct bearing with respect to the north–south axis, but select the wrong east–west sense of that angle (DeSante, 1983; Alerstam, 1991a, b). In these cases, the proportions of endogenous and environmentally-induced disorders cannot be estimated, and possible defective compass mechanisms remain obscure. Environmentally induced misorientation has been suggested by Alerstam and Högstedt (Alerstam, 1991a, b). The authors found relatively high degrees of vagrancy, i.e. abnormal migratory directions, in birds breeding in areas with magnetic anomalies in Russia and Siberia. This abnormal behaviour is thought to result from a disturbed magnetic compass. For more details of magnetic anomalies and their possible effects (e.g. the so-called 'K effect', see Keeton *et al.*, 1974).

Disorientation may occasionally lead to catastrophic mortality. Alerstam (1991a) reported recent occurrences of mass deaths in the Baltic Sea. At least 20 000 migrants perished in spring 1985 and autumn 1988. The disasters occurred after a complete overcast sky and dense fog which lasted for several days and which, above all, led to very restricted visibility. From surveillance radar film echos it is highly likely that a wide variety of diurnal, as well as nocturnal, migrants can suffer from disorientation. Although many studies have demonstrated that migrating birds are able to orientate accurately under a completely overcast sky (Martin, 1990; Alerstam, 1991a), others did not (e.g. Åkesson, 1993), and orientation appears to be restricted under conditions of extremely low visibility. This indicates, however, that magnetic orientation is limited, most probably due to sensory restrictions or because the compass information that derives from the earth's magnetic field is 'noisy' (Martin, 1990; Wallraff, 1991a). Disorientation expressed as attraction to lighthouses was found to be strongly correlated with the absence of the moon (Martin, 1990), again indicating that visibility appears to be involved.

There is a great deal of literature on wind effects on avian orientation which, fortunately, has recently been critically compiled by Richardson (1991). In this array of complex relationships 'we are beginning to find a few patterns in this variable behavior' after a long series of often contradictory publications. Two decades ago some workers thought that most passerines compensate almost totally for lateral drift. Now it is known that there are variable reactions to wind that may be selectively advantageous with respect to energy- and time-selected optimal migration strategies (section 2.8). Richardson (1991) summarized the reactions as follows. During the day, overland migrants at high altitudes may often allow crosswinds to drift their tracks laterally from the preferred heading. At low altitudes, however, many diurnal migrants adjust their headings to compensate for drift and may even overcompensate to allow for previous displacement by wind. At night, some overland migrants compensate fully for drift. Compensation may be more common where there are prominent topographic features. Over large waterbodies, compensation is rarely, if ever, complete. Wave patterns may only allow for partial compensation. In coastal areas, leading line effects and extensive coastal migration may play an important role. Other adaptations can include reduction of drift by flying at times and/or altitudes without strong crosswinds (Gauthreaux, 1991), as well as reorientation. During large parts of the migratory journey, migrants may, to an extent, tolerate uncorrected wind drift at certain times in order to require less time and energy for migration or to correct for drift accumulated during previous stages (Bruderer and Liechti, 1990). Wind drift may increase with strong winds, whereas in crosswinds with speeds of up to 2–3 m/s compensation may be complete (Liechti, 1992). During the terminal phase of migration, however, drift appears to be more reduced by various measures in order to reach a specific goal. Larger birds are more ready to migrate in light opposed winds, perhaps, because they generally fly faster (Evans and Davidson, 1990). Williams (1991) simulated autumnal migrant flights from central North America. The study demonstrated the general importance of wind selection as opposed to compensation for wind drift for migration strategies. Much remains to be done to understand the complex behavioural and orientation processes related to wind and other meteorological influences. In all analyses the phenomenon of 'pseudodrift' (section 1.5) has to be considered, i.e. whether birds at different altitudes follow different wind directions in various ways.

Vagrants are often displaced by strong winds over extreme distances (section 2.17). Interestingly, it has been suggested, that their displacement may not be entirely passive. Instead, it may be shown by pioneers, mainly immatures, taking advantage of wind to explore beyond their normal range at minimum cost. Thus, some vagrants would merely be individuals undertaking long-distance exploration (Elkins, 1988b).

# 4

# Current microevolutionary processes

Presumably, migratory behaviour is currently under substantial alteration and microevolutionary development in many bird species and populations, but can only be established in those cases for which sufficient data allow for comparisons between past and present tendencies. The view of microevolutionary changes is strongly supported by the extremely high evolutionary potential found in obligate partial migrants and in recent actually demonstrated microevolutionary processes of middle-distance migration in blackcaps (section 2.4). The most obvious recent changes in migratory behaviour can be summarized in five main categories (Berthold, 1995). (1) Reduction of migratoriness. For example, the European blackbird was an exclusively migratory species in Central Europe until the beginning of the 19th century and then became a partial migrant. Today, lowland populations consist of over 50% non-migrants. Some other European short-distance and partial migrants, like the black redstart, chiffchaff or goldfinch (*Carduelis carduelis*), have been found to winter in increasing numbers in their breeding areas, as do many other species in many areas of the world. (2) Extension of migratoriness. This rare phenomenon has been found in the serin (*Serinus serinus*). During the expansion from the Mediterranean area to Central Europe, the immigrants became increasingly migratory in contrast to the largely resident parent population. Similarly, the house finch became migratory in a transplanted population in eastern North America (Terrill, 1990). (3) Changes of migration periods, particularly later departures and earlier returns. Systematic sight observations and trapping activities of migrants have revealed that, at present, many species show trends or tendencies to leave their breeding areas in autumn later and, in part, also to return earlier in spring. Later departures have been found in the majority of short-distance migrants examined, and also in long-distance migrants. The opposite effect, an earlier departure, has rarely been found. (4) Changes in the migratory distance. In several species like the greylag goose or the cormorant (*Phalacrocorax carbo*) some populations show increasing tendencies to winter closer to the breeding areas. (5) Novel migratory directions and winter quarters. A number of species have extended their wintering areas by shifting their migratory directions. Although such alterations are often small, there is a very prominent recent example of this: the blackcap. Birds from Central Europe, which had migrated in southerly directions to the Mediterranean area and Africa have developed a novel northwestern migratory direction over the past 30 years (section 2.4). It thus provides the fortunate circumstances that a recent substantial alteration of migratory behaviour in Central Europe can immediately be analysed by field and experimental studies so that its cause, control mechanisms, selection forces and novel strategies can be examined in a stepwise manner.

## 4.1 CONTROL OF NOVEL MIGRATORY HABITS IN THE BLACKCAP

Over the past three decades, the number of blackcaps wintering in Britain and Ireland has increased from a few birds to a substantial population involving thousands of individuals. Surprisingly, the recruitment area of this novel wintering community appears to be entirely in Central Europe, mainly Germany and Austria. This is indicated by the steadily increasing number of recoveries of birds ringed in Europe and recovered in Britain, and vice versa (Berthold, 1995). This series was initiated by a spectacular recovery in 1961 of a male ringed in August near Linz, Austria, and found dying in a cat's claws in December in Ireland (Zink, 1962). No blackcaps of the breeding population in the British Isles have been found to winter there. Thus, continental blackcaps from Central Europe, which hitherto migrated exclusively in southern directions to Mediterranean and African winter quarters, have recently developed a novel migratory direction towards the northwest and have established new wintering areas (Figure 4.1).

The questions that have to be raised with respect to the development of this novel migratory trait are: why has it occurred in recent times? How is this novel habit controlled? Do continental blackcaps reach the British Isles for wintering at present accidentally, due to errors in their orientation system, or have they already developed specific control mechanisms? If adaptive microevolutionary processes are involved, what are the main selection forces? Do possible explanations for the origin of the migratory behaviour need to evoke major changes in existing migration patterns?

To start with the last question: Rice (1970) pointed out that no major

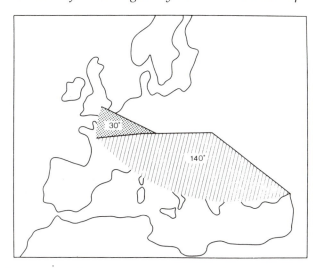

**Figure 4.1** "Fan-shaped" pattern of autumn migratory direction taken by blackcaps breeding in Europe. Note that the angle of "normal" migration (140°) need only be increased by about 30° for these birds to reach Britain and Ireland. (Source: Berthold and Terrill, 1988.)

changes and no recent substantial mutation or genetic rearrangement have to be postulated. Central European blackcaps normally migrate in directions spread over a broad fan-shaped angle of about 140° to their southern winter quarters. This angle need only be rotated northwest by about 30° for continental individuals to reach Britain in autumn (Figure 4.1). Supposing a continuous distribution of behaviour with respect to direction within that broad angle, it is likely that normal variation on the western side could account for individuals occasionally migrating towards the British Isles. They could then, under specific circumstances, prosper due to some selective advantage. In fact, although the phenomenal population growth in Britain's wintering blackcaps is very recent, their sporadic occurrence has been documented for well over a century (Berthold and Terrill, 1988). If this view is correct, the next question is why the broad fan of directions is strongly extended in its western tail end to the north?

A basic ultimate reason appears be the changed ecological situation for wintering blackcaps in the British Isles. Since the end of the Second World War, especially since the 1950s, when bird feeding started its dramatic increase, feeders became increasingly important for wintering birds in Britain, as in many other areas in Europe. Since blackcaps regularly successfully use a large variety of foodstuffs from feeders they have undoubtedly played a major role in the survival of many of the blackcaps wintering in Britain and Ireland (Rice, 1970; Hardy, 1978; Langslow, 1979; Leach, 1981; Simms, 1985; Bland, 1986). Feeders are especially important during late winter when berries, such as honeysuckle, blackberry, holly or seabuckthorn are depleted, on which blackcaps subsist to a large extent during autumn and early in the winter. These feeders appear to provide a solid nutritional basis for successful wintering of blackcaps in the British Isles which formerly, apart from exceptions, was impossible due to food shortage, above all in late winter. In addition to these anthropogenic factors, relatively mild winters in the recent past may also have contributed to establishing that new wintering tradition.

The next question is, which alterations in the control mechanisms of the normal migratory journeys of blackcaps may have occurred so that increasing numbers of individuals are able to reach the British Isles and to use the present favourable conditions for wintering? Theoretically, there are two possibilities: (1) individuals reach the British Isles purely accidentally due to an increasing phenotypical variation of migratory directions on the western tail of the migratory directions which now no longer has lethal effects. (2) Blackcaps wintering on the British Isles have developed, through high microevolutionary speed, specific control mechanisms in order to maintain or to stabilize that novel migration strategy. The fairly rapid growth of the wintering population, which is far above of any recent increase of a breeding population of blackcaps, favours the second idea. In the meantime, experimental analysis has clearly demonstrated that the novel migratory habit of continental blackcaps wintering in the British Isles is caused by microevolutionary processes of astounding evolutionary speed, and selection of a special inherited migratory direction (section 2.4). Further, it provides good reason to suppose that the novel migratory habit is selected on the basis of about seven, in part interrelated, selective advantages. (1) The formation of a new wintering area could by itself be a strong selection force for blackcaps since the density of the wintering populations in the Mediterranean, at least in favoured areas, can be extremely high and may cause substantial intraspecific competition. Shrinkage of Mediterranean wintering sites due to recent forest destruction may have made the situation more critical (Berthold, 1986). (2) The total distance of a round-trip migration to the British Isles is reduced by approximately one third in comparison to the normal journey to the Mediterranean. This may reduce costs associated with distance, which may be important in spring when the energetical situation can often be critical. More important, it appears that the strategy may allow birds wintering in the British Isles to reach the breeding areas after a shorter distance to cover earlier than conspecifics that have been wintering in Mediterranean areas. (3) Blackcaps migrating to Britain and Ireland migrate into shorter day lengths and they winter in considerably shorter day lengths than those in the Mediterranean region. Shorter day lengths are well known to terminate the so-called photorefractory period (i.e. the failure to respond physiologically to stimulating effects of long day lengths) earlier. Thus, blackcaps that winter in Britain and Ireland may respond to increased day length and come into spring migratory condition and/or reproductive state earlier than birds wintering to the south. (4) Blackcaps that winter in the British Isles may reach comparably earlier migratory and reproductive conditions not only due to the earlier termination of their refractory period, but, in addition, in comparison to birds from the southern wintering grounds, they are exposed to day lengths that increase in spring, before their homeward journey, at a faster rate. (5) An earlier stimulus for spring migration, combined with the shorter distance to cover, suggests that blackcaps wintering in the British Isles return to the breeding grounds often before conspecifics from the Mediterranean and in a more advanced reproductive state. This has been tested experimentally. When blackcaps from S Germany were exposed to photoperiodic conditions that simulated migration to the British Isles and to the Mediterranean region the following differences were found. For the first group, estimated mean arrival dates on the breeding grounds were 1 April (males 29 March, females 4 April), for the second group, however, 17 April (13 and 21 April, respectively) (Figure 4.2). In addition, mean testes size at the arrival time was about 25% higher in the first group than in the second. Presumably, circulating levels of testosterone and luteinizing hormone were also higher in the birds kept in conditions simulating migration to the British Isles (for more details see Terrill and Berthold, 1990). (6) Arriving on the breeding grounds earlier could be quite hazardous, especially in case of severe winter-like conditions in spring. However, here again, blackcaps that winter in the British Isles may be at an advantage relative to their southern wintering counterparts. They are most likely physiologically acclimatized to cooler weather than the southern birds. Further, they are more adapted to utilizing feeders in critical situations (whereas feeders are virtually nonexistent in the Mediterranean wintering area). With respect to the shorter migration distances they should arrive on their breeding grounds with greater fat reserves. Lastly, while blackcaps migrating from the Mediterranean area to the north are generally migrating into relatively harsher climatic conditions, birds coming from the British Isles are more likely to experience either steady or even milder climatic conditions. Thus, blackcaps wintering in the British Isles appear to be well prepared to profit from their early arrival on the breeding grounds. (7) Although not specifically demonstrated for the blackcap, it seems logical to propose that individuals that arrive before others gain a

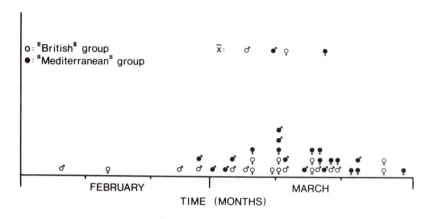

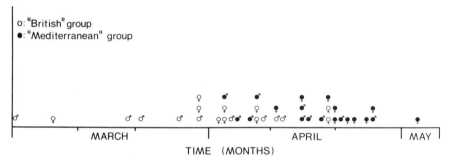

**Figure 4.2** Starting dates of spring migratory activity (above) and estimated arrival dates (below) for blackcaps held under British-winter photoperiod and Spanish-winter photoperiod. Five pairs of the 'Britishwintering' birds could have formed before the first 'Mediterranean-wintering' male paired. (Source: Terrill and Berthold, 1990.)

number of advantages with respect to breeding and fitness. Thus, birds returning from the British Isles may be more successful in obtaining and holding prime territories and getting an earlier start to breeding. Early breeding could enlarge clutch size (due to 'calendar effects', Berthold *et al.*, 1990b), increase the number of broods per season, or increase the likelihood of successful late or replacement broods. The relatively early arrival of both males and females from the British Isles in a more advanced reproductive state may also favour assortative mating within that group. This could lead to distributional concentrations of those birds, as probably occurs in Austria (Helbig, 1991d) – a sample of blackcaps from the contact zone (of SE- and SW-migrating individuals), near Linz, oriented SW to NW and was significantly different from both adjoining populations. It is suggested, therefore, that a distinct subpopulation with a large fraction of birds wintering in the British Isles has established itself in the contact zone. If such assortative mating occurs, the genetic transmission of the novel migratory direction may cause a substantial number of the offspring to migrate toward the new wintering grounds on the British Isles, and may then accelerate the new development. Thus, in all, a number of interrelated selective advantages appear to increase fitness in blackcaps wintering in the British Isles. This may easily make up for an eventually higher winter mortality in the British Isles compared to that in the Mediterranean wintering grounds. Increased fitness combined with the inheritance of the novel migratory direction appear, ultimately, to cause the rapid growth of that novel wintering population. Assortative mating and advantages in breeding remain to be established. The carrying capacity of the British Isles for wintering blackcaps, and thus the future development of the wintering population, is hard to judge. But if further climatic changes due to 'greenhouse' effects occur, potential wintering areas for blackcaps are most likely to spread over large parts of Europe, and have in fact already expanded to Scandinavia (Fransson and Stolt, 1993). For more details see Berthold and Terrill (1988), Terrill and Berthold (1990), Berthold and Helbig (1992) and Berthold (1995).

## 4.2 DEVELOPMENTS IN OTHER SPECIES AND PROSPECTS

There is no other bird species for which information about recent microevolutionary changes in migratory behaviour, including its causes and control mechanisms, would be as good as in the blackcap. But there are numerous cases of well-documented changes in migratory behaviour during recent centuries or decades of which a few were mentioned in the introduction of this section. In some cases, in addition to the blackcap, we have good reasons for making assumptions about the underlying control mechanisms and also to suppose microevolutionary processes to be involved. For instance, a number of European partial migrants have increased the non-migratory fractions of certain populations. This is the case in the Central European populations of the European blackbird (Heyder, 1955; Berthold, 1993), and in more local populations of the European starling (Merkel and Merkel, 1983) or the cormorant; Schmidt, 1989). In the European blackbird, according to results of field and experimental studies (section 2.4), it is most likely that the following has happened. About 200 years ago individuals of the former migratory populations started to take advantage of improved conditions with respect to feeding, microclimate and possibly predation associated with human settlements. Most likely, the 'civilizing' of the blackbird has helped it to lose its fear of humans (through mutation or learning processes; Berthold, 1993). First-resident individuals and then resident fractions were most likely selected for during the civilizing process and increased up to about 50% at present. As a result, partially-resident populations are now more concentrated in towns and villages, and here these 'urban blackbirds' may further increase their sedentariness, whereas 'forest blackbirds' breeding more distant from human settlements have remained more migratory (for more details see Schwabl, 1983). The situation in starling populations has also recently changed to partial sedentariness in a similar manner (for details see Merkel and Merkel, 1983). The widespread phenomenon of 'urbanization' (synanthropy), recently observed in many species of ducks, gulls and songbirds, like song thrush or European robin, in temperate-zone and subarctic areas (Luniak and Mulsow, 1988) may also, in part, be based on the selection of genetically-determined non-migratory fractions. A most interesting case has recently been reported from the great crested grebe. In The Netherlands, the non-migratory fraction has drastically increased since the 1960s, most likely due to changes in the genetic composition of the breeding population. This presumed microevolutionary change is thought to be caused by an increase in human made suitable habitats (lakes; Adriaensen *et al.*, 1993). In other cases, facultative migrants may no longer be triggered to move due to formerly common shortages. This appears to be the case in great and blue tits where reduced emigrations and increasing wintering in the breeding area are thought to be due to increasing human winter feeding (Winkel, 1993). In the great tit, Saitou (1991) showed that birds of resident flocks raised more surviving offspring than immigrants.

In partial migrants, not only has sedentariness increased, but also the distance to travel may have changed. Apart from a lot of suggestive evidence a few cases are now well documented, like the cormorant in the area of the Baltic Sea (Schmidt, 1989) or the greylag goose of eastern Germany (Rutschke, 1990). With a considerable rise of the breeding populations, increasingly more individuals tend to winter closer to the breeding area. According to results obtained in the blackcap on the genetic variation of the amount of programmed migratory activity related to the distance to cover (section 2.4), it is possible that in species like the cormorant shorter migratory distances are now being selected for. In the blackcap, it was recently demonstrated that under moderate selection intensities a middle-distance migrant population (from S Germany) could evolve into a short-distance migrant population within 10–20 generations (Berthold and Pulido, 1994). Corresponding microevolutionary changes in the migratory directions, as demonstrated for the blackcap, may have also occurred in other species but remain to be demonstrated.

In many areas in Central Europe white storks are currently kept and bred in special stork stations in order to prevent the possible extinction of the western European populations (Bloesch, 1989; Rheinwald *et al.*, 1989). The offspring of such semi-domesticated breeding stocks is often regularly prevented from migration until breeding maturity is gained after several years. Breeding pairs established in that way do not normally depart from the breeding area, due to domestication effects, and then have to be maintained during the winter period. It would be most interesting to examine if and how rapidly this domestication process, in an exclusive and long-distance migrant, could lead to the selection of less-migratory individuals. In addition, white storks have recently started to establish wintering populations in Israel and Spain (Leshem and Tortosa, pers. comm.) Here, it will be interesting to see whether parts of the stork populations are presently selected for shorter migration distances, as appears to be the case in other species like the cormorant (see above). Also, hybridization of European storks with introduced African, less-migratory conspecifics, may have played a role, especially for the western population now wintering in part in Spain. This possibility should be tested at the molecular genetic level, e.g. by studying mitochondrial DNA sequences.

At present, it is unknown how the relatively rare cases are controlled in which migratoriness has recently developed in birds descending from non-migratory populations, like serin or house finch as mentioned earlier. On the one hand, a novel migratory habit could be triggered by environmental factors, and thus could be some type of facultative migration without specifically evolved control mechanisms. On the other hand, specific genetic control mechanisms are not unlikely. In Europe, by far most of the species of the present avifauna are neither exclusively migratory nor fully resident. However, the degrees of migratoriness and sedentariness in these species vary extremely from (almost) nil to complete. Even in almost 'classical' resident species, like the crested tit, occasionally 'migratory' individuals are detected by ringing, and correspondingly in the most typical migratory species, like garden warbler or swallow, individuals are found from time to time that try to winter within the breeding area (section 2.17). Many species of the present European avifauna may have altered their migratory habits repeatedly and considerably along with multiple changes in environmental conditions within Europe and northern Africa, during the ice ages as well as during pre- and postglacial periods. During such changes many species may have passed through various stages of migratoriness and sedentariness. A classical case is represented by the serin. As mentioned above, this species advanced from the Mediterranean from about 1800 onwards and became migratory during its spread up to northern Europe (Mayr, 1926; Olsson, 1969; Schüz *et al.*, 1971). At present, in the course of increasingly mild winters, the species has started wintering in S Germany and has become resident again. Furthermore, occasional hybridization of typical and less-typical migrants may increase gene flow, genetic exchange and thus intra-specific genetic variation and the microevolutionary potential towards more migratory or more sedentary behaviour, as is observed in redstarts in captivity and in the wild (section 2.4). Under these circumstances it is questionable whether, among present European bird species, geno-typically pure migrants or non-migrants really exist, or whether species generally represent genotypically partial migrants with, however, extremely variable amounts of phenotypical fractions of migrants and residents. A variable phenotypic expression, as well as a great deal of genetic variation, may easily be maintained by genotype-environmental interactions, as well as by both temporal and spatial environmental heterogenity. If, for instance, in the gene pools of present phenotypical resident species or populations, remaining genes for migratoriness exist, it is conceivable that under new environmental conditions, as in the case of the serin during its northward-spread migratoriness could be relatively rapidly selected for. The opposite, however, increasing selection toward sedentariness, is much more likely. If 'global warming' proceeds as predicted by many meteorologists, many obligate partially migratory populations should rapid shift to increased residency up to full sedentariness through rapid microevolutionary processes (Berthold, 1995). In a recent review, Maynard Smith (1993) has pointed out that currently observed rapid evolutionary changes observed in creatures all concern adaptations to changes in the environment pro-duced by humans. If this holds for the future, proceeding 'global warming' and its side-effects, as well as worldwide human made habitat changes, for avian communities may provide a basis for wide-spread and rapid microevolutionary changes of avian migratory habits. Apart from the examples mentioned above, there are many more and increasing reports in the literature that species like, for example, swans, geese, gulls or songbirds, are changing migration routes and times, and wintering areas in recent decades in relation to weather and other factors (e.g. Erskine, 1988; Johnson and Herter, 1990; Berthold, 1995). The analysis of individual cases may well elucidate further microevo-lutionary processes. If substantial changes of migratory habits through microevolutionary processes due to increasing human made environ-mental alterations occur, as briefly outlined in section 2.17, the following scenario is highly likely. Improved wintering conditions due to 'global warming' leads to increasing sedentariness of many partial and short-distance migrants in higher latitudes. Well-adapted resident species also benefit from improved wintering and other conditions, mainly by reduced winter mortality. This leads to an overall increase in resident individuals and an increasing occupancy of ecological niches by this group. Long-distance migrants, with the slowest adaptation rate for various reasons, will then have increasing difficulties in obtaining ecological niches allowing for successful reproduction and are most likely to continue and accelerate their already widespread decline (Berthold, 1995). For students of avian migratory habits and their control, the coming decades may thus become a unique challenge, as they are likely to be for all those who are inclined to engage in conservation measures for our beautiful migratory birds.

# 5

# Synopsis: control of bird migration – the present view

The topic of the 'control' of bird migration deals with the endogenous mechanisms, environmental stimuli and physiological adaptations which regulate the onset of migration, and factors which generate its course and termination in goal areas. This chapter summarizes the main aspects, results and conclusions of the previous chapters.

Routes of long-, middle- and short-distance migrants cover our planet like a network, and there are no barriers such as deserts, mountains or oceans that cannot regularly be crossed by migrants (section 1.1). Most likely, avian migration is of polyphyletic origin. For this and for ecological reasons, an almost immeasurable amount of different migratory patterns have evolved, from dismigration to chain migration (p. 8). Originally, avian migration might simply have evolved from hunger (through 'hunger restlessness', section 2.1) or from juvenile dispersal and nomadism through facultative, to, finally, obligate migration (p. 10). The available facts demonstrate that avian migration in its present stage is a rather safe enterprise and a very successful proposition (section 1.1) – the result of reliable control mechanisms (chapters 2 and 3) and quick adaptations (chapter 4). However, occasionally mass deaths occur when numerous emergency measures, by no means satisfactorily known, fail to prevent disasters (sections 1.1, 2.13 and 3.7).

Although von Pernau in 1702 already recognized the urge to migrate as an innate behaviour, and Johann Andreas Naumann (1795–1817) started with first simple experiments, the period of experimental investigations of bird migration began with systematic experiments by Rowan (1925). Rowan was also first to develop a complex theory of the control of avian migration (section 1.2).

Among the novel methods in the study of bird migration a number are especially promising, e.g. satellite-tracking of individual migrants, the quantitative determination of hormones, comprehensive energetic and *in vitro* studies, recent developments in wind-tunnel technology for flight performance investigations, and large-scale breeding of migratory birds in captivity allowing for experimental genetic studies in migrants (section 1.3). What we are interested in when studying control mechanisms and strategies of avian migration is summarized in seven questions in section 1.4. Migratory birds clearly have morphological, physiological and behavioural prerequisites. Above all, their major flight muscles form a highly specialized 'aero-engine' which especially in long-distance migrants, has high oxidative capacities, mainly for fatty acid oxidation. Further, continuous trailing wing-tip vortices may be of crucial importance for the excellent flight performance of migrants, among numerous other prerequisites (section 1.6). However, despite these preadaptations, migratory performance is regularly improved during migratory disposition or 'zugdisposition', a state of readiness which comprises hypertrophy of flight muscles (sections 1.5 and 2.7), deposition of fat as a fuel (sections 2.1 and 2.7–2.9), integration of enzyme systems (section 2.7), the development of migratory behaviour (sections 2.1 and 2.13) and other dispositions (sections 2.6, 2.9 and 2.22). Among behavioural adjustments during the migratory period the development of nocturnality in many purely diurnal species during the non-migratory seasons is most conspicuous. Nocturnal migration is explained by seven hypotheses (daytime feeding, time gain, energy saving, increased physical safety, predator avoidance, forced flight and star compass allowance; section 2.1). The main forces in the evolution of nocturnal migration appear to be related to the first three hypotheses. In experimental birds, nocturnality is expressed as nocturnal 'migratory restlessness' or 'zugunruhe', a key factor for the study of bird migration. Zugunruhe, in diurnal migrants displayed during the daytime, is the expression of an instinct and is a fairly good reflection of the species- and population-specific migratory behaviour and the course of migration. It appears to be a cage-adapted migratory activity expressed as a reduced flight behaviour (wing whirring, migration in a sitting position). Five to six other kinds of restlessness ('acclimatization', 'winter', 'summer', 'roost-time', 'hunger' and possibly 'dispersal') can be distinguished from migratory restlessness when carefully analysed (section 2.1).

Two biological clocks are of basic importance for the control of avian migration: circadian and circannual rhythms. Although surprisingly few studies have concentrated on circadian aspects in migratory birds, there is no doubt that zugunruhe, metabolic events and hormonal mechanisms are based on circadian rhythms (sections 2.2 and 2.6). For the development of seasonal migratory activity a testosterone-controlled model with two coupled circadian oscillators has been proposed (section

2.2), however it appears not to be applicable to autumn migration (section 2.6).

Circannual rhythms involved in the control of migration have been demonstrated for 14 bird species from four different continents. They have been shown to control migratory activity, fattening, perhaps nutrition and orientation, related annual processes like moult and reproduction, and certainly control other events (sections 2.2, 2.4, 2.5, 2.9 and 2.22). There is evidence in a number of species that the endogenous migratory urge, based on circannual rhythmicity, can immediately trigger migrants to initiate their first departure (p. 56). Observations in other species also indicated rigid endogenous control of initiation and course of migration (p. 56, sections 2.3–2.5). In redstarts there is evidence for an immediate genetic control of the onset of autumn migration (p. 74). A substantial number of studies indicate that migratory activity functions to determine the course of migration. Activity patterns often appear to be endogenously- or, as shown in some cases, genetically-controlled patterns and seem to provide population-specific spatio-temporal programs to guide at least inexperienced migrants from their breeding area to unknown winter quarters (in combination with preprogrammed, genetically determined migratory directions, vector-navigation hypothesis; sections 2.3, 2.4 and 3.5). There is evidence from two species that differential migration can also be based on endogenous (sex-specific) time-programs (p. 65). The size of migratory fat reserves also often appears to be strictly controlled by an endogenously- or genetically-determined annual set-point mechanism. The programs for migratory activity and fattening may be largely independent of each other (sections 2.3 and 2.4). As to whether endogenous programs for avian dispersal exist remains to be demonstrated (p. 67).

Genetic control mechanisms of bird migration appear to be basic and widespread. Immediate genetic control has been demonstrated for the urge to migrate in exclusive migrants (probably polygenic control), the urge to migrate in partial migrants (with migratoriness and residency probably as threshold characters, determined by multiple loci), differential migration (sex-linked inheritance of migratoriness), for the dates of the onset and for temporal patterns of migratory activity during the autumn migratory period, the choice of migratory directions, including seasonal shifts, and most likely the patterns of body mass cycles (mainly due to fattening; see below). Genetic control could also be demonstrated for migration-related features: wing length, premigratory body mass, and time course and temporal pattern of juvenile moult (section 2.4).

Photoperiod, the most reliable environmental cue for the control of annual cycles, controls bird migration in at least two different ways: as the most important synchronizer or 'zeitgeber' of circannual and circadian rhythms, besides temperature, food supply, precipitation, social factors, etc., as other synchronizers, and by inhibiting or accelerating individual migratory processes. The dual system of endogenous circarhythms and photoperiod as a precise zeitgeber provides migratory birds with a viable basis for proper seasonal timing (section 2.5). From the second control mechanism, acceleration in the juvenile development of late-hatched individuals is conspicuous (short-day effect, calendar reaction; p. 86). Further, two functions, a primary stimulatory role and a role as an ultimate factor via day length with respect to increased or decreased feeding times, are still speculative (section 2.5).

Although physiologists have been studying bird migration for almost 70 years, our knowledge of how endocrine mechanisms are exactly involved in its control is still highly uncertain (p. 90). From gonadal hormones, it appears certain that androgens (testosterone and perhaps its metabolites) have a role in the development of events for spring migration (hyperphagia, fattening and to some extent activity). However, autumn migration proceeds quite similarly without a comparable hormone basis (p. 91). Also, in the regulation of partial migration, testosterone is apparently not involved (p. 93). With respect to thyroid hormones it remains open to what extent they may play a role in any primary control of migratory events or whether they, more likely, exert fine tuning or simply general metabolic effects (section 2.6).

Towards the spring migratory period, long days and circannual rhythms activate the hypothalamo–hypophysial system (p. 84). From the heterogeneous group of hypothalamic and hypophysial hormones, prolactin has been most intensively studied. Its administration can induce or accelerate fat deposition, but its actions on fattening are not fully understood. In the regulation of migratory fattening and activity, adrenocorticosteroids are also involved, but may primarily affect fine tuning. Corticosterone, as a stress indicator, was generally found to be low in fat birds but appears to increase in fat depleted individuals (section 2.6). Other hormones like epinephrine, norephrine, resorpine, growth hormone or melatonin showed some parallels to migratory processes but their possible control mechanisms are unclear. The latter appears to be involved in circadian pacemaker systems which are also related to migration (p. 99). With respect to neuronal mechanisms, there is evidence that migratory traits are controlled by the infundibular nucleus, median eminence and the ventral medial hypothalamus. But the knowledge of these mechanisms is still in its infancy (p. 100).

Fats have evolved as major fuel for bird migration due to five main reasons: (1) they have the highest concentration of metabolic energy stores; (2) they can be stored without water or protein (tank principle); (3) they can be handled most efficiently among energy carriers in intermediary metabolism; (4) most body tissues, and especially the flight muscles of migratory birds, are able to oxidize fatty acids efficiently and completely; and (5) muscle fibres relying on lipids are exhausted relatively slowly (sections 2.1 and 2.7.1). During migratory hyperlipogenesis, fatty acids are transported to 15 fat 'organs' (section 2.1) where they are stored mostly as triglycerides in the form of unsaturated fatty acids (mainly oleic and linoleic acid). The low melting points of these fatty acids may influence ease of mobilization and lipolysis. Fat deposition rate can lead to a daily body mass increase of up to about 10%. Maximum fat deposition rates decline with body mass and may be metabolically or environmentally limited (section 2.7.1). Many migrants have virtually no lipid reserves during summer, but many depart a few weeks later with fat deposits which can increase their body mass by slightly more than 100% (102–115%). Maximum fat deposits (above 60%) are reached in smaller long-distance migrants (passerines, hummingbirds). In waders they may lead to an 'adaptive overloading', especially in spring, see below. Short- and mid-range migrants accumulate smaller reserves, in partial migrants about 30%, and in many raptors, using gliding flight, they account for 10–20% (sections 2.1 and 2.8). Fat deposition can be accompanied by a substantial increase in fat-free body mass, primarily due to protein storage. Protein storage is often expressed in breast muscle hypertrophy. In other tissues it possibly occurs for repair 'to keep the engine going'. Carbohydrate (glycogen) storage appears to be suppressed during the migratory season in a number of species but there are exceptions which need further investigation. For some species, decrease in body water content during the premigratory period is reported and interpreted as a strategy of adaptive dehydration or ballast reduction. It is not found in many other species and also needs further research (section 2.7.2). Body composition of migrants can vary adaptively in both fat and fat-free mass in relation to specific requirements of different types or routes of migration, or even depend on the environmental conditions of the goal areas. In long-distance migrants, late arriving in the breeding grounds, maximum fat levels are similar in autumn and spring. In other species, higher energy reserves observed in the spring migratory period are thought to provide a safety margin for more rapid migration (which is known to often contain longer bouts in spring; sections 2.8 and 2.15) or an insurance against poor conditions at staging and breeding areas. There is considerable variation with respect to the size of fat deposits in age and sex classes and with regard to the beginning of lipid deposition in relation to the onset of migration. The majority of passerine migrants depart with fat loads amounting to 20–30% and reach maximum fat deposits before ecological barriers have to be crossed. Migratory fattening has to be considered under the selection pressure of at least four different constraints – time, energy, predation minimization and increasing human influences (section 2.8).

Numerous models and equations have been developed to estimate flight ranges of migrants with given fat deposits. Maximum ranges for waders account for up to 14 000 km, with an average of about 6000–7000 km (p. 116). Recently calculated ranges for trans-Sahara passerine migrants were only about 1100–1400 km indicating that they would need the aid of tail winds to get across the desert (p. 116). Overall energy budgets for migration and all annual activities are so far only available in a few species (p. 122).

Migratory birds have developed several nutritional strategies and adaptations that provide the necessary surplus energy for migration. These include hyperphagia – probably the most important mechanism, with an increase of metabolized energy levels by about 20–40% and dietary changes to easily accessible food, above all adaptive seasonal frugivory in many carnivorous/insectivorous species consuming fruits as an 'easy prey' which provide sufficient carbohydrate intake for fattening. Some fruits like black elder berries are clearly preferred and especially effective for fattening (p. 130). Further adaptations are: selection of energy- or nutrient-rich food (experimentally shown for diets with high fat levels in some species, not found in others, p. 137); increased food absorption and assimilation efficiency (in a warbler, p. 135); decreased rate of standard metabolism (one proof, p. 138); movements to more favourable feeding grounds (widespread, p. 125); devoting more time to actual feeding (a common but not general habit, p. 126); reduced locomotor activity during the daytime (widespread; section 2.13); and avoiding temporal overlap between fattening and other seasonal activities (section 2.22). Possibly, G3PD involved in fat metabolism is functionally linked to the display of migratory behaviour (section 2.4.2).

From the complex thermoregulation and water balance and the question whether dehydration or fuel exhaustion may limit long-distance migration the following emerged. The temperatures of birds measured during active flight are generally greater than 41°C. Thus, overheating may become a problem for migrants and heat dissipation should be efficient. Heat dissipation is closely linked to water metabolism so that overheating may cause dehydration. However, during migration, birds could actually obtain water homeostasis due to metabolic water production (from fat metabolism, p. 140) by ascending to altitudes where the air is cool enough to keep evaporative heat dissipation at the required level, or by flying at night using lower temperatures (up to about 10°C). Data gathered in the Sahara indicate that the main limiting factor for long-distance migration appears to be fuel rather than dehydration (section 2.10).

Birds, and especially migrants, possess extraordinary adaptations to high altitudes. Vultures, for instance, are able to soar to altitudes of more than 10 000 m, and many species routinely fly over the Himalayan mountains at 8000–9000 m. Those birds must be equipped with special mechanisms that prevent hypoxia and altitudinal sickness (sections 1.1 and 2.11). Indeed, pulmonary adaptations have led to an almost perfect gas exchange (section 2.11.1). Under hypoxic high-altitude conditions, tested trans-Himalayan migrants do not increase red blood cells nor change haematocrit or viscosity of their blood. Moreover, hypoxia may cause a greater increase in cerebral blood flow in birds than in mammals. This unique cardiovascular adaptation may essentially contribute to the exceptional tolerance of birds to extreme high altitude (section 2.11.2). As special blood adaptations in several species, two-stage and three-stage cascades of haemoglobins were found with graded oxygen affinities. Their different oxygen-binding capacities are adaptations for differing partial pressures of oxygen at different altitudes (section 2.11.3). Although birds are uniquely equipped for high-altitude flights, most migratory bird flight occurs below 3000 m (section 2.12). Great variance in the altitudes of migratory flights is related to atmospheric dynamics, especially wind speed and direction, but also cloud thickness, moisture content, etc. (sections 1.17 and 2.12). Little is known about how birds are able to find the optimal flight level (p. 152).

Migratory periods are characterized by numerous behavioural adaptations. Typical are the development of nocturnal migratory activity (see above), a commonly observed reduction of the amount of daytime activity (section 2.13), and a so-called 'einschlafpause'. This pause in the evening, accompanied by anorexia, is thought to be used for decision-making as to whether to migrate and in which direction (using sunset cues; see below). Another typical change of the locomotor activity pattern is the disappearance of a second activity peak in the afternoon. This shift from a bigeminus to a single-peaked pattern has been proposed as a mechanism to allow for the development of nocturnal migratory activity (by uncoupling several circadian oscillators; see above, and sections 2.2 and 2.6). It is typical for fat individuals, whereas lean birds often show hyperactivity and multi-peaked patterns. This may broaden the use of microhabitat in lean birds, accompanied by an expanded feeding repertoire (section 2.13.1). Flocking behaviour during the migratory season is a widespread phenomenon in many birds which live only in pairs during the breeding period. It is explained by at least six hypotheses. The energy saving explanation states that in typical formation flight the overall reduction in flight power demand is of the magnitude of 20%. The cultural transmission explanation is based on the observation that many large migratory birds use traditionally specific staging areas and tiny specific wintering areas where inexperienced migrants need to be guided by experienced conspecifics. The orientation/navigation hypothesis proposes that migration in flocks may improve the accuracy of orientation. Few studies with domestic pigeons yielded positive and negative results. The foraging efficiency hypothesis supposes that migration in flocks may guarantee a more consistent food supply and be an important basis for refuelling, with some support from observations. The lowering predation risk hypothesis predicts mechanisms to lower predation by the 'Trafalgar effect' (through quick information), the 'dilution effect' (especially in large groups) and the 'confusion effect' (due to difficulties of predators to concentrate on individuals). The thermal location and utilization hypothesis concerns species that move on by gliding, and where necessary, thermals may be more efficiently located by flocks than by individuals (section 2.13.2). A number of nocturnal migrants have characteristic call notes during migration. They may help to maintain contact in flocks and to improve orientation. Proofs, however, are lacking (section 2.13.3). There is little information on how many species establish temporary territories during migration. It is reported from several systematic groups so it may be widespread, and appears to promote refuelling during stopover periods (section 2.13.4).

From the various modes of locomotion during migration flight, and especially powered (flapping) flight, is by far the most important method of forward drive in migratory birds. Gliding, as the second basic type of avian flight, is of prime importance for migration in large birds. Its energetic costs are only about 15–30% of that of powered flight (p. 170). Wanderers of the oceans also cover large distances without flapping their wings, in so-called dynamic soaring (p. 171). Many small migrants use 'bounding' flight in which the bird introduces pauses between bouts of flapping. This intermittent flight is a unique method of balancing the mechanical power output by the flapping wings and the power input by the flight muscles (p. 172). Swimming, walking, hopping and sliding play a minor role in bird migration (p. 164). 'Migration during resting', i.e. hopping in the migratory direction during foraging trips, reported for several species and areas, was found to be of no significant importance in nocturnal passerine migrants in Central Europe (p. 164).

For migrants, two flight speeds are of prime importance: the maximum range speed which allows the maximum range to be covered on a given fuel reserve, and the optimal flight speed which minimizes the total duration of the migratory journey, incorporating time devoted to flight as well as to energy accumulation. At present, it cannot yet be readily answered as to whether migratory flight speed and overall migration speed are primarily selected with respect to time, energy or predation minimization, and possible other factors like accomplishment of endogenous programs (in naïve first-time migrants, section 2.14).

With respect to migratory episodes and stopover periods the following emerges. The majority of migrants proceed fairly slowly from the breeding areas to the winter quarters in short hops, often interrupted by longer stopovers. For instance, median migration speeds through Europe in passerines are only in the magnitude of 75 km/day in early departing long-distance migrants, 50 km/day in later migrating birds and about 30 km/day in partial migrants (sections 2.8 and 2.15). Until recently, stopover periods determined by retraps and sight observations were considerably underestimated. Several models now allow more realistic estimates. For a Central European staging area for passage migrants of 37 songbird species an overall mean stopover period of about 13 days during the autumn migratory period was calculated (p. 176). Stopover periods vary in different species, age and sex classes, and seasons (may be generally shorter during the normally more rapid spring migration). Stopover periods are often inversely related to the migrant's fat reserves, may decrease with progressing migratory state and premigratory moult, and can also depend on dominance status and locality. They also appear to be controlled by an internal time-program (as part of the endogenous migration scheme of many migrants; section 2.15). This may cause individual birds, as shown for some species, not only to migrate annually at the same dates but also to stage for similar periods each year (p. 174).

Migratory birds are the champions in overcoming ecological barriers like deserts, oceans, etc. Large deserts like the Sahara are transversed by an intermittent strategy by fairly large numbers of migrants. Birds migrate at night and rest during the daytime. Most likely, exclusive non-stop trans-desert migration also plays an important role. It remains to be shown what the proportions of the different strategies are (section 2.16.1). Other barriers are presumably mainly overcome by non-stop flights, some mountains like the Alps may be circumvented by a number of species due to directional programs (section 2.16.2).

Weather and climate exert primary stimulatory and modifying effects on migratory disposition and on actual migration. Meteorological factors can trigger departure (hard-weather movements, p. 186, or the cyclonic weather movements of swifts, p. 188), can delay or hasten migrants, deflect them from their headings, or ground or even kill them. Grounding occurs occasionally in spectacular massive falls (p. 198), and kills are obvious in occasional cases of mass mortality (p. 198). From the extremely complex relationships, the following general rules have emerged. There is a distinct tendency to avoid migrating in inclement weather and there are specific synoptic weather features that promote migration in different migratory seasons. For instance, mass autumn migration in higher latitudes of the northern hemisphere occurs when low pressure systems are being transposed by anticyclones. In spring, mass migration occurs as mild-weather migration when lows approach. Migrants presumably react to locally-measurable variables like wind or temperature rather than to synoptic features. Wind and precipitation are likely to be the two weather factors which affect migration the most (sections 2.17.1–2.17.3), whereas the role of temperature is less clear (section 2.17.4). Due to the extreme complexity of relationships between weather factors and migration, forecasting ability is limited (p. 197).

In many studies effects of population density and social rank on migration have been tested. Field studies yielded many suggestive correlations. Clear experimental evidence that social dominance proximately controls regular, partial or differential migration is still missing. In many cases of irregular migration it is extremely likely that social factors, deriving from overpopulation, are proximately involved in triggering movements. But, so far, it remains to be shown whether food shortage or an imbalance between food supply and population level is decisive (section 2.18).

Seasonal changes in resource availability are not only the main ultimate factors for avian migration but also seem to play a proximate role. While results from field studies remain as ambiguous as those related to social factors, experimental results appear more conclusive. Experimentally altered food availability can affect migratory disposition and migratory behaviour in captive birds in several ways. The pattern of migratory activity can be changed and facultative migratory activity can be induced after normal migratory periods. In several species, food deprivation increases migratory activity (and could lead migrants to better feeding places) whereas during the refeeding phase migratory activity decreases sharply (which could allow individuals to stay longer for sufficient refuelling in a suitable staging area). Induced facultative migratory activity can be directed towards the normal migratory direction and appears to guide migrants within the goal area to suitable wintering places (section 2.19).

With respect to choice of goal areas there is a broad behavioural spectrum from winter residency to almost continual movement which appears to reflect an environmental gradient in resource availability. In some species, winter-site fidelity and tenacity may be so strong that individuals of all sex and age groups occupy territories throughout the winter. In others, the discovery of large-scale midwinter migration has given a new dimension to the phenomenon of bird migration (section 2.20.1). Extended autumn migratory movements in many species may well be based on endogenous migration programs. In other cases, extended migration within the wintering grounds is clearly facultatively, i.e. exogenously, controlled. The possibility to evoke facultative migratory activity ends with the onset of winter moult (section 2.20.2).

Differential migration leads to considerable latitudinal variation in sex and age classes on wintering grounds (sections 2.18 and 2.10). Five hypotheses attempt to explain this variation: (1) behavioural dominance; (2) arrival time and sexual selection; (3) body size and related physiology; (4) prior residence effects; and (5) different endogenous programs. They are still intensely debated (section 2.20.2), for residence effects see section 2.18, for endogenous programs see above.

The available evidence suggests that, in general, actual competition appears not to be severe for migrants, neither among migrants in staging areas nor between local residents and immigrants. Instead, migrants have developed a number of mechanisms to avoid or reduce competition which are based on niche segregation: species-specific habitat selection and differences in main migratory seasons, diurnal activity patterns, in food choice and foraging strategies (section 2.21).

During the migratory periods, many migratory birds tend to rest in their 'typical' habitats which are used during the breeding season. Often, however, this is impossible, and migrants have to share different and even exotic habitats. But also in this case, many species tested showed clear-cut species-specific habitat selection, including those of distinct vertical layers, which act as a mechanism to prevent competition. Proper habitat choice appears to be based on search image ('habitat conception'), and its key factors seem to be morphological characteristics of the birds and structural properties of habitats which act as a lock-and-key mechanism (section 2.21.1).

During wintering, open-country species prefer quite similar habitats to those in the breeding areas. In migrants breeding in woodland the situation is quite variable. In Africa, the majority of these migrants occur in dry fairly open habitats. They are characterized by an opportunistic use of sporadically available resources in space and time (only seasonally available niches), and they are more eurytopic. In the Neotropics and parts of Asia, a higher percentage of the wintering migrants from northern woodlands occupy rainforests. Further, adaptive food selection and foraging structures (e.g. to switch food resources opportunistically) appear to be generally important for migrants to allow them to act as an integral part of the tropical avifaunas rather than as 'intruders' (section 2.21.2).

A migratory bird often has to fit up to five main seasonal processes into its annual schedule which requires careful coordination and tuning control. In particular, migratory periods have to be adapted strongly to the breeding season, and moult has to be adapted to migration (p. 232). As a rule, various processes of juvenile development including juvenile moult, start earlier, proceed more rapidly and, thus, are of shorter duration in typical migrants compared to resident or less typical migratory forms. Part of this adaptive acceleration of development appears to start already in the egg (p. 233), part in juvenile development is under genetic control and part based on photoperiodic control (short-day effect, sections 2.4 and 2.5, above).

As a result of these adaptations, migrants have more time to prepare for migration (section 2.22.1).

In adult birds, as a rule, wing-feather moult and migration are mutually exclusive in order to avoid travelling with 'gappy' wings (p. 236). To guarantee this rule, a number of moult adaptations have evolved: to moult before or after migration or twice, to moult during parts of the breeding season, to pass staggered moult (a special case in terns, p. 237) or suspended moult ('interrupted moult', arrested during migration, a widespread and variable adaptation, section 2.22.2).

Especially in long-distance migrants, the fitting of a successful reproductive cycle between two subsequent migratory periods often needs specific adaptations, above all in species which may be *en route* for up to 10–10.5 months per year (p. 240). Most migrants initiate gonadal development before the homeward migration on the wintering grounds, presumably often based on endogenous circannual rhythms. Populations wintering far away from the breeding area show a tendency to initiate gonadal development (and migratory activity) earlier than those wintering closer to the breeding grounds (controlled by circannual rhythms and photoperiod, p. 241). In migrants arriving late in breeding areas gonadal development at arrival is more advanced as in earlier migrants (p. 241). Further adaptations in late arriving species with short periods for breeding are pair formation in the winter quarters or *en route*, nest building and egg laying immediately after arrival, a shortened incubation and nestling period, accumulation of protein reserves for egg laying during premigratory feeding, etc. With respect to away migration, Arctic breeders adapted to large-scale breeding failure may depart extremely early, in other cases late termination of breeding may cause a late departure (section 2.22.3).

Orientation mechanisms are likely to be part of the basic equipment of birds in general (also of resident forms, for daily routine activities) and may essentially represent phylogenetically ancient mechanisms (p. 9). In migrants, they may be especially perfect. Compass orientation (flying innate courses) in unexperienced migrants and true navigation based on gradient maps in experienced birds appear to be the two basic orientation mechanisms of avian migration (section 3.1.1). Initial satellite-tracking studies demonstrated straight routes and directness over long distances. To what extent migrants follow orthodromes or loxodromes remains to be shown (section 3.1.2). Leading-line effects indicate that migrants make use of landmarks but the general role of landmarks is unknown (section 3.2).

In birds, three biological compass systems have been demonstrated. For the sun compass, possibly used by many diurnal migrants, the crucial reference quantity appears to be the sun azimuth ('sun azimuth compass', section 3.3.1). The magnetic compass (demonstrated in about 15 species) relies on the axial course of the field lines and their inclination in space ('inclination compass') and not on polarity or intensity. After crossing the equator, migrants appear to reverse their reaction with respect to the magnetic field lines (section 3.3.2). So far, neither the structure nor the mechanism of the magnetic receptor (which is possibly represented by the photoreceptor, section 3.4.2) are known. The star compass (demonstrated in about five species) is based on learning to use the axis of stellar rotation and the constellation of Ursa major (section 3.3.3). From other cues, 'sunset cues', the light polarization pattern during sunset, appears to be important. There is indication that visual access to natural skylight polarization patterns is necessary for calibration of magnetic orientation during daylight (section 3.3.4).

About seven hypotheses have been proposed to explain avian navigation, largely based on the map-and-compass concept (section 3.5.4). The vector-navigation hypothesis states that many first-time migrants, equipped with innate time-and-direction vectors (spatio-temporal programs, based on inherited patterns of migratory activity and genetically preprogrammed migratory directions), are more or less 'automatically' guided to their unknown winter quarters. This hypothesis is at present the best supported (section 3.5.1). Olfactory navigation is presently the most debated hypothesis. It is not certain whether olfaction plays a role in migration and whether there are olfactory cues in the environment that produce gradient fields over larger areas which could be used for olfactory navigation by migrants (sections 3.4.3 and 3.5.2). Whether true magnetic navigation occurs, possibly based on a magnetocline compass (connected with vector-navigation), is unknown (section 3.5.3), and the same holds true for other cues and mechanisms (section 3.5.4).

There is increasing evidence that the navigational portion of bird migration comprises a system integrating a number of redundant mechanisms. Most likely, the primary framework is a time-and-direction program, the magnetic compass acts as a primary compass (possibly with limited effectiveness under low visibility, p. 252), the star compass develops as a largely independent mechanism but later functions, like other mechanisms, as an integral part of one complex system (section 3.6).

Wind exerts numerous effects on avian orientation. Since migrants can possibly sense the wind while aloft (p. 258) they may rather behave adaptively to it than to be its plaything. Some rules of this adaptive behaviour begin to emerge. During large parts of the migratory journey migrants may, to an extent, tolerate uncorrected wind drift in order to require less time and energy for migration. During the terminal phase of migration, however, drift appears to be reduced by various measures. Drift compensation, common in low altitude diurnal migration or overland nocturnal migration, is by no means sufficiently understood and needs further research (section 3.7).

Migratory behaviour is currently under substantial alteration and presumably microevolutionary development in many bird populations. The most obvious recent changes are: reduction of migratoriness (widespread; a number of examples in section 4.2), extension of migratoriness (rare; two examples in chapter 4), changes in migration periods (very common; many examples in chapter 4), changes in the migratory distance and establishment of novel migratory directions and winter quarters (some cases documented in section 4.2 and the following). Continental blackcaps from Central Europe, which hitherto migrated in southern directions to Mediterranean and African winter quarters, have recently (within the last 30 years) developed a novel migratory direction towards the northwest and have established new wintering areas on the British Isles. This novel migratory habit has been shown to be based on rapid microevolutionary processes (section 4.1). In the blackcap it was also demonstrated that under moderate selection intensities a middle-distance migrant population (from Central Europe) could evolve into a short-distance migrant population within 10–20 generations (section 4.2).

With the recent findings on genetic control of, and rapid microevolutionary processes in, bird migration in mind, the following scenario appears likely with respect to future developments. Many species of the present avifauna, above all of higher latitudes, may have altered their migratory habits repeatedly and considerably along with multiple changes in environmental conditions. During such changes they may have passed through various changes of migratoriness and sedentariness. Furthermore, low but regular hybridization of typical and less typical migrants may increase gene flow, genetic exchange, genetic variation and consequently the microevolutionary potential. Therefore, many species possibly represent genotypically partial migrants with, however, extremely variable amounts of phenotypical fractions of migrants and residents. These fractions could be relatively rapidly selected for by genotype-environmental interactions. 'Global warming' could lead to (further) increasing sedentariness of partially migratory populations and that, along with an overall increase in resident individuals due to reduced winter mortality, could further affect long-distance migrants through increasing competition on breeding grounds (section 4.2).

I would like to close this chapter with a citation from Rowan's (1931) book on *The Riddle of Migration* were he concluded (in chapter II, on Environment, Past and Present): 'If the fully established migrations with which we are so familiar today in the northern hemisphere depend neither on a bird's personal experience nor on a conscious knowledge of the experience of its ancestors or the factors of its environment, their seasonableness, precision and accuracy, on the other hand, can leave little doubt that they hinge in some other way on the experience of past generations. Such experience has not been perpetuated by word of mouth or in writing and we must therefore assume that it has been handed on by another method, – genetically, by inheritance. We must assume that the habit has been acquired by individuals of the past, that it has somehow become inherent and that it has survived because it remains of value or is even essential, today. In short, it is now instinctive.'

With respect to the increasing evidence of the recent past that endogenous and genetic factors and mechanisms play a basic role in the control of all essential parts of avian migration, Rowan's conclusion should be recalled when future studies are conceived.

# References

Able, K.P. (1973) The role of weather variables and flight direction in determining the magnitude of nocturnal bird migration. *Ecology*, **54**, 1031–1041.

Able, K.P. (1980) Mechanisms of orientation, navigation, and homing. In *Animal Migration, Orientation, and Navigation* (ed. S.A. Gauthreaux), Academic Press, New York, San Francisco, London, pp. 284–373.

Able, K.P. (1991) The development of migratory orientation mechanisms. In *Orientation in Birds* (ed. P. Berthold), Birkhäuser, Basel, Boston, Berlin, pp. 166–179.

Able, K.P. and Able, M.A. (1993) Daytime calibration of magnetic orientation in a migratory bird requires a view of skylight polarization. *Nature*, **364**, 523–525.

Adejuwon, J.O., Balogun, E.E. and Adejuwon, S.A. (1990) On the annual and seasonal patterns of rainfall fluctuations in sub-Saharan West Africa. *Int. J. Climatol.*, **10**, 839–848.

Adriaensen, F. (1988) An analysis of robins (*Erithacus rubecula*) ringed or recovered in Belgium: age and mortality, speed, direction and distance of migration. *Gerfaut*, **78**, 3–24.

Adriaensen, F. and Dhondt, A.A. (1990) Population dynamics and partial migration of the European robin (*Erithacus rubecula*) in different habitats. *J. Anim. Ecology*, **59**, 1077–1090.

Adriaensen, F., Dhondt, A.A. and Matthysen, E. (1990) Bird migration. *Nature*, **347**, 23.

Adriaensen, F., Ulenaers, P. and Dhondt, A.A. (1993) Ringing recoveries and the increase in numbers of European great crested grebes (*Podiceps cristatus*). *Ardea*, **81**, 59–70.

Ainley, M.G. (1988) William Rowan and the experimental approach in ornithology. *Acta XIX Congr. Internat. Ornithol.*, Ottawa 1986, pp. 2737–2745.

Åkesson, S. (1993) Effect of geomagnetic field on orientation of the marsh warbler, *Acrocephalus palustris*, in Sweden and Kenya. *Anim. Behav.*, **46**, 1157–1167.

Åkesson, S., Karlsson, L., Pettersson, J. and Walinder, G. (1992) Body composition and migration strategies: a comparison between Robins (*Erithacus rubecula*) from two stop-over sites in Sweden. *Vogelwarte*, **36**, 188–195.

Alatalo, R.V., Gustafsson, L. and Lundberg, A. (1984) Why do young passerine birds have shorter wings than older birds? *Ibis*, **126**, 410–415.

Alerstam, T. (1979) Wind as selective agent in bird migration. *Ornis Scand.*, **10**, 76–93.

Alerstam, T. (1981) The course and timing of bird migration. In *Animal Migration* (ed. D.J. Aidley), Cambridge University Press, Cambridge, New York, Melbourne, pp. 9–54.

Alerstam, T. (1990) *Bird Migration*. Cambridge University Press, Cambridge, New York, Melbourne.

Alerstam, T. (1991a) Ecological causes and consequences of bird orientation. In *Orientation in Birds* (ed. P. Berthold), Birkhäuser, Basel, Boston, Berlin, pp. 202–225.

Alerstam, T. (1991b) Bird flight and optimal migration. *Tree*, **7**, 210–215.

Alerstam, T., Gudmundsson, G.A., Jönsson, P.E., Karlsson, J. and Lindström, Å. (1990) Orientation, migration routes and flight behaviour of knots, turnstones and brant geese departing from Iceland in spring. *Arctic*, **43**, 201–214.

Alerstam, T. and Lindström, Å. (1990) Optimal bird migration: the relative importance of time, energy and safety. In *Bird Migration: Physiology and Ecophysiology* (ed. E. Gwinner), Springer, Berlin, Heidelberg, New York, pp. 331–351.

Alerstam, T. and Pettersson, S.G. (1991) Orientation along great circles by migrating birds using a sun compass. *J. theor. Biol.*, **152**, 191–202.

Alonso, J.C., Alonso, J.A., Cantos, F.J. and Bautista, L.M. (1990) Spring crane *Grus grus* migration through Gallocanta, Spain. I. Daily variations in migration volume. *Ardea*, **78**, 365–378.

Altenburg, W. and van Spanje, T. (1989) Utilization of mangroves by birds in Guinea-Bissau. *Ardea*, **77**, 57–72.

Anon. (1968) Das Grasmücken-Programm des Max-Planck-Instituts für Verhaltensphysiologie. *Vogelwarte*, **24**, 320–323.

Arnold, T.W. (1991) Geographic variation in sex ratios of wintering American kestrels *Falco sparverius*. *Ornis Scand.*, **22**, 20–26.

Aschoff, J. (1960) Exogenous and endogenous components in circadian rhythms. *Cold Spring Harbor Symp. Quant. Biol.*, **25**, 11–28.

Aschoff, J. (1964) Biologische Periodik als selbsterregte Schwingung. *Veröff. Arbeitsgemeinsch. Forschung Nordrhein-Westfalen*, **138**, 51–79.

Aschoff, J. (1967) Circadian rhythms in birds. *Proc. XIV Internat. Ornithol. Congr.*, Oxford 1966, pp. 81–105.

Aschoff, J. (1979) Circadian rhythms: influences of internal and external factors on the period measured in constant conditions. *Z. Tierpsychol.*, **49**, 225–249.

Aschoff, J. (1987) Effects of periodic availability of food on circadian rhythms. In *Comparative Aspects of Circadian Clocks* (eds T. Hiroshige and K. Honma), Proc. 2. Sapporo Symp. Biol. Rhythm, pp. 19–41.

Astheimer, L.B., Buttemer, W.A. and Wingfield, J.C. (1992) Interactions of corticosterone with feeding, activity and metabolism in passerine birds. *Ornis Scand.*, **23**, 355–365.

Badgerow, J.P. (1988) An analysis of function in the formation flight of Canada geese. *Auk*, **105**, 749–755.

Bairlein, F. (1981) Ökosystemanalyse der Rastplätze von Zugvögeln: Beschreibung und Deutung der Verteilungsmuster von ziehenden Kleinvögeln in verschiedenen Biotopen der Stationen des 'Mettnau-Reit-Illmitz-Programmes'. *Ökol. Vögel*, **3**, 7–137.

Bairlein, F. (1985a) Dismigration und Sterblichkeit in Süddeutschland beringter Schleiereulen (*Tyto alba*). *Vogelwarte*, **33**, 81–108.

Bairlein, F. (1985b) Efficiency of food utilization during fat deposition in the long-distance migratory garden warbler, *Sylvia borin*. *Oecologia*, **68**, 118–125.

Bairlein, F. (1985c) Body weights and fat deposition of Palaearctic passerine migrants in the central Sahara. *Oecologia*, **68**, 141–146.

Bairlein, F. (1990) Nutrition and food selection in migratory birds. In *Bird Migration: Physiology and Ecophysiology* (ed. E. Gwinner), Springer, Berlin, Heidelberg, New York, pp. 198–213.

Bairlein, F. (1991a) Body mass of garden warblers (*Sylvia borin*) on migration: a review of field data. *Vogelwarte*, **36**, 48–61.

Bairlein, F. (1991b) Nutritional adaptations to fat deposition in the long-distance migratory garden warbler (*Sylvia borin*). *Acta XX Congr. Internat. Ornithol.*, Christchurch 1990, pp. 2149–2158.

Bairlein, F. (1992) Recent prospects on trans-Saharan migration of songbirds. *Ibis*, **134**, Suppl. 1, 41–46.

Bairlein, F. (1995) *Ökologie der Vögel*. Fischer, Stuttgart.

Bairlein, F., Beck, P., Feiler, W. and Querner, U. (1983) Autumn weights of some Palaearctic passerine migrants in the Sahara. *Ibis*, **125**, 404–407.

Bairlein, F., Leisler, B. and Winkler, H. (1986) Morphologische Aspekte der Habitatwahl von Zugvögeln in einem SW-deutschen Rastgebiet. *J. Ornithol.*, **127**, 463–473.

Bairlein, F. and Totzke, U. (1992) New aspects on migratory physiology of trans-Saharan passerine migrants. *Ornis Scand.*, **23**, 244–250.

Baker, J.R. (1938) The evolution of breeding seasons. In *Evolution: Essays on Aspects of Evolutionary Biology* (ed. G.R. de Beer), Oxford University Press, Oxford, London, New York, pp. 161–177.

Baker, R.R. (1978) *The Evolutionary Ecology of Animal Migration*. Hodder and Stoughton, London.

Baker, R.R. (1987) Integrated use of moon and magnetic compasses by the heart-and-dart moth, *Agrotis exclamationis*. *Anim. Behav.*, **35**, 94–101.

Bakken, G.S., Murphy, M.T. and Erskine, D.J. (1991) The effect of wind and air temperature on metabolism and evaporative water loss rates of dark-eyed juncos, *Junco hyemalis*: A standard operative temperature scale. *Physiol. Zool.*, **64**, 1023–1049.

Balciauskas, M. and Zalakevicius, M. (1991) Accuracy of methods for the study of nocturnal migration. *Acta Ornithol. Lituanica*, **4**, 39–51.

Balda, R.P. and Wiltschko, W. (1991) Caching and recovery in scrub jays: transfer of sun-compass directions from shaded to sunny areas. *Condor*, **93**, 1020–1023.

Baldaccini, N.E. and Bezzi, E.M. (1989) Orientational responses to different light stimuli by adult and young sedge warbler (*Acrocephalus schoenobaenus*) during autumn migration: A funnel technique study. *Behaviour*, **110**, 121–123.

Baldassarre, G.A. and Fischer, D.H. (1984) Food habits of fall migrant shorebirds on the Texas high plains. *J. Field Ornithol.*, **55**, 220–229.

Balen, J.H. van and Hage, F. (1989) The effect of environmental factors on tit movements. *Ornis Scand.*, **20**, 99–104.

Banzett, R.B., Nations, C.S., Wang, N., Butler, J.P. and Lehr, J.L. (1992) Mechanical independence of wingbeat and breathing in starlings. *Respir. Physiol.*, **89**, 27–36.

Barter, M. and Hou, W.T. (1990) Can waders fly non-stop from Australia to China? *Stilt*, **17**, 36–39.

Bastian, A. (1992) Mobilität von Kleinvögeln in einem süddeutschen Rastgebiet während der Wegzugperiode. *Ökol. Vögel*, **14**, 121–163.

Bastian, A. and Berthold, P. (1991) Wandern nachts ziehende Kleinvögel während ihrer Tagesrast in Mitteleuropa langsam in Zugrichtung weiter oder sind sie stationär? *J. Ornithol.*, **132**, 325–327.

Bauer, K.M. and Glutz von Blotzheim, U.N. (1968) *Handbuch der Vögel Mitteleuropas 2* (ed. G. Niethammer). Akademische Verlagsgesellsch., Frankfurt.

Beason, R.C. (1992) You can get there from here: responses to simulated magnetic equator crossing by the bobolink (*Dolichonyx oryzivorus*). *Ethology*, **91**, 75–80.

Beason, R.C. and Semm, P. (1991) Neuroethological aspects of avian orientation. In *Orientation in Birds* (ed. P. Berthold), Birkhäuser, Basel, Boston, Berlin, pp. 106–127.

Bech, C., Rautenberg, W. and May-Rautenberg, B. (1988) Thermoregulatory responses of the pigeon (*Columba livia*) to selective changes in the inspired air temperature. *J. Comp. Physiol. B*, **157**, 747–752.

Bélanger, L. and Bédard, J. (1990) Energetic cost of man-induced disturbance to staging snow geese. *J. Wildl. Manage.*, **54**, 36–41.

Bellrose, F.C. (1958) The orientation of displaced waterfowl. *Wilson Bull.*, **70**, 20–40.

Belthoff, J.R. and Gauthreaux, S.A. Jr (1991) Partial migration and differential winter distribution of house finches in the eastern United States. *Condor*, **93**, 374–382.

Bensch, S., Hasselquist D., Hedenström, A. and Ottosson, U. (1990) Rapid moult among palaearctic passerines in West Africa – an adaptation to the oncoming dry season? *Ibis*, **133**, 47–52.

Bergman, G. (1941) Der Frühlingszug von *Clangula hyemalis* (L.) und *Oidemia nigra* (L.) bei Helsingfors. *Ornis Fenn.*, **18**, 1–26.

Berndt, R. and Henß, M. (1967) Die Kohlmeise, *Parus major*, als Invasionsvogel. *Vogelwarte*, **24**, 17–37.

Bernstein, M.H. (1989) Respiration by birds at high altitude and in flight. In *Physiology of Cold Adaptation in Birds* (eds C. Bech and R.E. Reinertsen), Plenum Press, New York, London, pp. 197–206.

Berthold, P. (1969) Über Populationsunterschiede im Gonadenzyklus europäischer *Sturnus vulgaris*, *Fringilla coelebs*, *Erithacus rubecula* und *Phylloscopus collybita* und deren Ursachen. *Zool. Jb. Syst.*, **96**, 491–557.

Berthold, P. (1971) Dependence of timing and pattern of annual events on the date of birth in birds. *Abstr. XXV Internat. Congr. Union Physiol. Sci.*, München, IX, p. 623.

Berthold, P. (1973) Relationships between migratory restlessness and migration distance in six *Sylvia* species. *Ibis*, **115**, 594–599.

Berthold, P. (1974) Endogene Jahresperiodik – Innere Jahreskalender als Grundlage der jahreszeitlichen Orientierung bei Tieren und Pflanzen. Konstanzer Universitätsreden 69, Universitätsverlag, Konstanz.

Berthold, P. (1975) Migration: Control and metabolic physiology. In *Avian Biology 5* (eds D.S. Farner and J.R. King), Academic Press, New York, San Francisco, London, pp. 77–128.

Berthold, P. (1976) Animalische und vegetabilische Ernährung omnivorer Singvogelarten: Nahrungsbevorzugung, Jahresperiodik der Nahrungswahl, physiologische und ökologische Bedeutung. *J. Ornithol.*, **117**, 145–209.

Berthold, P. (1978a) Die quantitative Erfassung der Zugunruhe bei Tagziehern: Eine Pilotstudie an Ammern (*Emberiza*). *J. Ornithol.*, **119**, 334–336.

Berthold, P. (1978b) Circannuale Rhythmik: Freilaufende selbsterregte Periodik mit lebenslanger Wirksamkeit bei Vögeln. *Naturwiss.*, **65**, 546.

Berthold, P. (1979a) Über die photoperiodische Synchronisation circannualer Rhythmen bei Grasmücken (*Sylvia*). *Vogelwarte*, **30**, 7–10.

Berthold, P. (1979b) Beziehungen zwischen Zugunruhe und Zug bei der Sperbergrasmücke *Sylvia nisoria*: eine ökophysiologische Untersuchung. *Vogelwarte*, **30**, 77–84.

Berthold, P. (1980) Untersuchung der Nachtunruhe diesjähriger und adulter sowie handaufgezogener und gefangener *Sylvia atricapilla*. *Vogelwarte*, **30**, 255–259.

Berthold, P. (1984a) The control of partial migration in birds: a review. *Ring*, **10**, 253–265.

Berthold, P. (1984b) The endogenous control of bird migration: a survey of experimental evidence. *Bird Study*, **31**, 19–27.

Berthold, P. (1985a) Vergleichende Untersuchung von Jugendentwicklung und Zugverhalten bei Garten- und Hausrotschwanz, *Phoenicurus phoenicurus* und *Ph. ochruros*. *J. Ornithol.*, **126**, 383–392.

Berthold, P. (1985b) Endogenous components of annual cycles of migration and moult. *Acta XVIII Congr. Internat. Ornithol.*, Moskau 1982, pp. 922–929.

Berthold, P.(1986) Wintering in a Mediterranean blackcap (*Sylvia atricapilla*) population: strategy, control, and unanswered questions. *Proc. First Conf. Birds Wintering Mediterranean Region*, Aulla 1985, pp. 261–272.

Berthold, P. (1988a) The Control of Migration in European Warblers. *Acta XIX. Congr. Internat. Ornithol.*, Ottawa 1986, pp. 215–249.

Berthold, P.(1988b) Unruhe-Aktivität bei Vögeln: eine Übersicht. *Vogelwarte*, **34**, 249–259.

Berthold, P. (1988c) The biology of the genus *Sylvia* – a model and a challenge for Afro-European co-operation. *Tauraco*, **1**, 3–28.

Berthold, P. (1988d) Evolutionary aspects of migratory behavior in European warblers. *J. evol. Biol.*, **1**, 195–209.

Berthold, P. (1990a) Wegzugbeginn und Einsetzen der Zugunruhe bei 19 Vogelpopulationen – eine vergleichende Untersuchung. Proc. Int. 100. DO-G Meeting, Current Topics Avian Biol., Bonn 1988, *J. Ornithol.*, 131, Sonderh., pp. 217–222.

Berthold, P. (1990b) Genetics of Migration. In *Bird Migration: The Physiology and Ecophysiology* (ed. E. Gwinner), Springer, Berlin, Heidelberg, New York, pp. 269–280.

Berthold, P. (1991a) Genetic control of migratory behaviour in birds. *Tree*, **6**, 254–257.

Berthold, P. (ed.) (1991b) *Orientation in Birds*. Birkhäuser, Basel, Boston, Berlin.

Berthold, P. (1991c) Spatiotemporal programmes and genetics of orientation. In *Orientation in Birds* (ed. P. Berthold), Birkhäuser, Basel, Boston, Berlin, pp. 86–105.

Berthold, P. (1991d) Patterns of avian migration in light of current global 'greenhouse' effects: A Central European perspective. *Acta XX Congr. Internat. Ornithol.*, Christchurch 1990, pp. 780–786.

Berthold, P. (1993) *Bird Migration. A general survey*. Oxford University Press, Oxford, London, New York.

Berthold, P. (1995) Microevolution of migratory behaviour illustrated by the Blackcap *Sylvia atricapilla*: 1993 Witherby Lecture. *Bird Study*, **42**, 89–100.

Berthold, P., Gwinner, E. and Klein, H. (1970) Vergleichende Untersuchung der Jugendentwicklung eines ausgeprägten Zugvogels, *Sylvia borin*, und eines weniger ausgeprägten Zugvogels, *S. atricapilla*. *Vogelwarte*, **25**, 297–331.

Berthold, P., Gwinner, E. and Klein, H. (1971) Circannuale Periodik bei Grasmücken (*Sylvia*). *Experientia*, **27**, 399.

Berthold, P., Gwinner, E., Klein, H. and Westrich, P. (1972a) Beziehungen zwischen Zugunruhe und Zugablauf bei Garten- und Mönchsgrasmücke (*Sylvia borin* und *S. atricapilla*). *Z. Tierpsychol.*, **30**, 26–35.

Berthold, P., Gwinner, E. and Klein, H. (1972b) Circannuale Periodik bei

Grasmücken I. Periodik des Körpergewichtes, der Mauser und der Nachtunruhe bei *Sylvia atricapilla* und *S. borin* unter verschiedenen konstanten Bedingungen. *J. Ornithol.*, **113**, 170–190.

Berthold, P., Gwinner, E. and Querner, U. (1974) Vergleichende Untersuchung der Jugendentwicklung südfinnischer und südwestdeutscher Gartengrasmücken, *Sylvia borin*. *Ornis Fenn.*, **51**, 146–154.

Berthold, P., Bairlein, F. and Querner, U. (1976) Über die Verteilung von ziehenden Kleinvögeln in Rastbiotopen und den Fangerfolg von Fanganlagen. *Vogelwarte*, **28**, 267–273.

Berthold, P. and Gwinner, E. (1978) Jahresperiodik der Gonadengröße beim Fichtenkreuzschnabel (*Loxia curvirostra*). *J. Ornithol.*, **119**, 338–339.

Berthold, P. and Leisler, B. (1980) Migratory restlessness of the marsh warbler *Acrocephalus palustris*. *Naturwiss.*, **67**, 472.

Berthold, P. and Querner, U. (1981) Genetic basis of migratory behavior in European warblers. *Science*, **212**, 77–79.

Berthold, P. and Querner, U. (1982a) Partial migration in birds: experimental proof of polymorphism as a controlling system. *Experientia*, **38**, 805.

Berthold, P. and Querner, U. (1982b) Genetic basis of moult, wing length, and body weight in a migratory bird species, *Sylvia atricapilla*. *Experientia*, **38**, 801–802.

Berthold, P. and Querner, U. (1982c.) The annual cycle of the blackcap *Sylvia atricapilla* on the Cape Verde Islands: characteristics, environmental and genetic control. *Proc. Seventh Pan-African Ornithol. Congr.*, Nairobi 1988, pp. 327–336.

Berthold, P. and Querner, U. (1992) On the control of suspended moult in an European trans-Saharan migrant, the Orphean warbler. *J. Yamashina Inst. Ornithol.*, **14**, 157–165.

Berthold, P., Brensing, D. and Heine, G. (1986) Tageszeitliche 'Fangmuster' von Kleinvögeln und deren Bedeutung. *J. Ornithol.*, **127**, 515–517.

Berthold, P. and Querner, U. (1988) Was Zugunruhe wirklich ist – eine quantitative Bestimmung mit Hilfe von Video-Aufnahmen bei Infrarotlichtbeleuchtung. *J. Ornithol.*, **129**, 372–375.

Berthold, P. and Terrill, S.B. (1988) Migratory behaviour and population growth of blackcaps wintering in Britain and Ireland: some hypotheses. *Ringing Migration*, **9**, 153–159.

Berthold, P., Mohr, G. and Querner, U. (1990a) Steuerung und potentielle Evolutionsgeschwindigkeit des obligaten Teilzieherverhaltens: Ergebnisse eines Zweiweg-Selektionsexperiments mit der Mönchsgrasmücke (*Sylvia atricapilla*). *J. Ornithol.*, **131**, 33–45.

Berthold, P., Querner, U. and Schlenker, R. (1990b) *Die Mönchsgrasmücke*. Die Neue Brehm-Bücherei, Wittenberg Lutherstadt.

Berthold, P., Wiltschko, W., Miltenberger, H. and Querner, U. (1990c) Genetic transmission of migratory behavior into a nonmigratory bird population. *Experientia*, **46**, 107–108.

Berthold, P., Fliege G., Heine, G., Querner, U. and Schlenker, R. (1991) Autumn migration, resting behaviour, biometry and moult of small birds in Central Europe. *Vogelwarte*, **36**, Sonderh.

Berthold, P. and Helbig, A.J. (1992) The genetics of bird migration: stimulus, timing, and direction. *Ibis*, **134**, Suppl. 1, 9–14.

Berthold, P., Helbig, A.J., Mohr, G. and Querner, U. (1992a) Rapid microevolution of migratory behaviour in a wild bird species. *Nature*, **360**, 668–669.

Berthold, P., Nowak, E. and Querner, U. (1992b) Satelliten-Telemetrie beim

Weißstorch (*Ciconia ciconia*) auf dem Wegzug – eine Pilotstudie. *J. Ornithol.*, **133**, 155–163.

Berthold, P., Kaiser, A., Querner, U. and Schlenker, R. (1993) Analyse von Fangzahlen im Hinblick auf die Bestandsentwicklung von Kleinvögeln nach 20jährigem Betrieb der Station Mettnau, Süddeutschland. *J. Ornithol.*, **134**, 283–299.

Berthold, P. and Pulido, F. (1994) Heritability of migratory activity in a natural bird population. *Proc. R. Soc. Lond. B.*, **257**, 311–315.

Bibby, C.J. (1992) Conservation of migrants on their breeding grounds. *Ibis*, **134**, Suppl. 1, 29–34.

Bibby, C.J. and Green, R.E. (1981) Autumn migration strategies of reed and sedge warblers. *Ornis Scand.*, **12**, 1–12.

Biebach, H. (1977) Das Winterfett der Amsel (*Turdus merula*). *J. Ornithol.*, **118**, 117–133.

Biebach, H. (1983) Genetic determination of partial migration in the European robin (*Erithacus rubecula*). *Auk*, **100**, 601–606.

Biebach, H. (1985) Sahara stopover in migratory flycatchers: fat and food affect the time program. *Experientia*, **41**, 695–697.

Biebach, H. (1988) Ecophysiology of resting willow warblers (*Phylloscopus trochilus*) crossing the Sahara. *Acta XIX Congr. Internat. Ornithol.*, Ottawa 1986, pp. 2162–2168.

Biebach, H. (1990) Strategies of trans-Sahara migrants. In *Bird Migration: Physiology and Ecophysiology* (ed. E. Gwinner), Springer, Berlin, Heidelberg, New York, pp. 352–367.

Biebach, H. (1991) Is water or energy crucial for trans-Sahara migrants? *Acta XX Congr. Internat. Ornithol.*, Christchurch 1990, pp. 773–779.

Biebach, H. (1992) Flight-range estimates for small trans-Sahara migrants. *Ibis*, **134**, Suppl. 1, 47–54.

Biebach, H., Wegner, H. and Habersetzer, J. (1985) Measuring migratory restlessness in captive birds by an ultrasonic system. *Experientia*, **41**, 411–412.

Biebach, H., Friedrich, W. and Heine, G. (1986) Interactions of bodymass, fat, foraging and stopover period in trans-Sahara migrating passerine birds. *Oecologia*, **69**, 370–379.

Biebach, H., Friedrich, W., Heine, G., Jenni, L., Jenni-Eiermann, S. and Schmidl, D. (1991) The daily pattern of autumn bird migration in the northern Sahara. *Ibis*, **133**, 414–422.

Biewener, A.A., Dial, K.P. and Goslow, G.E. (1992) Pectoralis muscle force and power output during flight in the starling. *J. Exp. Biol.*, **164**, 1–18.

Bingman, V.P. (1988) The avian hippocampus: its role in the neural organization of the spatial behavior of homing pigeons. *Acta XIX Congr. Internat. Ornithol.*, Ottawa 1986, pp. 2075–2082.

Bingman, V.P. and Mench, J.A. (1990) Homing behavior of hippocampus and parahippocampus lesioned pigeons following short-distance releases. *Behav. Brain Res.*, **40**, 227–238.

Birjukow, W.J. and Netschwal, N.A. (1988) Algoritmy obnarushenija i rasposnawanija na osnowe statistitscheskowo analisa radiolokazionnych otrasheni ot ptiz. *Tes. Dokl. Pribalt. Ornitol. Konf.*, **12**, 20–21.

Black, C.P. and Tenney, S.M. (1980) Oxygen transport during progressive hypoxia in high-altitude and sea-level waterfowl. *Respir. Physiol.*, **39**, 217–239.

Black, J.M. and Owen, M. (1989) Parent–offspring relationships in wintering barnacle geese. *Anim. Behav.*, **37**, 187–198.

Blake, J.G. and Loiselle, B.A. (1992) Fruits in the diets of neotropical migrants birds in Costa Rica. *Biotropica*, **24**, 200–210.

Bland, R.L. (1986) Blackcap. In *The Atlas of Wintering Birds in Britain and Ireland* (ed. P. Lack), Poyser, Calton.

Blem, C.R. (1980) The energetics of migration. In *Animal Migration, Orientation, and Navigation* (ed. S.A. Gauthreaux Jr), Academic Press, New York, San Francisco, London, pp. 175–224.

Blem, C.R. (1990) Avian energy storage. In *Current Ornithology* (ed. M. Power), Plenum Press, New York, London, pp. 59–113.

Bloesch, M. (1989) Der Storchenansiedlungsversuch in Altreu (Schweiz). In *Weißstorch* (eds G.J. Rheinwald, H. Ogden and H. Schulz), *Schriftenr. Dachverb. Dt. Avifaunisten*, **10**, 437–444.

Bluhm, C.K., Schwabl, H., Schwabl, I., Perera, A., Follet, B.K., Goldsmith, A.R. and Gwinner, E. (1991) Variation in hypothalamic gonadotrophin-releasing hormone content, plasma and pituitary LH, and in-vitro testosterone release in a long-distance migratory bird, the garden warbler (*Sylvia borin*), under constant photoperiods. *J. Endocrinol.*, **128**, 339–345.

Boddy, M. (1991) Some aspects of frugivory by bird populations using coastal dune scrub in Lincolnshire. *Bird Study*, **38**, 188–199.

Boddy, M. (1992) Timing of whitethroat *Sylvia communis* arrival, breeding and moult at a coastal site in Lincolnshire. *Ringing Migration*, **13**, 65–72.

Bögel, R. (1990) Measuring flight altitude of griffon vultures by radio-telemetry. *Acta XX Congr. Internat. Ornithol.*, Christchurch 1990, Suppl., pp. 489–490.

Borowicz, V.A. (1988) Fruit consumption by birds in relation to fat content of pulp. *Am. Midl. Nat.*, **119**, 121–127.

Bowmaker, J.K. (1988) Avian colour vision and the environment. *Acta XIX Congr. Internat. Ornithol.*, Ottawa 1986, pp. 1284–1294.

Brackenbury, J.H. (1991) Ventilation, gas exchange and oxygen delivery in flying and flightless birds. *Soc. Exp. Biol.*, **41**, 125–147.

Braithwaite, V.A., Guilford, T.C., Dawkins, M.S. and Krebs, J.R. (1990) Visual landmarks in pigeon homing: An operant approach. *Acta XX Congr. Internat. Ornithol.*, Christchurch 1990, Suppl., pp. 382–383.

Braunitzer, G. and Hiebl, J. (1989) Molekulare Aspekte der Höhenatmung von Vögeln. Hämoglobine der Streifengans (*Anser indicus*), der Andengans (*Chloephaga melanoptera*) und des Sperbergeiers (*Gyps rueppellii*). *Naturwiss.*, **75**, 280–287.

Brensing, D. (1977) Nahrungsökologische Untersuchungen an Zugvögeln in einem südwestdeutschen Durchzugsgebiet. *Vogelwarte*, **29**, 44–56.

Brensing, D. (1989) Ökophysiologische Untersuchungen der Tagesperiodik von Kleinvögeln. *Ökol. Vögel*, **11**, 1–148.

Brooke, M. and Prince, P.A. (1990) Nocturnality in seabirds. *Acta XX Congr. Internat. Ornithol.*, Christchurch 1990, Suppl., p. 279.

Brown, R.E. and Fedde, M.R. (1993) Airflow sensors in the avian wing. *J. Exp. Biol.*, **179**, 13–30.

Bruderer, B. (1971) Radarbeobachtungen über den Frühlingszug im Schweizerischen Mittelland. *Ornithol. Beob.*, **68**, 89–158.

Bruderer, B. and Bruderer, H. (1993) Distribution and habitat preference of redbacked shrikes *Lanius collurio* in southern Africa. *Ostrich*, **64**, 141–147.

Bruderer, B. and Jacquat, B. (1972) Zur Bestimmung von Flügelschlagfrequenzen tag- und nachtziehender Vogelarten mit Radar. *Ornithol. Beob.*, **69**, 189–206.

Bruderer, B. and Jenni, L. (1990) Migration across the Alps. In *Bird Migration: Physiology and Ecophysiology* (ed. E. Gwinner), Springer, Berlin, Heidelberg, New York, pp. 60–77.

Bruderer, B. and Liechti, F. (1990) Richtungsverhalten nachtziehender Vögel in

Süddeutschland und der Schweiz unter besonderer Berücksichtigung des Windeinflusses. *Ornithol. Beob.*, **87**, 271–293.

Bruggers, R.L. and Elliott, C.C.H. (eds) (1989) *Quelea quelea: Africa's Bird Pest*. Oxford University Press, Oxford, London, New York.

Bruns, K. and Ten Thoren, B. (1990) Zugvorbereitung und Zugunruhe bei der Ringelgans *Branta bernicla bernicla*. Proc. Int. 100. DO-G Meeting, Current Topics Avian Biol., Bonn 1988, *J. Ornithol.*, **131**, Sonderh., pp. 223–229.

Bub, H. (1983) *Ornithologische Beringungsstationen in Europa*. Schriftenr. Dachverb. Dt. Avifaunisten, 7.

Bub, H. and Oelke, H. (1989) The history of bird marking till the inception of scientific bird ringing. *Ring*, **12**, 141–163.

Buchholz, R. and Levey, D.J. (1990) The evolutionary triad of microbes, fruits and seed dispersers: an experiment in fruit choice by cedar waxwings, *Bombycilla cedorum. Oikos*, **59**, 200–204.

Budeau, D.A., Ratti, J.T. and Ely, C.R. (1991) Energy dynamics, foraging ecology, and behavior of prenesting greater white-fronted geese. *J. Wildl. Manage.*, **55**, 556–563.

Bullough, W.S. (1942) The reproductive cycles of the British and continental races of the starling (*Sturnus vulgaris L.*). *Phil. Trans. Roy. Soc.*, **231**, 165–246.

Burger, J. and Gochfeld, M. (1991) Human distance and birds: tolerance and response distances of resident and migrant species in India. *Environ. Conserv.*, **18**, 158–165.

Butler, P.J. (1991) Exercise in birds. *J. Exp. Biol.*, **160**, 233–262.

Butler, P.J. and Woakes, A.J. (1990) The physiology of bird flight. In *Bird Migration: Physiology and Ecophysiology* (ed. E. Gwinner), Springer, Berlin, Heidelberg, New York, pp. 300–318.

Buurma, L.S., Lensink, R. and Linnartz, L.G. (1986) Altitude of diurnal broad front migration over Twente; a comparison of radar and visual observations in October 1984. *Limosa*, **59**, 169–182.

Calder, W.A. and King, J.R. (1974) Thermal and caloric relations of birds. In *Avian Biology 4* (eds D.S. Farner and J.R. King), Academic Press, New York, San Francisco, London, pp. 259–413.

Calvo, B. and Furness, R.W. (1992) A review of the use and the effects of marks and devices on birds. *Ringing Migration*, **13**, 129–151.

Campbell, R.R., Etches, R.J. and Leatherland, J.F. (1981) Seasonal changes in plasma prolactin concentration and carcass lipid levels in the lesser snow goose (*Anser caerulescens caerulescens*). *Comp. Biochem. Physiol.*, **68**, 653–658.

Carmi, N., Pinshow, B., Porter, W.P. and Jaeger, J. (1992) Water and energy limitations on flight duration in small migrating birds. *Auk*, **109**, 268–276.

Carmi, N., Pinshow, B., Horowitz, M. and Bernstein, M.H. (1993) Birds conserve plasma volume during thermal and flight-incurred dehydration. *Physiol. Zool.*, **66**, 829–846.

Carpenter, F.L. and Hixon, M.A. (1988) A new function for torpor: fat conservation in a wild migrant hummingbird. *Condor*, **90**, 373–378.

Carpenter, F.L., Paton, D.C. and Hixon, M.A. (1983) Weight gain and adjustment of feeding territory size in migrant hummingbirds. *Proc. Natl. Acad. Sci.*, **80**, 7259–7263.

Cassone, V.M. (1990) The pineal organ and other components of avian circadian organization. *Acta XX Congr. Internat. Ornithol.*, Christchurch 1990, Suppl., pp. 311–312.

Castro, G. and Myers, J.P. (1989) Flight range estimates for shorebirds. *Auk*, **106**, 474–476.

Castro, G. and Myers, J.P. (1990) Validity of predictive equations for total body fat in sanderlings from different nonbreeding areas. *Condor*, **92**, 205–209.

Castro, G., Wunder B.A. and Knopf, F.L. (1991) Temperature-dependent loss of mass by shorebirds following capture. *J. Field Ornithol.*, **62**, 314–318.

Castro, G., Myers, J.P. and Ricklefs, R.E. (1992) Ecology and energetics of sanderlings migrating to four latitudes. *Ecology*, **73**, 833–844.

Catterall, C.P. and Kikkawa, J. (1989) Habitat learning and its relationship with site fidelity and reproduction in silvereyes. *Acta XX Congr. Internat. Ornithol.*, Christchurch 1990, Suppl., p. 409.

Child, G.I. and Marshall, S.G. (1970) A method of estimating carcass fat and fat-free weights in migrant birds from water content of specimens. *Condor*, **72**, 116–119.

Choudhury, S. and Black, J.M. (1991) Testing the behavioural dominance and dispersal hypothesis in Pochard. *Ornis Scand.*, **22**, 155–159.

Clementi, M.E., Sanna, M.T., Giardina, B. and Brunori, M. (1991) Flight and heat dissipation in migratory birds. *Rend. Fis. Acc. Lincei*, **2**, 315–321.

Coates–Estrada, R. and Estrada, A. (1989) Avian attendance and foraging at army-ant swarms in the tropical rain forest of Los Tuxtlas, Veracruz, Mexico. *J. Trop. Ecol.*, **5**, 281–292.

Cochran, W.W. and Kjos, C.G. (1985) Wind drift and migration of thrushes: A telemetry study. *Illinois Nat. Hist. Survey Bull.*, **33**, 297–330.

Coemans, M. and Vos, J. (1992) *On the Perception of Polarized Light by the Homing Pigeon*. Koninklijke Bibliotheek, Den Haag.

Cooke, W. W. (1905) Routes of bird migration. *Auk*, **22**, 1–11.

Cornwallis, R.K. and Townsend, A.D. (1968) Waxwings in Britain and Europe during 1965/66. *Brit. Birds*, **61**, 97–118.

Craig, A. and Larochelle, J. (1991) The cooling power of pigeon wings. *J. Exp. Biol.*, **155**, 193–202.

Creutz, G. (1987) *Geheimnisse des Vogelzugs*. Die Neue Brehm-Bücherei, Wittenberg Lutherstadt.

Cristol, D.A., Nolan, V. Jr and Ketterson, E.D. (1990) Effect of prior residence on dominance status of dark-eyed juncos, *Junco hyemalis*. *Anim. Behav.*, **40**, 580–586.

Cristol, D.A. and Evers, D.C. (1992) Dominance status and latitude are unrelated in wintering dark-eyed juncos. *Condor*, **94**, 539–542.

Curry-Lindahl, K. (1981) *Bird Migration in Africa 1 and 2*. Academic Press, New York, San Francisco, London.

Czeschlik, D. (1976) Der Einfluß des Wetters auf die Zugunruhe von Garten- und Mönchsgrasmücke (*Sylvia borin* und *S. atricapilla*). Dissertation University of Innsbruck.

Dall'Antonia, P., Ioalè, P., Mango, F. and Papi, F. (1990) Reconstruction of the flight paths of homing pigeons by means of a flight-path recorder. *Ethol. Ecol. Evol.*, **3**, 304–305.

Daum-Benz, P., Munro, U. and Wiltschko, W. (1988) Ontogenie der Zugorientierung bei Gartengrasmücken. *Verh. Dtsch. Zool. Ges.*, **81**, 227.

Davidson, N.C. (1984) How valid are flight range estimates for waders? *Ringing Migration*, **5**, 49–64.

Davidson, N.C. and Evans, P.R. (1988) Prebreeding accumulation of fat and muscle protein by Arctic-breeding shorebirds. *Acta XIX Congr. Internat. Ornithol.*, Ottawa 1986, pp. 342–352.

Debussche, M. and Isenmann, P. (1984) Origine et nomadisme des Fauvettes à tête noire (*Sylvia atricapilla*) hivernant en zone méditerranéenne française. *Oiseau Rev. Franç. Ornithol.*, **54**, 101–107.

Degen, T. and Jenni, L. (1990) Biotopnutzung von Kleinvögeln in einem

Naturschutzgebiet und im umliegenden Kulturland während der Herbstzugzeit. *Ornithol. Beob.*, **87**, 295–325.

DeSante, D.F. (1983) Vagrants: when orientation or navigation goes wrong. *Point Reyes Bird Observ. Newsletter*, **61**, 12–16.

Deviche, P. (1991) Androgen-opioid interaction on feeding in a migratory songbird. *Amer. Zool.*, **31**, 5.

Deviche, P. (1992) Regulation of food intake in a migratory songbird (*Junco hyemalis*): participation of endorphinergic mechanisms. *Ornis Scand.*, **23**, 260–263.

Dhondt, A.A. (1983) Variations in the number of overwintering stonechats possibly caused by natural selection. *Ringing Migration*, **4**, 155–158.

Dierschke, V. (1989) Automatisch-akustische Erfassung des nächtlichen Vogelzuges bei Helgoland im Sommer 1987. *Vogelwarte*, **35**, 115–131.

Dingle, H. (1980) Ecology and evolution of migration. In *Animal Migration, Orientation, and Navigation* (ed. S.A. Gauthreaux, Jr), Academic Press, New York, San Francisco, London, pp. 1–101.

Dingle, H. (1991) Evolutionary genetics of animal migration. *Amer. Zool.*, **31**, 253–264.

Dingle, H. (1994) Genetic Analyses of Animal Migration. In *Quantitative Genetic Studies of Behavioral Evolution* (ed. C.R.B. Boake), University of Chicago Press, Chicago, London, pp. 145–164.

Dingle, H. (1995) *The Bioloy of Migration. 2. Migration: a definition.* In press.

Dinse, V. (1991) Über den Heimzug von Kleinvögeln in Hamburg. Eine Auswertung von Fangdaten im Rahmen des Mettnau–Reit–Illmitz–Programms. *Hamburger Avifaun. Beitr.*, **23**, 1–125.

Dittami, J. (1981) Seasonal changes in the behavior and plasma titers of various hormones in barheaded geese, *Anser indicus*. *Z. Tierpsychol.*, **55**, 289–324.

Dolnik, V.R. (1975) *Migracionnoe Sostojanie Ptic.* Isdatelstwo Nauka, Moskau.

Dolnik, V.R. (1985) Nocturnal bird migration over arid and mountainous regions of middle Asia and Kazakhstan. *Proc. Zool. Inst. USSR Acad. Sci., Leningrad*, **169**, 1–147.

Dolnik, V.R. (1990) Bird migration across arid and mountainous regions of Middle Asia and Kazakhstan. In *Bird Migration: Physiology and Ecophysiology* (ed. E. Gwinner), Springer, Berlin, Heidelberg, New York, pp. 368–386.

Dolnik, V.R. and Blyumental, T.I. (1964) The bioenergetics of bird migrations. *Succ. Mod. Biol.*, **58**, 280–301.

Dolnik, V.R. and Bolshakov, K.V. (1985) Preliminary results of vernal nocturnal bird passage study over arid and mountain areas of central Asia: latitudinal crossing. In *Spring Nocturnal Bird Passage Over Arid and Mountain Areas of Asia Middle and Kazakhstan* (ed. V.R. Dolnik), USSR Acad. Sci. Zool. Inst., Leningrad, pp. 260–294.

Dorka, V. (1966) Das jahres- und tageszeitliche Zugmuster von Kurz- und Langstreckenziehern nach Beobachtungen auf den Alpenpässen Cou/Bretolet. *Ornithol. Beob.*, **63**, 165–223.

Doughty, R.W. (1989) *Return of the Whooping Crane.* University of Texas Press, Austin, p. 192.

Dowsett-Lemaire, F. and Dowsett, R.J. (1987) European reed and marsh warblers in Africa: migration patterns, moult and habitat. *Ostrich*, **58**, 65–85.

Drent, R. and Piersma, T. (1990) An exploration of the energetics of leap-frog migration in arctic breeding waders. In *Bird Migration: Physiology and Ecophysiology* (ed. E. Gwinner), Springer, Berlin, Heidelberg, New York, pp. 399–412.

Drent, R., Ebbinge, B. and Weijand, B. (1981) Balancing the energy budgets of arctic-breeding geese throughout the annual cycle: a progress report. *Verh. Ornithol. Ges. Bayern*, **23**, 239–264.

Driedzic, W.R. and Crowe, H.L. (1993) Adaptations in pectoralis muscle, heart mass, and energy metabolism during premigratory fattening in semipalmated sandpipers (*Calidris pusilla*). *Can. J. Zool.*, **71**, 1602–1608.

Dunn, E.H. and Nol, E. (1980) Age-related migratory behavior of warblers. *J. Field Ornithol.*, **51**, 254–269.

Dunn, P.O., May, T.A. and McCollough, M.A. (1988) Length of stay and fat content of migrant semipalmated sandpipers in eastern Maine. *Condor*, **90**, 824–835.

Dyachenko, V.P. and Dolnik, T.V. (1990) The period of circadian rhythm of activity in chaffinches with different level of migratory fat. *Baltic Birds*, **5**, 102–107.

Eastwood, E. (1967) *Radar Ornithology*, Methuen, London.

Eaton, S.W., O'Connor, P.D., Osterhaus, M.B. and Anicette, B.Z. (1963) Some osteological adaptations in parulidae. *Proc. XIII Internat. Ornithol. Congr.*, Ithaca 1962., pp. 71–83.

Edmonds, D.T. (1992) A magnetite null detector as the migrating bird's compass. *Proc. R. Soc. Lond. B*, **249**, 27–31.

Edwards, H.H., Schnell, G.D., DuBois, R.L. and Hutchison, V.H. (1992) Natural and induced remanent magnetism in birds. *Auk*, **109**, 43–56.

Eiserer, L.A. (1979) Roosttime restlessness in captive American robins (*Turdus migratorius*). *Animal Learning Behavior*, **7**, 406–412.

Ekman, J. and Hake, M. (1988) Avian flocking reduces starvation risk: an experimental demonstration. *Behav. Ecol. Sociobiol.*, **22**, 91–94.

Elkins, N. (1988a) Can high-altitude migrants recognize optimum flight levels? *Ibis*, **130**, 562–563.

Elkins, N. (1988b) *Weather and Bird Behaviour*. Poyser, Calton.

Ellegren, H. (1990a) Timing of autumn migration in bluethroats *Luscinia s. svecica* depends on timing of breeding. *Ornis Fenn.*, **67**, 13–17.

Ellegren, H. (1990b) Autumn migration speed in Scandinavian bluethroats *Luscinia s. svecica*. *Ringing Migration*, **11**, 121–131.

Ellegren, H. (1991) Stopover ecology of autumn migrating bluethroats *Luscinia s. svecica* in relation to age and sex. *Ornis Scand.*, **22**, 340–348.

Ellegren, H. (1993) Speed of migration and migratory flight lengths of passerine birds ringed during autumn migration in Sweden. *Ornis Scand.*, **24**, 220–228.

Ellegren, H. and Staav, R. (1990) Ruggningsflyttning hos blåhaken *Luscinia s. svecica*. *Vår Fågelvärld*, **49**, 80–86.

Elmore, J.B. Sr and Palmer-Ball, B. Jr (1991) Mortality of migrant birds at two central Kentucky tv towers. *Kentucky Warbler*, **67**, 67–71.

Emlen, S.T. (1967a) Migratory orientation in the indigo bunting, *Passerina cyanea*. Part I: Evidence for use of celestial cues. *Auk*, **84**, 309–342.

Emlen, S.T. (1967b) Migratory orientation in the indigo bunting, *Passerina cyanea*. Part II: Mechanism of celestial orientation. *Auk*, **84**, 463–489.

Emlen, S.T., Wiltschko, W., Demong, N.J., Wiltschko, R. and Bergman, S. (1976) Magnetic direction finding: Evidence of its use in migratory indigo buntings. *Science*, **193**, 505–508.

Erskine, A.J. (1988) The changing patterns of Brant migration in eastern North America. *J. Field Ornithol.*, **59**, 110–119.

Evans, P.R. (1970) Timing mechanisms and the physiology of bird migration. *Sci. Progr. (London)*, **58**, 263–275.

Evans, P.R. (1991) Seasonal and annual patterns of mortality in migratory shorebirds: some conservation implications. In *Bird Population Studies* (eds C.M. Perrins, J.-D. Lebreton and G.J.M. Hirons), Oxford University Press, Oxford, London, New York, pp. 346–359.

Evans, P.R. and Townshend, D.J. (1988) Site faithfulness of waders away from the breeding grounds: How individual migration patterns are established. *Acta XIX Congr. Internat. Ornithol.*, Ottawa 1986, pp. 594–603.

Evans, P.R. and Davidson, N.C. (1990) Migration strategies and tactics of waders breeding in arctic and north temperate latitudes. In *Bird Migration: Physiology and Ecophysiology* (ed. E. Gwinner), Springer, Berlin, Heidelberg, New York, pp. 387–398.

Evans, P.R., Davidson, N.C., Uttley, J.D. and Evans, R.D. (1992) Premigratory hypertrophy of flight muscles: an ultrastructural study. *Ornis Scand.*, **23**, 238–243.

Ewert, D.N. and Askins, R.A. (1991) Flocking behavior of migratory warblers in winter in the Virgin Islands. *Condor*, **93**, 864–868.

Faaborg, J. and Chaplin, S.B. (1988) *Ornithology: An Ecological Approach*. Prentice Hall, Englewood Cliffs, New Jersey.

Falconer, D.S. (1981) *Introduction to Quantitative Genetics*. Longman, London, New York.

Faraci, F.M. (1991) Adaptations to hypoxia in birds: How to fly high. *Annu. Rev. Physiol.*, **53**, 59–70.

Farner, D.S. (1955) The annual stimulus for migration. Experimental and physiologic aspects. In *Recent Studies of Avian Biology* (ed. A. Wolfson), University of Illinois Press, Urbana, pp. 198–237.

Farner, D.S. (1966) Über die photoperiodische Steuerung der Jahreszyklen bei Zugvögeln. *Biol. Rundsch.*, **4**, 228–241.

Farner, D.S. and Serventy, D.L. (1960) The timing of reproduction in birds in the arid regions of Australia. *Anat. Rec.*, **137**, 354.

Farner, D.S., King, J.R. and Stetson, M.H. (1969) The control of fat metabolism in migratory birds. In *Progr. Endocrinol., Proc. Int. Congr. Endocrinol.*, 1968 Int. Congr. Ser. No. 184, pp. 152–157.

Fasola, M. and Fraticelli, F. (1990) Non-competitive habitat use by foraging passerine birds during spring migrations. *Ethol. Ecol. Evol.*, **2**, 363–371.

Fedde, M.R. (1990) High-altitude bird flight: Exercise in a hostile environment. *NIPS*, **5**, 191–193.

Ferns, P.N. (1975) Feeding behaviour of autumn passage migrants in north east Portugal. *Ringing Migration*, **1**, 3–11.

Finch, D.M. (1991) Population ecology, habitat requirements, and conservation of neo-tropical migratory birds. *US For. Serv. Gen. Tech. Rep.*, **205**, 1–26.

Fitzgerald, J.A. (1990) Effects of long and short photoperiods on nutrient preferences of migratory dark-eyed juncos and non-migratory house sparrows. *Bird Behav.*, **8**, 87–94.

Fitzgerald, S.D., Sullivan, J.M. and Everson, R.J. (1990) Suspected ethanol toxicosis in two wild cedar waxwings. *Avian Diseases*, **34**, 488–490.

Fliege, G. (1984) Das Zugverhalten des Stars (*Sturnus vulgaris*) in Europa: Eine Analyse der Ringfunde. *J. Ornithol.*, **125**, 393–446.

Flousek, J. and Smrcek, M. (1984) Daily activity of birds based on results of the Operation Baltic in the Krkonose Mountains. *Opera Corcontica*, **21**, 103–126.

Folland, C.K., Palmer, T.N. and Parker, D.E. (1986) Sahel rainfall and world-wide sea temperatures, 1901–85. *Nature*, **320**, 602–607.

Fransson, T. and Stolt, B.-O. (1993) Is there an autumn migration of continental Blackcaps (*Sylvia atricapilla*) into northern Europe? *Vogelwarte*, **37**, 89–95.

Frederick II (1964) *Über die Kunst mit Vögeln zu jagen*. Insel, Frankfurt.

Fry, C.H. (1992) The Moreau ecological overview. *Ibis*, **134**, Suppl. 1, 3–6.

Fuller, R.J. and Crick, H.Q.P. (1992) Broad-scale patterns in geographical and habitat distribution of migrant and resident passerines in Britain and Ireland. *Ibis*, **134**, Suppl. 1, 14–20.

Ganzhorn, J.U. and Burkhardt, J.F. (1991) Pigeon homing: new airbag experiments to assess the role of olfactory information for pigeon navigation. *Behav. Ecol. Sociobiol.*, **29**, 69–75.

Gardiazabal y Pastor, A. (1990) Untersuchungen zur Ökologie rastender Kleinvögel im Nationalpark von Doñana, (Spanien): Ernährung, Fettdeposition, Zugstrategie. Dissertation Univ. Köln.

Gatter, W. (1992) Zugzeiten und Zugmuster im Herbst: Einfluß des Treibhauseffekts auf den Vogelzug? *J. Ornithol.*, **133**, 427–436.

Gatter, W. and Behrndt, M. (1985) Unterschiedliche tageszeitliche Zugmuster alter und junger Vögel am Beispiel der Rauchschwalbe (*Hirundo rustica*). *Vogelwarte*, **33**, 115–120.

Gaunt, A.S., Hikida, R.S, Jehl, J.R. Jr and Fenbert, L. (1990) Rapid atrophy and hypertrophy of an avian flight muscle. *Auk*, **107**, 649–659.

Gauthier, M. and Thomas, D.W. (1990) Evaluation of the accuracy of $^{22}$Na and tritiated water for the estimation of food consumption and fat reserves in passerine birds. *Can. J. Zool.*, **68**, 1590–1594.

Gauthier-Hion, A. and Michaloud, G. (1989) Are figs always keystone resources for tropical frugivorous vertebrates? A test in Gabon. *Ecology*, **70**, 1826–1833.

Gauthreaux, S.A. (1969) A portable ceilometer technique for studying low–level nocturnal migration. *Bird Banding*, **40**, 309–320.

Gauthreaux, S.A. (1978) The ecological significance of behavioural dominance. In *Perspectives in Ethology* (eds P.P.G. Bateson and P.H. Klopfer), Plenum, New York, pp. 17–54.

Gauthreaux, S.A. (1982) The ecology and evolution of avian migration systems. In *Avian Biology 6* (eds D.S. Farner, J.R. King and K.C. Parkes), Academic Press, New York, San Francisco, London, pp. 93–168.

Gauthreaux, S.A. (1988) Age effects on migration and habitat selection. *Acta XIX Congr. Internat. Ornithol.*, Ottawa 1986, pp. 1106–1115.

Gauthreaux, S.A. (1991) The flight behavior of migrating birds in changing wind fields: radar and visual analyses. *Amer. Zool.*, **31**, 187–204.

Geil, S., Noer, H. and Rabøl, J. (1974) *Forecast models for bird migration in Denmark*. Bird Strike Committee, Denmark.

George, J.C., John, T.M. and Minhas, K.J. (1987) Seasonal degradative, reparative and regenerative ultrastructural changes in the breast muscle of the migratory Canada goose. *Cytobios*, **52**, 109–126.

Gessaman, J.A. (1990) Body temperatures of migrant accipiter hawks just after flight. *Wilson Bull.*, **102**, 133–137.

Giardina, B., Corda, M., Pellegrini, M.G., Sanna, M.T., Brix, O., Clementi, M.E. and Gondo, S.G. (1990) Flight and heat dissipation in birds: A possible molecular mechanism. *Febs Letters*, **270**, 173–176.

Gifford, C.E. and Odum, E.P. (1965) Bioenergetics of lipid deposition in the Bobolink, a transequatorial migrant. *Condor*, **67**, 383–403.

Ginn, H.B. and Melville, D.S. (1983) *Moult in Birds*. Brit. Trust Ornithol., Guide 19, Tring.

Glück, E. (1978) Aktivitätsuntersuchungen an Tagziehern (*Carduelis carduelis*). *J. Ornithol.*, **119**, 336–338.

Glutz v. Blotzheim, U.N. and Bauer, K.M. (1980) *Handbuch der Vögel Mitteleuropas 9*. Aula, Wiesbaden.

Gorney, E. and Yom-Tov, Y. (1990) Timing of spring migration in relation to body condition in the steppe buzzard *Buteo buteo vulpinus*. *Acta XX Congr. Internat. Ornithol.*, Christchurch 1990, Suppl., p. 336.

Goslow, G.E., Dial, K.P. and Jenkins, F.A. (1990) Bird flight: insights and complications. *BioScience*, **40**, 108–115.

Graczyk, R. (1963) Badania dosåwiadczalne nad etologia gatunków rodzaju *Turdus L. Roczn. WSR, Poznan*, **17**, 21–69.

Gray, J.M., Yarian, D. and Ramenofsky, M. (1990) Corticosterone, foraging behavior, and metabolism in dark-eyed juncos, *Junco hyemalis*. *Gen. Comp. Endocrinol.*, **79**, 375–384.

Grazulevicius, G. and Petraitis, A. (1990) Dependence of low altitude migration on weather conditions. *Baltic Birds*, **5**, 109–113.

Greenberg, R. (1986) Competition in migrant birds in the nonbreeding season. In *Current Ornithology 3* (ed. R.F. Johnston), Plenum Press, New York, pp. 281–307.

Greenewalt, C.H. (1975) The flight of birds. *Trans. Amer. Philos. Soc.*, **65**, 1–67.

Gregoire, P.E. and Davison Ankney, C. (1990) Agonistic behavior and dominance relationships among lesser snow geese during winter and spring migration. *Auk*, **107**, 550–560.

Gudmundsson, G.A. (1992) *Flight and Migration Strategies of Birds at Polar Latitudes*. Lund University, Lund.

Gudmundsson, G.A., Alerstam, T. and Larsson, B. (1992) Radar observations of northbound migration of the Arctic tern, *Sterna paradisaea*, at the Antarctic Peninsula. *Antarctic Sci.*, **4**, 163–170.

Gudmundsson, G.A., Lindström, Å. and Alerstam, T. (1991) Optimal fat loads and long–distance flights by migrating knots *Calidris canutus*, sanderlings *C. alba* and turnstones *Arenaria interpres*. *Ibis*, **133**, 140–152.

Gwinner, E. (1967a) Circannuale Periodik der Mauser und der Zugunruhe bei einem Vogel. *Naturwiss.*, **54**, 447.

Gwinner, E. (1967b) Wirkung des Mondlichtes auf die Nachtaktivität von Zugvögeln. Lotsenversuch an Rotkehlchen (*Erithacus rubecula*) und Gartenrotschwänzen (*Phoenicurus phoenicurus*). *Experientia*, **23**, 227.

Gwinner, E. (1968) Artspezifische Muster der Zugunruhe bei Laubsängern und ihre mögliche Bedeutung für die Beendigung des Zuges im Winterquartier. *Z. Tierpsychol.*, **25**, 843–853.

Gwinner, E. (1971) Orientierung. In *Grundriß der Vogelzugskunde* (ed. E. Schüz), Parey, Berlin, Hamburg, pp. 299–348.

Gwinner, E. (1972) Endogenous timing factors in bird migration. In *Animal Orientation and Navigation* (eds S.R. Galler, K. Schmidt-Koenig, G.J. Jacobs and R.E. Belleville), NASA, Washington D.C., pp. 321–338.

Gwinner, E. (1974) Testosterone induces 'splitting' of circadian locomotor activity rhythms in birds. *Science*, **185**, 72–74.

Gwinner, E. (1975) Effects of season and external testosterone on the freerunning circadian activity rhythm of European starlings (*Sturnus vulgaris*). *J. Comp. Physiol.*, **103**, 315–328.

Gwinner, E. (1986) *Circannual Rhythms*. Springer, Berlin, Heidelberg, New York.

Gwinner, E. (1987) Photoperiodic synchronization of circannual rhythms in gonadal activity, migratory restlessness, body weight, and molt in the garden warbler (*Sylvia borin*). In *Comparative Physiology of Environmental Adaptations III* (ed. Pévet), Karger, Basel, pp. 30–44.

Gwinner, E. (1989) Melatonin in the circadian system of birds: Model of internal resonance. In *Circadian Clocks and Ecology* (eds T. Hiroshige and K. Honma),

Hokkaido University Press, Sapporo, pp. 127–153.

Gwinner, E. (1990) Circannual rhythms in bird migration: Control of temporal patterns and interactions with photoperiod. In *Bird Migration: Physiology and Ecophysiology* (ed. E. Gwinner), Springer, Berlin, Heidelberg, New York, pp. 257–268.

Gwinner, E. and Czeschlik, D. (1978) On the significance of spring migratory restlessness in caged birds. *Oikos*, **30**, 364–372.

Gwinner, E., Biebach, H. and Kries, I. v. (1985) Food availability affects migratory restlessness in caged garden warblers (*Sylvia borin*). *Naturwiss.*, **72**, 51.

Gwinner, E. and Neusser, V. (1985) Die Jugendmauser europäischer und afrikanischer Schwarzkehlchen (*Saxicola torquata rubicula und axillaris*) sowie von F1–Hybriden. *J. Ornithol.*, **126**, 219–220.

Gwinner, E. and Wiltschko, W. (1978) Endogenously controlled changes in migratory direction of the garden warbler, *Sylvia borin*. *J. Comp. Physiol.*, **125**, 267–273.

Gwinner, E., Schwabl, H. and Schwabl-Benzinger, I. (1988) Effects of food-deprivation on migratory restlessness and diurnal activity in the garden warbler (*Sylvia borin*). *Oecologia*, **77**, 321–326.

Gwinner, E. and Dittami, J. (1990) Endogenous reproductive rhythms in a tropical bird. *Science*, **249**, 906–908.

Gwinner, E., Schwabl, H. and Schwabl-Benzinger, I. (1992a) The migratory time program of the garden warbler: is there compensation for interruptions? *Ornis Scand.*, **23**, 264–270.

Gwinner, E., Zeman, M., Schwabl-Benzinger, I., Jenni-Eiermann, S., Jenni, L. and Schwabl, H. (1992b) Corticosterone levels of passerine birds during migratory flight. *Naturwiss.*, **79**, 276–278.

Haartman, L. von (1969) The nesting habits of Finnish birds, I. Passeriformes. *Commentationes Biol.*, **32**, 3–187.

Hagan, J.M., Lloyd-Evans, L. and Altwood, J.L. (1991) The relationship between latitude and the timing of spring migration of North American landbirds. *Ornis Scand.*, **22**, 129–136.

Hainsworth, F.R. (1989) Wing movements and positioning for aerodynamic benefit by Canada geese flying in formation. *Can. J. Zool.*, **67**, 585–589.

Hall, J.C. (1990) Genetics of circadian rhythms. *Annu. Rev. Genet.*, **24**, 659–697.

Hamilton, W.J. (1962) Evidence concerning the function of nocturnal call notes of migratory birds. *Condor*, **64**, 390–401.

Hands, H.M., Ryan, M.R. and Smith, J.W. (1991) Migrant shorebird use of marsh, moist-soil, and flooded agricultural habitats. *Wildl. Soc. Bull.*, **19**, 457–464.

Hardy, E. (1978) Winter foods of Blackcaps in Britain. *Bird Study*, **25**, 60–61.

Harengerd, M., Pölking, F., Prünte, W. and Speckmann, M. (1972) *Die Tundra ist mitten in Deutschland*. Kilda, Greven.

Harrington, B.A., Leeuwenberg, F.J., Resende, S.L., McNeil, R., Thomas, B.T., Grear, J.S. and Martinez, E.F. (1990) Migration and weight change of white-rumped sandpiper *Calidris fuscicollis* during non-breeding seasons. *Acta XX Congr. Internat. Ornithol.*, Christchurch 1990, Suppl., p. 527.

Harris, M.P. (1970) Abnormal migration and hybridization of *Larus argentatus* and *L. fuscus* after interspecies fostering experiments. *Ibis*, **112**, 488–498.

Hashmi, D. (1990) *Soziales Zugverhalten: Untersuchungen am Baßtölpel.* Programm Kurzfassungen Vorträge 123. Jahresversammlung DO-G, p. 45.

Hayes, F.E., Goodman, S.M., Fox, J.A., Tamayo, T.G. and Lopez, N.E. (1990)

North American bird migrants in Paraguay. *Condor*, **92**, 947–960.

Healy, S.D. and Krebs, J.R. (1991) Hippocampal volume and migration in passerine birds. *Naturwiss.*, **78**, 424–426.

Hedenström, A. (1992) Flight performance in relation to fuel load in birds. *J. theor. Biol.*, **158**, 535–537.

Hedenström, A. and Alerstam, T. (1992) Climbing performance of migrating birds as a basis for estimating limits for fuel-carrying capacity and muscle-work. *J. Exp. Biol.*, **164**, 19–38.

Hedenström, A. and Pettersson, J. (1986) Differences in fat deposits and wing pointedness between male and female willow warblers caught on spring migration at Ottenby, SE Sweden. *Ornis Scand.*, **17**, 182–185.

Heinemann, D. (1992) Resource use, energetic profitability, and behavioral decisions in migrant rufous hummingbirds. *Oecologia*, **90**, 137–149.

Heitmeyer, M.E. and Fredrickson, L.H. (1990) Fatty acid composition of wintering female mallards in relation to nutrient use. *J. Wildl. Manage.*, **54**, 54–61.

Helbig, A.J. (1990) Genetics of migratory orientation in the blackcap *Sylvia atricapilla. Acta XX Congr. Internat. Ornithol.*, Christchurch 1990, Suppl., p. 383.

Helbig, A.J. (1991a) Inheritance of migratory direction in a bird species: a cross-breeding experiment with SE- and SW-migrating blackcaps (*Sylvia atricapilla*). *Behav. Ecol. Sociobiol.*, **28**, 9–12.

Helbig, A.J. (1991b) Experimental and analytical techniques used in bird orientation research. In *Orientation in Birds* (ed. P. Berthold), Birkhäuser, Basel, Boston, Berlin, pp. 270–306.

Helbig, A.J. (1991c) Genetische Grundlagen der Zugorientierung bei Vögeln. *Verh. Dt. Zool. Ges.*, **84**, 343–344.

Helbig, A.J. (1991d) SE- and SW-migrating blackcap (*Sylvia atricapilla*) populations in Central Europe: Orientation of birds in the contact zone. *J. evol. Biol.*, **4**, 657–670.

Helbig, A.J. (1992) Ontogenetic stability of inherited migratory directions in a nocturnal bird migrant: comparison between the first and second year of life. *Ethol. Ecol. Evol.*, **4**, 375–388.

Helbig, A.J. and Franz, D. (1990) Einflug der Eiderente *Somateria mollissima* nach Mittel- und Südeuropa im Herbst 1988. *Limicola*, **4**, 229–249.

Helbig, A.J., Berthold, P., Mohr, G. and Querner, U. (1994) Inheritance of a novel migratory direction in central European blackcaps. *Naturwiss.*, **81**, 184–186.

Helms, C.W. (1963) Annual cycle and Zugunruhe in birds. *Proc. XIII Internat. Ornithol. Congr.*, Ithaca 1962, pp. 925–939.

Helms, C.W. and Drury, W.H. (1960) Winter and migratory weight and fat field studies on some North American buntings. *Bird Banding*, **31**, 1–40.

Hepp, G.R. and Hines, J.E. (1991) Factors affecting winter distribution and migration distance of wood ducks from southern breeding populations. *Condor*, **93**, 884–891.

Herremans, M. (1989) Habitat and sampling related bias in sex-ratio of trapped blackcaps *Sylvia atricapilla. Ringing Migration*, **10**, 31–34.

Herremans, M. (1990a) Body-moult and migration overlap in reed warblers (*Acrocephalus scirpaceus*) trapped during nocturnal migration. *Gerfaut*, **80**, 149–158.

Herremans, M. (1990b) Can night migrants use interspecific song recognition to assess habitat? *Gerfaut*, **80**, 141–148.

Herremans, M. (1991) Viewpoint trans-Saharan migration strategies. *Ringing Migration*, **12**, 55.

Herrera, C.M. and Jordano, P. (1981) *Prunus mahaleb* and birds: the high-efficiency seed dispersal system of a temperate fruiting tree. *Ecol. Monogr.*, **51**, 203–218.

Heyder, R. (1955) Hundert Jahre Gartenamsel. *Beitr. Vogelk.*, **4**, 64–81.

Hiebl, I. and Braunitzer, G. (1988) Anpassungen der Hämoglobine von Streifengans (*Anser indicus*), Andengans (*Chloephaga melanoptera*) und Sperbergeier (*Gyps rueppellii*) an hypoxische Bedingungen. *J. Ornithol.*, **129**, 217–226.

Hildén, O. (1979) The timing of arrival and departure of the spotted redshank *Tringa erythropus* in Finland. *Ornis Fenn.*, **56**, 18–23.

Hildén, O. (1982) Winter ecology and partial migration of the goldcrest *Regulus* in Finland. *Ornis Fenn.*, **59**, 99–122.

Hildén, O. and Saurola, P. (1982) Speed of autumn migration of birds ringed in Finland. *Ornis Fenn.*, **59**, 140–143.

Hilgerloh, G. and Bingman, V.P. (1992) Radar observations of passerine migration over Frankfurt and Hannover, Germany. *J. Ornithol.*, **133**, 23–31.

Hilgerloh, G., Laty, M. and Wiltschko, W. (1992) Are the Pyrenees and the western Mediterranean barriers for trans-Saharan migrants in spring? *Ardea*, **80**, 375–381.

Hill, W.L. (1989) Reply to Briggs: The roles of endogenous and exogenous nutrient supplies. *Condor*, **91**, 494–495.

Hissa, R. (1988) Controlling mechanisms in avian temperature regulation: a review. *Acta Physiol. Scand.*, **132**, 1–147.

Holberton, R.L. (1993) An endogenous basis for differential migration in the dark-eyed junco. *Condor*, **95**, 580–587.

Holberton, R.L., Hanano, R. and Able, K.P. (1990) Age-related dominance in male dark-eyed juncos: effects of plumage and prior residence. *Anim. Behav.*, **40**, 573–579.

Holberton, R.L. and Able, K.P. (1992) Persistence of circannual cycles in a migratory bird held in constant dim light. *J. Comp. Physiol.*, **171**, 477–481.

Holmes, R.T., Sherry, T.W. and Reitsma, L. (1989) Population structure, territoriality and overwinter survival of two migrant warbler species in Jamaica. *Condor*, **91**, 545–561.

Holmgren, N. and Lundberg, S. (1990) Evolution of migration patterns in dominance structured populations. *Ring*, **13**, 219–220.

Holmgren, N. and Lundberg, S. (1993) Despotic behaviour and the evolution of migration patterns in birds. *Ornis Scand.*, **24**, 103–109.

Houston, A.I. (1990) Parent birds feeding young – to walk or to fly? *J. theor. Biol.*, **142**, 141–147.

Hummel, D. (1973) Die Leistungsersparnis beim Verbandsflug. *J. Ornithol.*, **114**, 259–282.

Hummel, D. and Beukenberg, M. (1989) Aerodynamische Interferenzeffekte beim Formationsflug von Vögeln. *J. Ornithol.*, **130**, 15–24.

Humphrey, P.S. and Parkes, K.C. (1959) An approach to the study of molts and plumages. *Auk*, **76**, 1–31.

Hutto, R.L. (1989) The effect of habitat alteration on migratory landbirds in a west Mexican tropical deciduous forest: a conservation perspective. *Conserv. Biol.*, **3**, 138–148.

Ieromnimon, V. (1977) Beobachtungen über die Wirkung von Hormonen auf das Zugverhalten bei Rotkehlchen (*Erithacus rubecula*). I. Die Wirkung von

Prolaktin (LtH=Hpr) im Jahreszyklus. *Vogelwarte*, **29**, 126–134.

Immelmann, K. (1963) Tierische Jahresperiodik in ökologischer Sicht. *Zool. Jb. Syst.*, **91**, 91–200.

Immelmann, K. (1971) Ecological aspects of periodic reproduction. In *Avian Biology 1* (eds D.S. Farner and J.R. King), Academic Press, New York, San Francisco, London, pp. 341–389.

Izhaki, I. and Saffriel, U.N. (1990) Weight losses due to exclusive fruit diet – interpretation and evolutionary implications: a reply to Mack and Sedinger. *Oikos*, **57**, 140–142.

Jackson, S. and Place, A.R. (1990) Gastrointestinal transit and lipid assimilation efficiencies in three species of sub-Antarctic seabird. *J. Exp. Zool.*, **255**, 141–154.

Jahnel, M., Balda, R.P. and Wiltschko, W. (1990) The influence of clockshifting on the recovery behaviour of seed caching corvids. *Acta XX Congr. Internat. Ornithol.*, Christchurch 1990, Suppl., p. 404.

Jakobsen, B. (1991) Premature fall migration of ducks apparently caused by hunting. *Dan. Ornithol. Foren. Tidsskr.*, **85**, 174–175.

James, D.A. (1990) Does early autumnal foliar color change target avian frugivory? *Acta XX Congr. Internat. Ornithol.*, Christchurch 1990, Suppl., pp. 421–422.

Jehl, J.R. (1990) Aspects of the molt migration. In *Bird Migration: Physiology and Ecophysiology* (ed. E. Gwinner), Springer, Berlin, Heidelberg, New York, pp. 102–116.

Jehl, J.R. (1994) Field estimates of energetics in migrating and downed black-necked grebes. *J. Avian Biol.*, **25**, 63–68.

Jellmann, J. (1989) Radarmessungen zur Höhe des nächtlichen Vogelzuges über Nordwestdeutschland im Frühjahr und im Hochsommer. *Vogelwarte*, **35**, 59–63.

Jenner, G.C. (1824) Some observations on the migration of birds. *Phil. Trans. Roy. Soc. Lond.*, **1**, 11–44.

Jenni, L. (1984) Herbstzugmuster von Vögeln auf dem Col de Bretolet unter besonderer Berücksichtigung nachbrutzeitlicher Bewegungen. *Ornithol. Beob.*, **81**, 183–213.

Jenni, L. and Jenni-Eiermann, S. (1987) Der Herbstzug der Gartengrasmücke *Sylvia borin* in der Schweiz. *Ornithol. Beob.*, **84**, 173–206.

Jenni, L., Reutimann, P. and Jenni-Eiermann, S. (1990) Recognizability of different food types in faeces and in alimentary flushes of *Sylvia* warblers. *Ibis*, **132**, 445–453.

Jenni-Eiermann, S. and Jenni, L. (1991) Metabolic responses to flight and fasting in night-migrating passerines. *J. Comp. Physiol. B*, **161**, 465–474.

Jenni-Eiermann, S. and Jenni, L. (1992) High plasma triglyceride levels in small birds during migratory flight: A new pathway for fuel supply during endurance locomotion at very high mass-specific metabolic rates? *Physiol. Zool.*, **65**, 112–123.

John, T.M. and George, J.C. (1986) Arginine vasotocin induces free fatty acid release from avian adipose tissue *in vitro*. *Arch. Int. Physiol. Biochem.*, **94**, 85–89.

John, T.M. and George, J.C. (1989) Seasonal ultrastructural changes in the pineal gland of the migratory Canada goose. *Cytobios*, **58**, 179–204.

John, T.M., George, J.C. and Scanes, C.G. (1983) Seasonal changes in circulating levels of luteinizing hormone and growth hormone in the migratory Canada goose. *Gen. Comp. Endocrinol.*, **51**, 44–49.

318     *References*

Johnson, S.R. and Herter, D.R. (1990) Bird migration in the Arctic: a review. In *Bird Migration: Physiology and Ecophysiology* (ed. E. Gwinner), Springer, Berlin, Heidelberg, New York, pp. 33–36.

Johnson, O.W., Morton, M.L., Bruner, P.L. and Johnson, P.M. (1989) Fat cyclicity, predicted migratory flight ranges, and features of wintering behavior in Pacific golden-plovers. *Condor*, **91**, 156–177.

Jones, P.J. and Ward, P. (1977) Evidence of pre-migratory fattening in three tropical granivorous birds. *Ibis*, **119**, 200–203.

Jordano, P. (1985) El ciclo anual de los paseriformes frugivoros en el matorral mediterraneo del sur de España: importancia de su invernada y variaciones interanuales. *Ardeola*, **32**, 69–94.

Jordano, P. (1989) Variacion de la dieta frugivora otoño-invernal del petirrojo (*Erithacus rubecula*): efectos sobre la condicion corporal. *Ardeola*, **36**, 161–183.

Jouventin, P. and Weimerskirch, H. (1990) Satellite tracking of wandering albatrosses. *Nature*, **343**, 746–748.

Jung, R.E. (1992) Individual variation in fruit choice by American robins (*Turdus migratorius*). *Auk*, **109**, 98–111.

Kaiser, A. (1992) Fat deposition and theoretical flight range of small autumn migrants in southern Germany. *Bird Study*, **39**, 96–110.

Kaiser A. (1993a) Rast- und Durchzugsstrategien mitteleuropäischer Singvögel. Analysen von Fang- und Wiederfangdaten von Fanganlagen zur Beschreibung der Ökophysiologie und des Verhaltens rastender Populationen. PhD thesis, University of Konstanz.

Kaiser, A. (1993b) A new multi-category classification of subcutaneous fat deposits of songbirds. *J. Field Ornithol.*, **64**, 246–255.

Kalela, O. (1954) Populationsökologische Gesichtspunkte zur Entstehung des Vogelzuges. *Ann. Zool. Soc. Zool. Bot. Fennicae Vanamo*, **16**, 1–30.

Karasov, W.H. and Levey, D.J. (1990) Digestive system trade-offs and adaptations of frugivorous passerine birds. *Physiol. Zool.*, **63**, 1248–1270.

Karlsson, L., Persson, K., Pettersson, J. and Walinder, G. (1988) Fat-weight relationships and migratory strategies in the robin *Erithacus rubecula* at two stop-over sites in south Sweden. *Ringing Migration*, **9**, 160–168.

Kasparek, M. (1981) *Die Mauser der Singvögel Europas – ein Feldführer*. Dachverb. Dt. Avifaunisten.

Keast, A. (1980) Spatial relationships between migratory parulid warblers and their ecological counterparts in the Neotropics. In *Migrant Birds in the Neotropics: ecology, behavior, distribution, and conservation* (eds A. Keast and E.S. Morton), Smithsonian Institution Press, Washington DC.

Keeton, W.T. (1970) Comparative orientational and homing performances of single pigeons and small flocks. *Auk*, **87**, 797–799.

Keeton, W.T., Larkin, T.S. and Windsor, D.M. (1974) Normal fluctuations in the earth's magnetic field influence pigeon orientation. *J. Comp. Physiol.*, **95**, 95–103.

Kelsey, M.G. (1989) A comparison of the song and territorial behaviour of a long-distance migrant, the marsh warbler *Acrocephalus palustris*, in summer and winter. *Ibis*, **131**, 403–414.

Kelsey, M.G. (1992) Conservation of migrants on their wintering grounds: an overview. *Ibis*, **134**, Suppl. 1, 109–112.

Kelsey, M.G., Backhurst, G.C. and Pearson, D.J. (1989) Age differences in the timing and biometrics of migrating marsh warblers in Kenya. *Ringing Migration*, **10**, 41–47.

Kemlers, E., Kemlers, A. and Petershof, E. (1990) Ornitofenologiskie noverojumi kuldigas apkartne. *Putni daba*, **3**, 112–125.

Kennedy, J.S. (1985) Migration, Behavioral and Ecological. In *Migration: Mechanisms and Adaptive Significance* (ed. M.A. Rankin), Contr. Marine Sci., **27**, Suppl., 5–26.

Kerlinger, P. (1989) *Flight Strategies of Migrating Hawks*. University of Chicago Press, Chicago, London.

Kerlinger, P. and Moore, F.R. (1989) Atmospheric structure and avian migration. *Curr. Ornithol.*, **6**, 109–142.

Kersten, M. and Piersma, T. (1987) High levels of energy expenditure in shorebirds, metabolic adaptations to an energetically expensive way of life. *Ardea*, **75**, 175–187.

Ketterson, E.D. and Nolan, V. Jr (1985) Intraspecific variation in avian migration: evolutionary and regulatory aspects. In *Migration: Mechanisms and Adaptive Significance* (ed. M.A. Rankin), Univ. Texas, Austin, pp. 553–579.

Ketterson, E.D. and Nolan, V. Jr (1990) Site attachment and site fidelity in migratory birds: experimental evidence from the field and analogies from neurobiology. In *Bird Migration: Physiology and Ecophysiology* (ed. E. Gwinner), Springer, Berlin, Heidelberg, New York, pp. 117–129.

Ketterson, E.D., Nolan, V., Ziegenfus, C., Cullen, D.P., Cawthorn, M. and Wolf L. (1991) Non-breeding season attributes of male dark-eyed junco that acquired breeding territories in their first year. *Acta XX Congr. Internat. Ornithol.*, Christchurch 1990, pp. 1229–1239.

Kiepenheuer, J. (1984) The magnetic compass mechanism of birds and its possible association with the shifting course directions of migrants. *Behav. Ecol. Sociobiol.*, **14**, 81–99.

King, J.R. (1961a) On the regulation of vernal premigratory fattening in the white-crowned sparrow. *Physiol. Zool.*, **34**, 145–157.

King, J.R. (1961b) The energetics of migration. *Condor*, **63**, 128–142.

King, J.R. (1963) Autumnal migratory-fat deposition in the white-crowned sparrow. *Proc. XIII Internat. Ornithol. Congr.*, Ithaca 1962, pp. 940–949.

King, J.R. (1968) Cycles of fat deposition and molt in white-crowned sparrows in constant environmental conditions. *Comp. Biochem. Physiol.*, **24**, 827–837.

King, J.R. (1970) Photoregulation of food intake and fat metabolism in relation to avian sexual cycles. *Colloq. Int. Cent. Nat. Rech. Sci.*, **172**, 365–385.

King, J.R. and Farner, D.S. (1965) Studies of fat deposition in migratory birds. *Ann. NY Acad. Sci.*, **131**, 422–440.

King, J.R., Follett, B.K., Farner, D.S. and Morton, M.L. (1966) Annual gonadal cycles and pituitary gonadotropins in *Zonotrichia leucophrys gambelii*. *Condor*, **68**, 476–487.

Kipp, F. (1943) Beziehungen zwischen dem Zug und der Brutbiologie der Vögel. *J. Ornithol.*, **91**, 144–153.

Kjellén, N. (1992) Differential timing of autumn migration between sex and age groups in raptors at Falsterbo, Sweden. *Ornis Scand.*, **23**, 420–434.

Klaassen, M., Kersten, M. and Ens, B.J. (1990) Energetic requirements for maintenance and premigratory body mass gain of waders wintering in Africa. *Ardea*, **78**, 209–220.

Klein, H., Berthold, P. and Gwinner, E. (1973) Der Zug europäischer Garten- und Mönchsgrasmücken (*Sylvia borin* und *S. atricapilla*). *Vogelwarte*, **27**, 73–134.

Kok, O.B., Ee C.A. van and Nel, D.G. (1991) Daylength determines departure date of the spotted flycatcher *Muscicapa striata* from its winter quarters. *Ardea*, **79**, 63–66.

Kolunen, H. and Peiponen, V.A. (1991) Delayed autumn migration of the swift

*Apus apus* from Finland in 1986. *Ornis Fenn.*, **68**, 81–92.

Kramer, G. (1949) Über Richtungstendenzen bei der nächtlichen Zugunruhe gekäfigter Vögel. In *Ornithologie als Biologische Wissenschaft* (eds E. Mayr and E. Schüz), Winter-Universitätsverlag, Heidelberg, pp. 269–283.

Kramer, G. (1953) Wird die Sonnenhöhe bei der Heimfindeorientierung verwertet? *J. Ornithol.*, **94**, 201–219.

Krapu, G.L. and Johnson, D.H. (1990) Conditioning of sandhill cranes during fall migration. *J. Wildl. Manage.*, **54**, 234–238.

Kreithen, M.L. (1978) Sensory mechanisms for animal orientation – can any new ones be discovered? In *Animal Migration, Navigation, and Homing* (eds K. Schmidt-Koenig and W.T. Keeton), Springer, Berlin, Heidelberg, New York, pp. 25–34.

Kreithen, M. L. and Keeton, W.T. (1974) Detection of changes in atmospheric pressure by the homing pigeon, *Columba livia*. *J. Comp. Physiol.*, **89**, 73–82.

Kriner, E. and Schwabl, H. (1991) Control of winter song and territorial aggression of female robins (*Erithacus rubecula*) by testosterone. *Ethology*, **87**, 37–44.

Kshatriya, M. and Blake, R.W. (1992) Theoretical model of the optimum flock size of bird flying in formation. *J. theor. Biol.*, **157**, 135–174.

Kuenzel, W.J. and Helms, C.W. (1967) Obesity produced in a migratory bird by hypothalamic lesions. *BioScience*, **17**, 395–396.

Kumar, V., Kumar, B.S., Singh, B.P. and Sarkar, A. (1991) A common functional basis for the photoperiodic mechanism regulating reproductive and metabolic responses in the migratory redheaded bunting. *Period. Biol.*, **93**, 169–174.

Kuroda, N. (1964) Analysis of variations by sex, age, and season of body weight, fat and some body parts in the dusky thrush, wintering in Japan: A preliminary study. *Misc. Rep. Yamashina Inst. Ornithol.*, **4**, 91–104.

Kvist, A., Lindström, A. and Tulp, I. (1993) Excessive migratory fattening in a captive bluethroat *Luscinia s. svecica*. *Ornis Svecica*, **3**, 161–164.

Lack, D. (1943/44) The problem of partial migration. *Brit. Birds*, **37**, 122–130, 143–150.

Lack, D. (1954) *The Natural Regulation of Animal Numbers*. Clarendon Press, Oxford.

Lack, D. (1968) Bird migration and natural selection. *Oikos*, **19**, 1–9.

Landsborough-Thomson, A. (1964) *A New Dictionary of Birds*. Nelson, London, Edinburgh.

Lane, S.G. (1972) A review of the co-operative silvereye project. *Australian Bird Bander*, **10**, 3–6.

Langslow, D.R. (1979) Movements of Blackcaps ringed in Britain and Ireland. *Bird Study*, **26**, 239–252.

Larkin, R.P. (1991) Flight speeds observed with radar, a correction: slow 'birds' are insects. *Behav. Ecol. Sociobiol.*, **29**, 221–224.

Laursen, K. (1978) Interspecific relationship between some insectivorous passerine species, illustrated by their diet during spring migration. *Ornis Scand.*, **9**, 178–192.

Lavée, D. and Safriel, U. (1973) Utilization of an oasis by desert-crossing migrant birds. *Israel J. Zool.*, **23**, 219.

Lavée, D. and Safriel, U. (1989) The dilemma for crossing-desert migrants – stopover or skip a small oasis? *J. Arid Environ.*, **17**, 69–81.

Lavée, D., Safriel, U.N. and Meilijson, I. (1991) For how long do trans-Saharan migrants stop over at an oasis? *Ornis Scand.*, **22**, 33–44.

Leach, I.H. (1981) Wintering blackcaps in Britain and Ireland. *Bird Study*, **28**, 5–14.

Legg, J. (1780a) *A Discourse on the Emigration of British Birds 1.* Collins and Johnson, Salisbury.

Legg, J. (1780b) *A Discourse on the Emigration of British Birds 2.* J. Brumby Bookseller, London.

Lehikoinen, E. and Lindström, J. (1988) Chernobyl radiation not the cause of fatally delayed autumn migration of the swift *Apus apus* in 1986. *Ornis Fenn.*, **65**, 37–38.

Leibel, R.L. (1992) Fat as fuel and metabolic signal. *Nutrition Reviews*, **50**, 12–26.

Leisler, B. (1990) Selection and use of habitat of wintering migrants. In *Bird Migration: Physiology and Ecophysiology* (ed. E. Gwinner), Springer, Berlin, Heidelberg, New York, pp. 156–174.

Leisler, B. (1992) Habitat selection and coexistence of migrants and Afrotropical residents. *Ibis*, **134**, Suppl. 1, 77–82.

Leshem, Y. (1989) Following raptor migration from the ground, motorized glider and radar at a junction of three continents. In *Raptors in the Modern World* (eds B.-U. Meyburg and R.D. Chancellor), WWGBP, Berlin, London, Paris, pp. 43–52.

Levey, D.J. and Duke, G.E. (1992) How do frugivores process fruit? Gastrointestinal transit and glucose absorption in cedar waxwings (*Bombycilla cedrorum*). *Auk*, **109**, 722–730.

Levey, D.J. and Karasov, W.H. (1992) Digestive modulation in a seasonal frugivore, the American robin (*Turdus migratorius*). *Am. J. Physiol.*, **262**, 711–718.

Levey, D.J. and Stiles, F.G. (1992) Evolutionary precursors of long-distance migration: resource availability and movement patterns in neotropical landbirds. *Amer. Nat.*, **140**, 447–476.

Ley, H.W. (1985) Vergleichende Untersuchungen zur Ontologie des Habitatverhaltens der Heckenbraunelle (*Prunella modularis*) und des Teichrohrsängers (*Acrocephalus scirpaceus*). Dissertation Univ. Berlin.

Liechti, F. (1992) Flugverhalten nächtlich ziehender Vögel in Abhängigkeit von Wind und Topographie. Dissertation Univ. Basel.

Liechti, F. and Bruderer, B. (1986) Einfluß der lokalen Topographie auf nächtlich ziehende Vögel nach Radarstudien am Alpenrand. *Ornithol. Beob.*, **83**, 35–66.

Lindström, Å. (1989) Finch flock size and risk of hawk predation at a migratory stopover site. *Auk*, **106**, 225–232.

Lindström, Å. (1990) The role of predation risk in stopover habitat selection in migrating bramblings, *Fringilla montifringilla. Behav. Ecol.*, **2**, 102–105.

Lindström, A., (1991) Maximum fat deposition rates in migrating birds. *Ornis Scand.*, **22**, 12–19.

Lindström, Å., Hasselquist, D., Bensch, S. and Grahn, M. (1990) Asymmetric contests over resources for survival and migration: a field experiment with bluethroats. *Anim. Behav.*, **40**, 453–461.

Lindström, Å. and Alerstam, T. (1992) Optimal fat loads in migrating birds: a test of the time-minimization hypothesis. *Amer. Nat.*, **140**, 477–491.

Lindström, Å. and Piersma, T. (1993) Mass changes in migrating birds: the evidence for fat and protein storage re-examined. *Ibis*, **135**, 70–78.

Löhrl, H. (1959) Zur Frage des Zeitpunktes einer Prägung auf die Heimatregion beim Halsbandschnäpper (*Ficedula albicollis*). *J. Ornithol*, **100**, 132–140.

Lövei, G.L. (1989) Passerine migration between the Palearctic and Africa. In *Current Ornithology 6* (ed. D.M. Power), Plenum Press, New York, pp. 143–174.

Lofts, B. (1962) Cyclical changes in the interstitial and spermatogenetic tissue of migratory waders 'wintering' in Africa. *Proc. Zool. Soc. London*, **138**, 405–413.

Lofts, B., Murton, R.K. and Westwood, N.J. (1967) Experimental demonstration of a post-nuptial refractor period in the turtle dove *Streptopelia turtur*. *Ibis*, **109**, 352–358.

Long, M.E. (1991) Secrets of animal navigation. *J. Nat. Geogr. Soc.*, **179**, 70–99.

Loria, D.E. and Moore, F.R. (1990) Energy demands of migration on red-eyed vireos, *Vireo olivaceus*. *Behav. Ecol.*, **1**, 24–35.

Lucanus, F. von (1923) *Die Rätsel des Vogelzuges*. Beyer und Söhne, Langensalza.

Lundberg, P. (1988) The evolution of partial migration in birds. *Tree*, **3**, 172–175.

Lundgren, B.O. (1988) Catabolic enzyme activities in the pectoralis muscle of migratory and non-migratory goldcrests, great tits, and yellowhammers. *Ornis Scand.*, **19**, 190–194.

Lundgren, B.O. and Kiessling, K.-H. (1985) Seasonal variation in catabolic enzyme activities in breast muscle of some migratory birds. *Oecologia*, **66**, 468–471.

Lundgren, B.O. and Kiessling, K.-H. (1986) Catabolic enzyme activities in the pectoralis muscle of premigratory and migratory juvenile reed warblers *Acrocephalus scirpaceus* (Herm). *Oecologia*, **68**, 529–532.

Lundgren, B.O. and Kiessling, K.-H. (1988) Comparative aspects of fibre types, and capillary supply in the pectoralis muscle of some passerine birds with differing migratory behaviour. *J. Comp. Physiol. B. Biochem. Syst. Environ. Physiol.*, **158**, 165–173.

Luniak, M. and Mulsow, R. (1988) Ecological parameters in urbanization of the European blackbird. *Acta XIX Congr. Internat. Ornithol.*, Ottawa 1988, pp. 1787–1793.

Luniak, M., Mulsow, R. and Walasz, K. (1990) Urbanization of the European blackbird – expansion and adaptations of urban population. *Proc. Intern. Symp. Warszawa*, Jablona, 1986, pp. 187–198.

Lynch, J.F. (1989) Distribution of overwintering nearctic migrants in the Yucatan peninsula, I: General patterns of occurrence. *Condor*, **91**, 515–544.

MacNally, R.C. (1990) The roles of floristics and physiognomy in avian community composition. *Austral. J. Ecol.*, **15**, 321–327.

Mädlow, W. (1992) Zur Habitatwahl auf dem Wegzug rastender Kleinvögel in einer norddeutschen Uferzone. Diplomarbeit, Univ. Berlin.

Maho le, Y. (1990) Hypometabolism as an adaptation to live and breed in the cold. *Acta XX Congr. Internat. Ornithol.*, Christchurch 1990, Suppl., p. 303.

Maisonneuve, C. and Bédard, J. (1992) Chronology of autumn migration by greater snow geese. *J. Wildl. Manage.*, **56**, 55–62.

Marsh, R.L. (1981) Catabolic enzyme activities in relation to premigratory fattening and muscle hypertrophy in the gray catbird (*Dumetella carolinensis*). *J. Comp. Physiol.*, **141**, 417–423.

Marsh, R.L. (1983) Adaptations of the gray catbird *Dumetella carolinensis* to long distance migration: energy stores and substrate concentrations in plasma. *Auk*, **100**, 170–179.

Marsh, R.L. and Dawson, W.R. (1982) Substrate metabolism in seasonally acclimatized American Goldfinches. *Am. J. Physiol.*, **242**, 563–569.

Marshall, A.J. (1952a) The condition of the interstitial and spermatogenetic tissue of migratory birds on arrival in England in April and May. *Proc. Zool. Soc. London*, **122**, 287–295.

Marshall, A.J. (1952b) Non-breeding among arctic birds. *Ibis*, **94**, 310–333.

Marshall, A.J. and Serventy, D.L. (1957) On the post-nuptial rehabilitation of the avian testis tunic. *Emu*, **57**, 59–63.

Marshall, A.J. and Williams M.C. (1959) The pre-nuptial migration of the yellow wagtail (*Motacilla flava*) from latitude 0°04'N. *Proc. Zool. Soc. London*, **132**, 313–320.

Martin, B.P. (1987) *World Birds*. Guinness, Enfield.

Martin, G. (1990) *Birds by Night*. Poyser, London.

Martin, G.R. (1990) The visual problems of nocturnal migration. In *Bird Migration: Physiology and Ecophysiology* (ed. E. Gwinner), Springer, Berlin, Heidelberg, New York, pp. 185–197.

Martin, G.R. (1991) The sensory bases of nocturnal foraging in birds. *Acta XX Congr. Internat. Ornithol.*, Christchurch 1990, pp. 1130–1135.

Martineau, L. and Larochelle, J. (1988) The cooling power of pigeon legs. *J. Exp. Biol.*, **136**, 193–208.

Mascher, J.W. and Stolt, B.O. (1961) Lufttryckets inverkan på ortolana lansparvens (*Emberiza hortulana* L.) aktivitet under vårflyttningsdperioden. *Vår Fågelvärld*, **20**, 97–111.

Masman, D. and Klaassen, M. (1987) Energy expenditure during free flight in trained and free-living Eurasian kestrels (*Falco tinnunculus*). *Auk*, **104**, 603–616.

Masman, D., Dijkstra, C., Daan, S. and Bult, A. (1989) Energetic limitation of avian parental effort: field experiments in the kestrel (*Falco tinnunculus*). *J. evol. Biol.*, **2**, 435–455.

Mattocks, P.W. Jr (1976) The role of gonadal hormones in the regulation of the premigratory fat deposition in the white-crowned sparrow, *Zonotrichia leucophrys gambelii*. MS Thesis, Univ. Washington, Seattle.

Maynard Smith, J. (1993) *The Theory of Evolution*. Cambridge University Press, Cambridge, New York, Melbourne.

Mayr, E. (1926) Die Ausbreitung des Girlitz. *J. Ornithol.*, **74**, 571–671.

McGreal, R.D. and Farner, D.S. (1956) Premigratory fat deposition in the Gambel white-crowned sparrow: some morphologic and chemical observations. *North-West Sci.*, **30**, 12–23.

McMillan, J.P., Gauthreaux, S.A. and Helms, C.W. (1970) Spring migratory restlessness in caged birds: a circadian rhythm. *BioScience*, **20,** 1259–1260.

McNamara, J.M. and Houston, A.I. (1990) The value of fat reserves and the tradeoff between starvation and predation. *Acta Biotheor.*, **38**, 37–61.

McNeil, R. (1969) La détermination du contenu lipidique et la capacité de vol chez quelques espèces d'oiseaux de rivage (*Charadriidae* et *Scolopadicae*). *Can. J. Zool.*, **47**, 525–536.

McNeil, R. (1991) Nocturnality in shorebirds. *Acta XX Congr. Internat. Ornithol.*, Christchurch 1990, pp. 1098–1104.

McNeil, R. and Caideux, F. (1972) Numerical formulae to estimate flight range of some North American shorebirds from fresh weight and wing length. *Bird Banding*, **43**, 107–113.

Mead, C.J. (1983) *Bird Migration*. Newness, Feltham.

Mead, C.J. and Watmough, B.R. (1976) Suspended moult of trans-Saharan migrants in Iberia. *Bird Study*, **23**, 187–196.

Mees, G.F. (1974) The migration of the Tasmanian race of the silvereye. *Australian Bird Bander*, **12**, 51–54.

Meier, A.H., Ferrell, B.R. and Miller, L.J. (1980) Circadian components of the circannual mechanisms in the white-throated sparrow. *Acta XVII Congr. Internat. Ornithol.*, Berlin 1978, pp. 458–462.

Mench, J.A. (1991) Stress in birds. *Acta XX Congr. Internat. Ornithol.*, Christchurch 1990, pp. 1905–1914.

Merkel, F.W. (1956) Untersuchungen über tages- und jahresperiodische

Aktivitätsänderungen bei gekäfigten Zugvögeln. *Z. Tierpsychol.*, **13**, 278–301.

Merkel, F.W. (1958) Untersuchungen über tages- und jahresperiodische Änderungen im Energiehaushalt gekäfigter Zugvögel. *Z. vergl. Physiol.*, **41**, 154–178.

Merkel, F.W. (1966) The sequence of events leading to migratory restlessness. *Ostrich*, **Suppl. 6**, 239–248.

Merkel, F.W. (1978) Angle sense in painted quails: a parameter of geodetic orientation. In *Animal Migration, Navigation, and Homing* (eds K. Schmidt-Koenig and W.T. Keeton), Springer, Berlin, Heidelberg, New York, pp. 269–274.

Merkel, I. and Merkel, F.W. (1983) Zum Wandertrieb der Stare. *Luscinia*, **45**, 63–74.

Milinski, M. (1986) Vorteile des Lebens im Verband. *Ornithol. Beob.*, **83**, 257–265.

Miller, A.H. (1931) Systematic revision and natural history of the American shrikes. *Univ. California Publ. Zool.*, **38**, 11–242.

Mills, E.D. and Rogers, D.T. (1992) Ratios of neotropical migrant and neotropical resident birds in winter in a citrus plantation in central Belize. *J. Field Ornithol.*, **63**, 109–116.

Milne, H. and Robertson, F.W. (1965) Polymorphisms in egg albumen protein and behaviour in the Eider Duck. *Nature*, **205**, 367–369.

Möhring, F.J. (1992) Untersuchung über die saisonalen Veränderungen im Gewicht und Fett beim europäischen Star (*Sturnus vulgaris*). Diplomarbeit, Univ. Bielefeld.

Mönkkönen, M. (1992) Life history traits of palaearctic and nearctic migrant passerines. *Ornis Fenn.*, **69**, 161–172.

Mönkkönen, M., Helle, P. and Welsh, D. (1992) Perspectives on palaearctic and nearctic bird migration: comparisons and overview of life-history and ecology of migrant passerines. *Ibis*, **134**, 7–13.

Monge, C. and León-Velarde, F. (1991) Physiological adaptation to high altitude: oxygen transport in mammals and birds. *Physiol. Reviews*, **71**, 1135–1171.

Moore, F.R. (1990) Prothonotary warblers cross the Gulf of Mexico together. *J. Field Ornithol.*, **61**, 285–287.

Moore, F.R. (1991) Ecophysiological and behavioral response to energy demand during migration. *Acta XX Congr. Internat. Ornithol.*, Christchurch 1990, pp. 753–760.

Moore, F.R. and Kerlinger, P. (1987) Stopover and fat deposition by North American wood-warblers (*Parulinae*) following spring migration over the Gulf of Mexico. *Oecologia*, **74**, 47–54.

Moore, F.R., Kerlinger, P. and Simons, T.R. (1990) Stopover on a gulf coast barrier island by spring trans-gulf migrants. *Wilson Bull.*, **102**, 487–500.

Moore, F.R. and Simm, P.A. (1986) Risk-sensitive foraging by a migratory bird *Dendroica coronata*. *Experientia*, **42**, 1054–1056.

Moore, F.R. and Yong, W. (1991) Evidence of food-based competition among passerine migrants during stopover. *Behav. Ecol. Sociobiol.*, **28**, 85–90.

Moreau, R.E. (1961) Problems of Mediterranean-Sahara migration. *Ibis*, **103a**, 373–427, 580–623.

Moreau, R.E. (1972) *The Palaearctic-African Bird Migration Systems*. Academic Press, New York, San Francisco, London.

Morel, G.J. and Morel, M.-Y. (1992) Habitat use by Palaearctic migrant passerine birds in West Africa. *Ibis*, **134**, Suppl. 1, 83–88.

Morse, D.H. (1989) *American Warblers. An ecological and behavioral perspective.* Harvard University Press, Cambridge, London.

Morton, M.L. (1992) Effects of sex and birth date on premigration biology, migration schedules, return rates and natal dispersal in the mountain white-crowned sparrow. *Condor*, **94**, 117–133.

Morton, M.L., Wakamatsu, M.W., Pereyra, M.E. and Morton, G.A. (1991) Postfledging dispersal, habitat imprinting, and philopatry in a montane, migratory sparrow. *Ornis Scand.*, **22**, 98–106.

Mueller, H.C. and Berger, D.D. (1992) Behavior of migrating raptors: differences between spring and fall. *J. Raptor Res.*, **26**, 136–145.

Muller, R.E. (1976) Effects of weather on the nocturnal activity of whitethroated sparrows. *Condor*, **78**, 186–194.

Nachtigall, W. (1987) *Vogelflug und Vogelzug*. Rasch and Röhring, Hamburg, Zürich.

Nachtigall, W. (1990) Wind tunnel measurements of long time flights in relation to the energetics and water economy of migrating birds. In *Bird Migration: Physiology and Ecophysiology* (ed. E. Gwinner), Springer, Berlin, Heidelberg, New York, pp. 319–330.

Nachtigall, W. (1993) Biophysik des Vogelflugs. Aspekte der Energetik beim Langstreckenflug. *Z. angew. Math. Mech.*, **73**, 191–202.

Nachtigall, W. and Hirth, K.D. (1990) Core temperature relations of pigeons during prolonged wind tunnel flight. *Acta XX Congr. Internat. Ornithol.*, Christchurch 1990, Suppl., p. 265.

Naik, D.V. and George J.C. (1964) Certain cyclic histological changes in the testis of the migratory starling, *Sturnus roseus* (Linnaeus). *Pavo*, **2**, 48–54.

Napolitano, G.E. and Ackman, R.G. (1990) Anatomical distribution of lipids and their fatty acids in the semipalmated sandpiper *Calidris pusilla* L. from Shepody Bay, New Brunswick, Canada. *J. Exp. Mar. Biol. Ecol.*, **144**, 113–124.

Naumann, J.A. (1795–1817) *Naturgeschichte der Land- und Wasser-Vögel des nördlichen Deutschlands und angrenzender Länder*. Osterloh und Aue, Köthen.

Naumann, J.F. (1897–1905) *Naturgeschichte der Vögel Mitteleuropas*. Köhler, Gera.

Necker, R. (1991) Die zentralnervöse Verarbeitung der Information von Federn-rezeptoren bei der Taube: Anpassungen an den Vogelflug. *Verh. Dt. Zool. Ges.*, 356.

Newton, I. (1975) *Finches*. Collins, London.

Nice, M.M. (1933) Zur Naturgeschichte des Singammers. *J. Ornithol.*, **81**, 552–595.

Nice, M.M. (1934) Zur Naturgeschichte des Singammers. *J. Ornithol.*, **82**, 1–96.

Nice, M.M. (1937) Studies in the life history of the song sparrow. I. *Trans. Linnaean Soc. New York*, **4**, 1–247.

Niemi, G.J. (1985) Patterns of morphological evolution in bird genera of New World and Old World peatlands. *Ecology*, **66**, 1215–1228.

Nikolaus, G. and Pearson, D. (1991) The seasonal separation of primary and secondary moult in Palaearctic passerine migrants on the Sudan coast. *Ringing Migration*, **12**, 46–47.

Nilsson, L. and Pirkola, M.K. (1991) Migration pattern of Finnish bean geese *Anser fabalis*. *Ornis Svecica*, **1**, 69–80.

Nisbet, I.C.T. and Drury, W.H. (1968) Short-term effects of weather on bird migration: a field study using multivariate statistics. *Anim. Behav.*, **16**, 496–530.

Nisbet, I.C.T. and Lord Medway (1972) Dispersion, population ecology and migration of Eastern Great Reed Warblers *Acrocephalus orientalis* wintering in Malaysia. *Ibis*, **114**, 451–494.

Nolan, V. Jr and Ketterson, E.D. (1990a) Effect of long days on molt and autumn migratory state of site-faithful dark-eyed juncos held at their winter sites. *Wilson Bull.*, **102**, 469–479.

Nolan, V. Jr and Ketterson, E.D. (1990b) Timing of autumn migration and its relation to winter distribution in dark-eyed juncos. *Ecology*, **71**, 1267–1268.

Nolan, V. Jr and Ketterson, E.D. (1991) Experiments on winter-site attachment in young dark-eyed juncos. *Ethology*, **87**, 123–133.

Norberg, U.M. (1990) *Vertebrate Flight*. Springer, Berlin, Heidelberg, New York.

Norgren, R.B. Jr (1990) Neural basis of avian circadian rhythms. *Bird Behaviour*, **8**, 57–66.

Norman, S.C. (1987) Body weights of willow warblers during autumn migration within Britain. *Ringing Migration*, **8**, 73–82.

Norman, S.C. (1990) Factors influencing the onset of post-nuptial moult in willow warblers *Phylloscopus trochilus*. *Ringing Migration*, **11**, 90–100.

Norman, S.C. (1991) Suspended split-moult systems – an alternative explanation for some species of Palearctic migrants. *Ringing Migration*, **12**, 135–138.

North, P.M. (1988) A brief review of the (lack of) statistics of bird dispersal. *Acta Ornithol.*, **24**, 63–74.

Novoa, F.F., Rosenmann, M. and Bozinovic, F. (1991) Physiological responses of four passerine species to stimulated altitudes. *Comp. Biochem. Physiol.*, **99A**, 179–183.

Nowak, E. and Berthold, P. (1991) Satellite-tracking: a new method in orientation research. In *Orientation in Birds* (ed. P. Berthold), Birkhäuser, Boston, Berlin, pp. 86–104.

Nowak, M. (1990) An evolutionarily stable strategy may be inaccessible. *J. theor. Biol.*, **142**, 237–241.

Obst, B.S. (1990) Intestinal nutrient absorption in birds. *Acta XX Congr. Internat. Ornithol.*, Christchurch 1990, Suppl., pp. 272–273.

O'Connor, R.J. (1990) Some ecological aspects of migrants and residents. In *Bird Migration: Physiology and Ecophysiology* (ed. E. Gwinner), Springer, Berlin, Heidelberg, New York, pp. 175–182.

Odum, E.P. and Major, J.C. (1956) The effect of diet on photoperiod-induced lipid deposition in the white-throated sparrow. *Condor*, **58**, 222–228.

Oehme, H. (1990a) Time pattern of wing motion in birds during cruising flight and its importance for flight energetics. *Acta XX Congr. Internat. Ornithol.*, Christchurch 1990, Suppl., p. 266.

Oehme, H. (1990b) Habitatwahl und Lokomotion: Anpassungen im Bereich der Vorderextremität. Proc. Int. 100. DO-G Meeting, Current Topics Avian Biol., Bonn 1988, *J. Ornithol.*, Sonderh., pp. 359–364.

Oehme, H. (1991) Zur Energetik des Streckenfluges kleiner Vögel. *Mitt. Zool. Mus. Berlin*, **67**, Suppl. Ann. Ornithol., 15, 109–120.

Olsson, V. (1969) Die Expansion des Girlitzes (*Serinus serinus*) in Nordeuropa in den letzten Jahrzehnten. *Vogelwarte*, **25**, 147–156.

Ormerod, S.J. (1989) The influence of weather on the body mass of migrating swallows *Hirundo rustica* in South Wales. *Ringing Migration*, **10**, 65–74.

Ormerod, S.J. (1990a) Time of passage, habitat use and mass change of *Acrocephalus* warblers in a South Wales reedswamp. *Ringing Migration*, **11**, 1–11.

Ormerod, S.J. (1990b) Possible resource partitioning in pairs of *Phylloscopus* and *Acrocephalus* warblers during autumn migration through a South Wales reedswamp. *Ringing Migration*, **11**, 76–85.

Ornat, A.L. and Greenberg, R. (1990) Sexual segregation by habitat in migratory warblers in Quintana Roo, Mexico. *Auk*, **107**, 539–543.

Orr, J.L. (1986) Canada geese flying in formation with Sandhill cranes. *Bull. Oklahoma Ornithol. Soc.*, **19**, 12.

Owen, M. and Black, J.M. (1991) The importance of migration mortality in non-passerine birds. In *Bird Population Studies* (eds C.M. Perrins, J.-D. Lebreton and G.J.M. Hirons), Oxford University Press, Oxford, London, New York, pp. 360–372.

Owen, M. and Ogilvie M.A. (1979) Wing molt and weights of barnacle geese in Spitsbergen. *Condor*, **81**, 42–52.

Owen, M., Wells, R.L. and Black, J.M. (1992) Energy budgets of wintering barnacle geese: the effects of declining food resources. *Ornis Scand.*, **23**, 451–458.

Palmgren, P. (1944) Studien über die Tagesrhythmik gekäfigter Zugvögel. *Z. Tierpsychol.*, **6**, 44–86.

Pambour, B. (1990) Vertical and horizontal distribution of five wetland passerine birds during the postbreeding migration period in a reed-bed of the Camargue, France. *Ringing Migration*, **11**, 52–56.

Pankoke, S. and Holländer, R. (1991) Der Einfluß von Cellulose und symbiontischen Bakterien bei der prämigratorischen Fettzunahme der Ringelgänse (*Branta bernicla bernicla*). *Ökol. Vögel*, **13**, 227–236.

Papi, F. (1990) Olfactory navigation in birds. In *Orientation in Birds* (ed. P. Berthold), Experientia, **46**, 352–363.

Papi, F. (1991) Olfactory navigation. In *Orientation in Birds* (ed. P. Berthold), Birkhäuser, Basel, Boston, Berlin, pp. 52–85.

Parnell, J.F. (1969) Habitat relations of the *Parulidae* during spring migration. *Auk*, **86**, 505–521.

Pathak, V.K. and Binosana, R.K. (1989) Iopanoic acid prevents spring premigratory increase in body weight of redheaded bunting *Emberiza bruniceps*. *Indian J. Exp. Biol.*, **27**, 598–601.

Pathak, V.K. and Chandola, A. (1982a) Involvement of thyroid gland in the development of migratory disposition in the red-headed bunting *Emberiza bruniceps*. *Horm. Behav.*, **16**, 46–59.

Pathak, V.K. and Chandola, A. (1982b) Seasonal variations in extrathyroidal conversion of thyroxine to triiodthyronine and migratory disposition in red-headed bunting. *Gen. Comp. Endocrinol.*, **47**, 433–440.

Patten, K.C. van and Price, T. (1990) Habitat choice in captive arctic warblers. *Auk*, **107**, 434–437.

Payevsky, V.A. (1990) Age and sex patterns in the territorial distribution of migrating birds. *Comm. Baltic Comm. Study Bird Migr.*, **23**, 46–54.

Payne, R.B. (1972) Mechanisms and control of molt. In *Avian Biology 2* (eds D.S. Farner and J.R. King), Academic Press, New York, San Francisco, London, pp. 103–155.

Payne, R.B. (1991) Natal dispersal and population structure in a migratory songbird, the indigo bunting. *Evolution*, **45**, 49–62.

Pearson, D.J. and Lack, P.C. (1992) Migration patterns and habitat use by passerine and near-passerine migrant birds in eastern Africa. *Ibis*, **134**, Suppl. 1, 89–98.

Péczely, P. (1976) Etude circanuelle de la fonction corticosurrènalienne chez les espèces de passereaux migrants et non-migrants. *Gen. Comp. Endocrinol.*, **30**, 1–11.

Pengelly, E.T. and Asmundson, S.M. (1971) Annual biological clocks. *Scientific American*, **224**, 72–78.

Pennycuick, C.J. (1969) The mechanics of bird migration. *Ibis*, **111**, 525–556.

Pennycuick, C.J. (1975) Mechanics of flight. In *Avian Biology 5* (eds D.S. Farner and J.R. King), Academic Press, New York, San Francisco, London, pp. 1–75.

Pennycuick, C.J. (1978) Fifteen testable predictions about bird flight. *Oikos*, **30**, 165–166.

Pennycuick, C. J.(1989) *Bird Flight Performance*. Oxford University Press, Oxford, London, New York.

Pennycuick, C.J., Fuller, M.R. and McAllister, L. (1989) Climbing performance of Harris' hawks (*Parabuteo unicinctus*) with added load: Implications for muscle mechanics and for radiotracking. *J. Exp. Biol.*, **142**, 17–29.

Perdeck, A.C. (1958) Two types of orientation in migrating starlings, *Sturnus vulgaris* L., and chaffinches, *Fringilla coelebs* L., as revealed by displacement experiments. *Ardea*, **46**, 1–37.

Perdeck, A.C. (1964) An experiment on the ending of autumn migration in starlings. *Ardea*, **52**, 133–139.

Pernau, F.A. von (1702) *Unterricht. Was mit dem lieblichen Geschöpff, denen Vögeln, auch ausser den Fang, nur durch die Ergründung deren Eigenschafften und Zahmmachung oder anderer Abrichtung man sich vor Lust und Zeit-Vertreib machen könne.* Neue Presse, Coburg, 1982.

Perrins, C.M. (1990) *The Illustrated Encyclopaedia of Birds. The definitive guide to birds of the world.* Headline, London.

Phillips, J.B. (1986) Two magnetoreception pathways in a migratory salamander. *Science*, **233**, 765–767.

Phillips, J.B. and Moore, F.R. (1992) Calibration of the sun compass by sunset polarized light patterns in a migratory bird. *Behav. Ecol. Sociobiol.*, **31**, 189–193.

Piersma, T. (1987) Hink, stap or sprong? Reisbeperkingen van arctische steltlopers door voedselzoeken, vetopbouw en vliegsnelheid. *Limosa*, **60**, 185–194.

Piersma, T. (1988) Breast muscle atrophy and constraints on foraging during the flightless period of wing moulting great crested grebes. *Ardea*, **76**, 96–106.

Piersma, T. (1990) Premigratory 'fattening' usually involves more than the deposition of fat alone. *Ringing Migration*, **11**, 113–115.

Piersma, T. (1994) Close to the Edge: *Energetic Bottlenecks and the Evolution of Migratory Pathways in Knots*. De Volharding, Amsterdam.

Piersma, T. and Davidson, N. (1992) The migrations and annual cycles of five subspecies of knots in perspective. *Wader Study Group Bull.*, **64**, Suppl., 187–197.

Piersma, T., Drent, R. and Wiersma, P. (1991) Temperate versus tropical wintering in the world's northernmost breeder, the knot: metabolic scope and resource levels restrict subspecific options. *Acta XX Congr. Internat. Ornithol.*, Christchurch 1990, pp. 761–772.

Piersma, T. and Jukema J. (1990) Budgeting the flight of a long-distance migrant: changes in nutrient reserve levels of bar-tailed godwits at successive spring staging sites. *Ardea*, **78**, 315–337.

Piersma, T. and Poot, M. (1993) Where waders may parallel penguins: spontaneous increase in locomotor activity triggered by fat depletion in a voluntarily fasting knot. *Ardea*, **81**, 1–8.

Piersma, T., Zwarts, L. and Bruggemann, J.H. (1990) Behavioural aspects of the departure of waders before long-distance flights: flocking, vocalizations, flight paths and diurnal timing. *Ardea*, **78**, 157–184.

Piper, W.H. (1990) Site tenacity and dominance in wintering white-throated sparrows *Zonotrichia albicollis* (Passeriformes: Emberizidae). *Ethology*, **85**, 114–122.

Piper, W.H. and Wiley, R.H. (1990) Correlates of range size in wintering white-throated sparrows, *Zonotrichia albicollis*. *Anim. Behav.*, **40**, 545–552.

Pittendrigh, C.S. (1960) Circadian rhythms and the circadian organization of living systems. *Cold Spring Harbor Symp. Quant. Biol.*, **25**, 159–184.

Pittendrigh, C.S. (1981) Circadian systems: entrainment. In *Handbook of Behavioral Neurobiology 4* (ed. J. Aschoff), Plenum, New York, pp. 95–124.

Place, A.R. (1990) The avian digestive system–an optimally designed plug-flow chemical reactor with recycle? *Acta XX Congr. Internat. Ornithol.*, Christchurch 1990, Suppl., p. 272.

Place, A.R. and Stiles, E.W. (1992) Living off the wax of the land: bayberries and yellow-rumped warblers. *Auk*, **109**, 334–345.

Pohl, H. (1971) Seasonal variation in metabolic functions of bramblings. *Ibis,* **113**, 185–193.

Pohl, H. (1992) How can a bird measure changes in photoperiod at 10 degrees or less north or south of the equator? *Abstr. Soc. Res. Biol. Rhythms Congr.*, Amelio Island, p. 35.

Pond, C.M. (1978) Morphological aspects and the ecological and mechanical consequences of fat deposition in wild vertebrates. *Ann. Rev. Ecol. Syst.*, **9**, 519–570.

Prescott, D.R.C. (1991) Winter distribution of age and sex classes in an irruptive migrant, the evening grosbeak (*Coccothraustes vespertinus*). *Condor*, **93**, 694–700.

Prince, P.A., Wood, A.G., Barton, T. and Croxall, J.P. (1992) Satellite tracking of wandering albatrosses (*Diomedea exulans*) in the South Atlantic. *Antarctic Science*, **4**, 31–36.

Prinzinger, R. (1990) Temperaturregulation bei Vögeln. I. Thermoregulatorische Verhaltensweisen. *Luscinia*, **46**, 255–302.

Prinzinger, R., Preßmar, A. and Schleucher, E. (1991) Body temperature in birds. *Comp. Biochem. Physiol.*, **99A**, 499–506.

Prop, J. and Vulink, T. (1992) Digestion by barnacle geese in the annual cycle: the interplay between retention time and food quality. *Funct. Ecol.*, **6**, 180–189.

Pruitt, N.L., Taylor, S.L. and Fernst, P.N. (1990) Membrane lipid composition and overwintering site of the wading bird *Calidris alpina*. *Comp. Biochem. Physiol.*, **97B**, 849–853.

Prys-Jones, R.P. (1991) The occurrence of biannual primary moult in passerines. *Bull. B.O.C.*, **111**, 150–152.

Pulido, F. and Berthold, P. (1991) Migratory behavior and its consequences for the genetic structure of blackcap (*Sylvia atricapilla*) populations. *Abstr. 3rd Congr. ESEB*, Debrecen, p. 88.

Putzig, P. (1938) Beobachtungen über Zugunruhe beim Rotkehlchen (*Erithacus rubecula*). *Vogelzug*, **9**, 10–14.

Quay, W.B. (1989) Insemination of Tennessee warblers during spring migration. *Condor*, **91**, 660–670.

Quine, D.B. and Kreithen, M.L. (1981) Frequency shift discrimination: can homing pigeons locate infrasounds by doppler shifts? *J. Comp. Physiol.*, **141**, 153–155.

Rabøl, J. (1990) Influence of weather, especially wind (direction), on the outdoor migratory activity of caged redstarts *Phoenicurus phoenicurus*, robins *Erithacus rubecula,* and garden warblers *Sylvia borin*. *Rep. Nat. Forest Nature Agency*, pp. 2–49.

Rabøl, J. (1993) The orientation systems of long-distance passerine migrants displaced in autumn from Denmark to Kenya. *Ornis Scand.*, **24**, 183–196.

Ralph, C.J. (1978) Disorientation and possible fate of young passerine coastal migrants. *Bird Banding*, **49**, 237–247.

Ramenofsky, M. (1990) Fat storage and fat metabolism in relation to migration. In *Bird Migration: Physiology and Ecophysiology* (ed. E. Gwinner), Springer, Berlin, Heidelberg, New York, pp. 214–231.

Ramenofsky, M., Wilson, W.H. and O'Reilly, K.M. (1990) Spring and autumnal migration: similar energetic requirements but separate physiological and behavioral strategies? *Acta XX Congr. Internat. Ornithol.*, Christchurch 1990, Suppl., p. 381.

Rankin, M.A. (1991) Endocrine effects on migration. *Amer. Zool.*, **31**, 217–230.

Rappole, J.H. (1988) Intra- and intersexual competition in migratory passerine birds during the non-breeding season. *Acta XIX Congr. Internat. Ornithol.*, Ottawa 1986, pp. 2308–2317.

Rappole, J.H., Morton, E.S., Lovejoy, T.E. and Ruos, J.L. (1983) Nearctic avian migrants in the neotropics. *US Dept. Interior, Fish Wildl. Service, Washington*.

Rappole, J.H. and Warner, D.W. (1976) Relationships between behavior, physiology and weather in avian transients at migration stopover sites. *Oecologia*, **26**, 193–212.

Rappole, J.H., Ramos, M.A. and Winker, K. (1989) Wintering wood thrush movements and mortality in southern Veracruz. *Auk*, **106**, 402–410.

Raveling, D.G. and Lefebre, E.A. (1967) Energy metabolism and theoretical flight range of birds. *Bird Banding*, **38**, 97–113.

Rayner, J.M.V. (1988) Form and function in avian flight. In *Current Ornithology* 5 (ed. R.F. Johnston), Plenum Press, New York, London, pp. 1–66.

Rayner, J.M.V. (1990) The mechanisms of flight and bird migration performance. In *Bird Migration: Physiology and Ecophysiology* (ed. E. Gwinner), Springer, Berlin, Heidelberg, New York, pp. 283–299.

Rees, E.C. (1987) Conflict of choice within pairs of Bewick's swans regarding their migratory movement to and from the wintering grounds. *Anim. Behav.*, **35**, 1685–1693.

Rees, E.C. (1989) Consistency in the timing of migration for individual Bewick's swans. *Anim. Behav.*, **38**, 384–393.

Reinikainen, A. (1937) The irregular migrations of the crossbill, *Loxia c. curvirostra*, and their relation to the cone-crop of the conifers. *Ornis Fenn.*, **14**, 55–64.

Repasky, R.R. (1991) Temperature and the northern distributions of wintering birds. *Ecology*, **72**, 2274–2285.

Rheinwald, G., Ogden, J. and Schulz, H. (eds) (1989) *Weißstorch*. Schriftenr. Dachverb. Dt. Avifaunisten, **10**.

Rice, C. (1970) Wintering blackcaps in Wiltshire. *Wiltshire Archeol. Nat. Hist. Mag.*, **65**, 12–15.

Richardson, W.J. (1990) Timing of bird migration in relation to weather: updated review. In *Bird Migration: Physiology and Ecophysiology* (ed. E. Gwinner), Springer, Berlin, Heidelberg, New York, pp. 78–101.

Richardson, W.J. (1991) Wind and orientation of migrating birds: a review. In *Orientation in Birds* (ed. P. Berthold), Birkhäuser, Basel, Boston, Berlin, pp. 226–249.

Richardson, R.D., Boswell, T., Weatherford, S.C., Wingfield, J.C. and Woods, S.C. (1992) Central and peripheral regulators of food intake in the white-crowned sparrow. *Appetite*, **19**, 212.

Riddiford, N. (1990) Tree pipit with suspended or arrested moult. *Ringing Migration*, **11**, 104.

Ridgill, S.C. and Fox, A.D. (1990) Cold weather movements of waterfowl in western Europe. *Internat. Waterfowl and Wetlands Research Bureau, IWRB News, Special publ. No. 13.*

Ringleben, H. (1956) Kanadagänse (*Branta canadensis*) in Deutschland. *Ornithol. Mitt.*, **8**, 185–187.

Rio, C.M. del, Baker, H.G. and Baker, I. (1992) Ecological and evolutionary implications of digestive processes: bird preferences and the sugar constituents of floral nectar and fruit pulp. *Experientia*, **48**, 544–551.

Rio, C.M. del and Karasov, W.H. (1990) Digestion strategies in nectar and fruit-eating birds and the sugar composition of plan rewards. *Amer. Nat.*, **136**, 618–637.

Ritchison, G., Belthoff, J.R. and Sparks, E.J. (1992) Dispersal restlessness: evidence for innate dispersal by juvenile eastern screech-owls? *Anim. Behav.*, **4**, 57–65.

Roby, D.D. (1991) A comparison of two noninvasive techniques to measure total body lipid in live birds. *Auk*, **108**, 509–518.

Rodda, G.H. and Phillips, J.B. (1992) Navigational systems develop along similar lines in amphibians, reptiles, and birds. *Ethol. Ecol. Evol.*, **4**, 43–51.

Rogers, D.T. Jr and Odum, E.P. (1966) A study of autumnal postmigrant weights and vernal fattening of North American migrants in the tropics. *Wilson Bull.*, **78**, 415–433.

Rogers, C.M., Theimer, T.L., Nolan, V. Jr and Ketterson, E.D. (1989) Does dominance determine how far dark-eyed juncos, *Junco hyemalis*, migrate into their winter range? *Anim. Behav.*, **37**, 498–506.

Rohwer, S. and Johnson, M.S. (1992) Scheduling differences of molt and migration for Baltimore and Bullock's orioles persist in a common environment. *Condor*, **94**, 992–994.

Root, T. (1988) Energy constraints on avian distributions and abundances. *Ecology*, **69**, 330–339.

Root, T. (1989) Energy constraints on avian distributions: a reply to Castro. *Ecology*, **70**, 1183–1185.

Rösner, H.-U. (1990) Sind Zugmuster und Rastplatzansiedlung des Alpenstrandläufers (*Calidris alpina alpina*) abhängig vom Alter? *J. Ornithol.*, **131**, 121–139.

Rowan, W. (1925) Relation of light to bird migration and developmental changes. *Nature*, **115**, 494–495.

Rowan, W. (1926) On photoperiodism, reproductive periodicity, and the annual migrations of birds and certain fishes. *Proc. Boston Soc. Nat. Hist.*, **38**, 147–189.

Rowan, W. (1931) *The Riddle of Migration.* Williams and Wilkins, Baltimore.

Rowan, W. (1946) Experiments in bird migration. *Trans. Roy. Soc. Can.*, **Sect. 5**, 123–135.

Rowan, W. and Batrawi, A.M. (1939) Comments on the gonads of some European migrants collected in East Africa immediately before their spring departure. *Ibis*, **14**, 58–65.

Rüttiger, L. and Schmidt-Koenig, K. (1991) Radiopeilung und Flugweganalyse heimkehrender Brieftauben: Umstimmung der inneren Uhr und Einfluß der Erfahrung. *Verh. Dt. Zool. Ges.*, 357–358.

Ruiz, G.M., Connors, P.G., Griffin, S.E. and Pitelka, F.A. (1989) Structure of a wintering dunlin population. *Condor*, **91**, 562–570.

Russell, R.B. (1991) Nocturnal flight by migrant 'diurnal' raptors. *J. Field Ornithol.*, **62**, 505–508.

Rutschke, E. (1990) Zur Etho-Ökologie einer Wildpopulation der Graugans (*Anser anser*). Proc. Int. 100. DO-G Meeting, Current Topics Avian Biol.,

# 332    References

Bonn 1988, *J. Ornithol.*, Sonderh., pp. 365–371.

Rutschke, E. (1992) *Die Wildschwäne Europas*. Deutscher Landwirtschaftsverlag Berlin GmbH.

Ryzhanovskii, V.N. (1987) Relationship of the postnuptial molt and breeding and migration in passerines in the subarctic. *Ekologiya*, **3**, 31–36.

Saarela, S., Keith, J.S., Hohtola, E. and Trayhurn, P. (1991) Is the 'mammalian' brown fat-specific mitochondrial uncoupling protein present in adipose tissues of birds? *Comp. Biochem. Physiol.*, **100B**, 45–49.

Safriel, U.N. and Lavée, D. (1988) Weight changes of cross-desert migrants at an oasis – do energetic considerations alone determine the length of stopover? *Oecologia*, **76**, 611–619.

Saitu, T. (1991) Comparison of breeding success between residents and immigrants in the great tit. *Acta XX Congr. Internat. Ornithol.*, Christchurch 1990, pp. 1196–1203.

Salomonsen, F. (1955) Evolution and bird migration. *Acta XI Congr. Internat. Ornithol.*, Basel 1954, pp. 337–339.

Salomonsen, F. (1967) *Fugletraekket ogdets gader*. Munksgaard, Kopenhagen.

Sandberg, R. (1991) Sunset orientation of robins, *Erithacus rubecula*, with different fields of sky vision. *Behav. Ecol. Sociobiol.*, **28**, 77–83.

Sandell, M. and Smith, H.G. (1991) Dominance, prior occupancy, and winter residency in the great tit (*Parus major*). *Behav. Ecol. Sociobiol.*, **29**, 147–152.

Savard, R., Ramenofsky, M. and Greenwood M.R.C. (1991) A north-temperate migratory bird: a model for the fate of lipids during exercise of long duration. *Can. J. Physiol. Pharmacol.*, **69**, 1443–1447.

Scanes, C.G., Jallageas, M. and Assenmacher, M. (1980) Seasonal variations in the circulating concentrations of growth hormone in male Pekin ducks (*Anas platyrhynchos*) and teal (*Anas crecca*): correlations with thyroidal function. *Gen. Comp. Endocrinol.*, **41**, 76–89.

Schaefer, G.W. (1968) Energy requirements of migratory flight. *Ibis*, **110**, 413–414.

Schindler, J., Berthold, P. and Bairlein, F. (1981) Über den Einflußr simulierter Wetterbedingungen auf das endogene Zugzeitprogramm der Gartengrasmücke *Sylvia borin*. *Vogelwarte*, **31**, 14–32.

Schmidt, R. (1989) Änderungen im Zugverhalten des Kormorans (*Phalacrocorax carbo*) im Zusammenhang mit seinem Bestandsanstieg. *Beitr. Vogelk.*, **35**, 219–221.

Schmidt-Koenig, K. (1975) *Migration and Homing in Animals*. Springer, Berlin, Heidelberg, New York.

Schmidt-Koenig, K. (1980) *Das Rätsel des Vogelzugs*. Hoffmann and Campe, Hamburg.

Schmidt-Koenig, K. (1991) On maps and compasses in homing pigeons. *Verh. Dt. Zool. Ges.*, **84**, 125–133.

Schmidt-Koenig, K., Ganzhorn, J.U. and Ranvaud, R. (1991) The suncompass. In *Orientation in Birds* (ed. P. Berthold), Birkhäuser, Basel, Boston, Berlin, pp. 1–15.

Schöps, M. (1991) Der Einfluß von Infraschall auf das Orientierungssystem von Brieftauben. *Verh. Dt. Zool. Ges.*, 361–362.

Schüz, E., Berthold, P., Gwinner, E. and Oelke, H. (1971) *Grundriß der Vogelzugskunde*. Parey, Berlin, Hamburg.

Schwabl, H. (1983) Ausprägung und Bedeutung des Teilzugverhaltens einer südwestdeutschen Population der Amsel *Turdus merula*. *J. Ornithol.*, **124**, 101–116.

Schwabl, H., Bairlein, F. and Gwinner, E. (1991) Basal and stress-induced corticosterone levels of garden warblers, *Sylvia borin*, during migration. *J. Comp. Physiol. B*, **161,** 576–580.

Schwabl, H. and Farner, D.S. (1989a) Dependency on testosterone of photo-periodically-induced vernal fat deposition in female white-crowned sparrows. *Condor*, **91**, 108–112.

Schwabl, H. and Farner, D.S. (1989b) Endocrine and environmental control of vernal migration in male white-crowned sparrows, *Zonotrichia leucophrys gambelii. Physiol. Zool.*, **62**, 1–10.

Schwabl, H., Gwinner, E., Benvenuti, S. and Ioalè, P. (1991) Exposure of dunnocks (*Prunella modularis*) to their previous wintering site modifies autumnal activity pattern: Evidence for site recognition? *Ethology*, **88**, 35–45.

Schwabl, H., Schwabl-Benzinger, I. and Farner, D.S. (1984a) Relationship between migratory disposition and plasma levels of gonadotrophins and steroid hormones in the European blackbird. *Naturwiss.*, **71**, 329–330.

Schwabl, H., Wingfield, J.C. and Farner, D.S. (1984b) Endocrine correlates of autumnal behavior in sedentary and migratory individuals of a partially migratory population of the European blackbird (*Turdus merula*). *Auk*, **101**, 499–507.

Schwabl, H., Schwabl-Benzinger, I., Goldsmith, A.R. and Farner, D.S. (1988) Effects of ovariectomy on long day-induced premigratory fat deposition, plasma levels of luteinizing hormone and prolactin, and molt in white-crowned sparrows, *Zonotrichia leucophrys gambelii. Gen. Comp. Endocrinol.*, **71**, 398–405.

Schwabl, H. and Silverin, B. (1990) Control of partial migration and autumnal behavior. In *Bird Migration: Physiology and Ecophysiology* (ed. E. Gwinner), Springer, Berlin, Heidelberg, New York, pp. 144–155.

Schwanke, W. and Rutschke, E. (1990) Strukturunterschiede in Flugrufen der Saatgans (*Anser fabalis*) bei Nebel und bei guter Sicht. *Beitr. Vogelk.*, **36**, 171–172.

Seibert, H.C. (1949) Differences between migrant and non-migrant birds in food and water intake at various temperatures and photoperiods. *Auk*, **66**, 128–153.

Senar, J.C., Burton, P.J.K. and Metcalfe, N.B. (1992) Variation in the nomadic tendency of a wintering finch *Carduelis spinus* and its relationship with body condition. *Ornis Scand.*, **23**, 63–72.

Senar, J.C., Copete, J.L. and Metcalfe, N.B. (1990) Dominance relationships between resident and transient wintering siskins. *Ornis Scand.*, **21**, 129–132.

Sharrock, J.T.R. and Sharrock, E.M. (1976) *Rare birds in Britain and Ireland*. Poyser, Berkhamsted.

Sherry, T.W. (1990) When are birds dietarily specialized? Distinguishing eco-logical from evolutionary approaches. *Std. Avian Biol.*, **13**, 337–352.

Silverin, B., Viebke, P.A. and Westin, J. (1989) Hormonal correlates of migration and territorial behavior in juvenile willow tits during autumn. *Gen. Comp. Endocrinol.*, **75**, 148–156.

Simms, E. (1985) *British Warblers*. Collins, London.

Skagen, S.K., Knopf, F.L. and Cade, B.S. (1993) Estimation of lipids and lean mass of migrating sandpipers. *Condor*, **95**, 944–956.

Smith, H.G. and Nilsson, J.-Å. (1987) Intraspecific variation in migratory pat-tern of a partial migrant, the blue tit (*Parus caeruleus*): an evaluation of different hypotheses. *Auk*, **104**, 109–115.

Sniegowski, P.D., Ketterson, E.D. and Nolan, V. Jr (1988) Can experience alter the avian annual cycle? Results of migration experiments with indigo buntings. *Ethology*, **79**, 333–341.

Snow, B. and Snow, D. (1988) *Birds and Berries*. Poyser, Calton.

Sokolov, L.V. (1990) Territorial control of stopping migratory flight. *Comm. Baltic Comm. Study Bird Migr.*, **23**, 30–45.

Sotthibandhu, S. and Baker, R.R. (1979) Celestial orientation by the large yellow underwing moth, *Noctua pronuba* L. *Anim. Behav.*, **27**, 786–800.

Spina, F. and Massi, A. (1992) Post-nuptial moult and fat accumulation of the ashy-headed wagtail (*Motacilla flava cinereocapilla*) in northern Italy. *Vogelwarte*, **36**, 211–220.

Spina, F., Piacentini, D. and Frugis, S. (1985) Vertical distribution of blackcap (*Sylvia atricapilla*) and garden warbler (*Sylvia borin*) within the vegetation. *J. Ornithol.*, **126**, 431–434.

Stetson, M.H. and Erickson, J.E. (1972) Hormonal control of photoperiodically induced fat deposition in white-crowned sparrows. *Gen. Comp. Endocrinol.*, **19**, 355–362.

Stewart, A.G. (1978) Swans flying at 8000 metres. *Brit. Birds*, **71**, 459–460.

Stiles, E.W. (1982) Fruit flags: two hypotheses. *Amer. Nat.*, **120**, 500–509.

Stiles, E.W. (1993) The influence of pulp lipids on fruit preference by birds. *Vegetatio*, **107/108**, 227–235.

Stolt, B.-O. (1969) Temperature and air pressure experiments on activity in passerine birds with notes on seasonal and circadian rhythms. *Zool. Bidrag, Uppsala*, **38**, 175–201.

Stolt, B.-O. (1977) Activity patterns and migration in some passerine birds. *Acta Univ. Upps. Abstr. Upps. Diss. Sci.*, **397**, 1–12.

Streif, M. (1991) Analyse der Biotoppräferenzen auf dem Wegzug in Süddeutschland rastender Kleinvögel. *Ornithol. Jahresb. Bad.-Württemb.*, **7**, 1–132.

Streng, A. and Wallraff, H.G. (1992) Attempts to determine the roles of visual and olfactory inputs in initial orientation and homing of pigeons over familiar terrain. *Ethology*, **91**, 203–219.

Stresemann, E. (1934) Aves. In *Handbuch der Zoologie 7* (eds W. Kükenthal and T. Krumbach), De Gruyter, Berlin, Leipzig.

Stresemann, E. (1951) *Die Entwicklung der Ornithologie von Aristoteles bis zur Gegenwart*. Peters, Berlin.

Stresemann, E. (1967) Inheritance and adaptation in moult. *Proc. XIV Internat. Ornithol. Congr.*, Oxford 1966, pp. 75–80.

Stresemann, E. and Stresemann, V. (1966) Die Mauser der Vögel. *J. Ornithol.*, **107**, Sonderh.

Strikwerda, T.S., Fuller, M.R., Seegar, W.S., Howey, P.W. and Black, H.D. (1986) Bird-borne satellite transmitter and location program. *J. Hopkins APL Tech. Digest*, **7**, 203–208.

Suarez, R.K. (1992) Hummingbird flight: sustaining the highest mass-specific metabolic rates among vertebrates. *Experientia*, **48**, 565–570.

Summers, R.W. and Waltner, M. (1979) Seasonal variations in the mass of waders in southern Africa with special reference to migration. *Ostrich*, **50**, 21–37.

Suter, W. and Eerden, M.R. van (1992) Simultaneous mass starvation of wintering diving ducks in Switzerland and The Netherlands: a wrong decision in the right strategy? *Ardea*, **80**, 229–242.

Sutherland, W.J. (1988) The heritability of migration. *Nature*, **334**, 471–472.

Svazas, S. (1990) Flocking behaviour of nocturnal migrants. *Acta Ornithol. Lituanica, Vilnius*, **2**, 36–55.

Swain, S.D. (1992) Flight muscle catabolism during overnight fasting in a passerine bird, *Erymophila alpestris*. *J. Comp. Physiol. B*, **162**, 383–392.

Swann, R.L. and Baillie, S.R. (1979) The suspension of moult by trans-Saharan migrants in Crete. *Bird Study*, **26**, 55–58.

Swanson, D.L. (1990) Seasonal variations of vascular oxygen transport in the dark-eyed junco. *Condor*, **92**, 62–66.

Swingland, I.R. (1984) Intraspecific differences in movement. In *The Ecology of Animal Movement* (eds I.R. Swingland and P.J. Greenwood), Clarendon, Oxford, pp. 102–115.

Szekely, T., Sozou, P.D. and Houston, A.I. (1991) Flocking behaviour of passerines: a dynamic model for the non-productive season. *Behav. Ecol. Sociobiol.*, **28**, 203–213.

Tamm, S. (1980) Bird orientation: single homing pigeons compared with small flocks. *Behav. Ecol. Sociobiol.*, **7**, 319–322.

Terborgh, J. (1989) *Where Have All the Birds Gone?* Princeton University Press, Princeton.

Terrill, S.B. (1987) Social dominance and migratory restlessness in the dark-eyed junco (*Junco hyemalis*). *Behav. Ecol. Sociobiol.*, **21**, 1–11.

Terrill, S.B. (1990) Ecophysiological aspects of movements by migrants in the wintering quarters. In *Bird Migration: Physiology and Ecophysiology* (ed. E. Gwinner), Springer, Berlin, Heidelberg, New York, pp. 130–143.

Terrill, S.B. (1991) Evolutionary aspects of orientation and migration in birds. In *Orientation in Birds* (ed. P. Berthold), Birkhäuser, Basel, Boston, Berlin, pp. 180–201.

Terrill, S.B. and Able, K.P. (1988) Bird migration terminology. *Auk*, **105**, 205–206.

Terrill, S.B. and Berthold, P. (1989) Experimental evidence for endogenously programmed differential migration in the blackcap (*Sylvia atricapilla*). *Experientia*, **45**, 207–209.

Terrill, S.B. and Berthold, P. (1990) Ecophysiological aspects of rapid population growth in a novel migratory blackcap (*Sylvia atricapilla*) population: an experimental approach. *Oecologia*, **85**, 266–270.

Thalau, H.-P. and Wiltschko, W. (1987) Einflüsse des Futterangebots auf die Tagesaktivität von Trauerschnäppern (*Ficedula hypoleuca*) auf dem Herbstzug. *Cour. Forsch.-Inst. Senckenberg*, **97**, 67–73.

Thaler-Kottek, E. (1990) *Die Goldhähnchen. Winter- und Sommergoldhähnchen* Regulus regulus, Regulus ignicapillus. Die Neue Brehm-Bücherei, Ziemsen, Wittenberg Lutherstadt.

Thienemann, J. (1927) *Rossitten*. Neumann, Neudamm.

Thomson, A.L. (1950) Factors determining the breeding seasons of birds. *Ibis*, **92**, 173–184.

Ticehurst, C.B. (1922) The Ibis. *Ibis*, **4**, 605–662.

Tögel, A. and Wiltschko, R. (1992) Detour experiments with homing pigeons: information obtained during the outward journey is included in the navigational process. *Behav. Ecol. Sociobiol.*, **31**, 73–79.

Torre-Bueno, J.R. (1978) Evaporative cooling and water balance during flight in birds. *J. Exp. Biol.*, **75**, 231–236.

Tsvelykh, A.N. (1990) Effect of the head and tail wind on the bird flight. *Zool. Zhurnal*, **69**, 82–92.

Tsvelykh, A.N. and Goroshko, O.A. (1991) Age dimorphism in the wing shape of *Hirundo rustica*. *Zool. Zhurnal*, **70**, 87–90.

Tucker, V.A. (1973) Bird metabolism during flight: Evaluation of a theory. *J. Exp. Biol.*, **58**, 689–709.

Tucker, V.A. (1993) Gliding birds: reduction of induced drag by wing tip slots between the primary feathers. *J. Exp. Biol.*, **180**, 285–310.

Tucker, V.A. and Heine, C. (1990) Aerodynamics of gliding flight in a harris' hawk, *Parabuteo unicinctus*. *J. Exp. Biol.*, **149**, 469–489.

Turek, F.W. and Gwinner, E. (1982) Role of hormones in the circadian organization of vertebrates. In *Vertebrate Circadian Systems* (eds J. Aschoff, S. Daan and G.A. Groos), Springer, Berlin, Heidelberg, New York, pp. 173–182.

Tyrberg, T. (1986) Cretaceous birds – a short review of the first half of avian history. *Verh. Ornithol. Ges. Bayern*, **24**, 249–275.

Underhill, L.G., Prys-Jones, R.P., Dowsett, R.J., Herroelen, P., Johnson, D.N., Lawn, M.R., Norman, S.C., Pearson, D.J. and Tree, A.J. (1992) The biannual primary moult of willow warblers *Phylloscopus trochilus* in Europe and Africa. *Ibis*, **134**, Suppl. 1, 286–297.

Vallyathan, N.V. and George, J.C. (1969) Effect of exercise on lipid levels in the pigeon. *Arch. Int. Physiol. Biochem.*, **77**, 863–868.

Viehmann, W. (1982) Orientierungsverhalten von Mönchsgrasmücken (*Sylvia atricapilla*) im Frühjahr in Abhängigkeit der Wetterlage. *Vogelwarte*, **31**, 452–457.

Wade, A.J., Marbut, M.M. and Round, J.M. (1990) Muscle fibre type and aetiology of obesity. *Lancet*, **335**, 805–808.

Wagner, H.O. (1930) Über Jahres- und Tagesrhythmus bei Zugvögeln. I. Mitteilung. *Z. Vgl. Physiol.*, **12**, 703–724.

Wagner, H.O. (1961) Beziehungen zwischen dem Keimdrüsenhormon Testosteron und dem Verhalten von Vögeln in Zugstimmung. *Z. Tierpsychol.*, **18**, 302–319.

Wagner, H.O. and Schildmacher, H. (1937) Über die Abhängigkeit des Einsetzens der nächtlichen Zugunruhe verfrachteter Vögel von der geographischen Breite. *Vogelzug*, **8**, 18–19.

Walasz, K. (1990) Experimental investigations on the behavioural differences between urban and forest blackbirds. *Acta Zool. Cracov.*, **33**, 235–271.

Walcott, C. (1991) Magnetic maps in pigeons. In *Orientation in Birds* (ed. P. Berthold), Birkhäuser, Basel, Boston, Berlin, pp. 38–51.

Waldvogel, J.A. (1989) Olfactory orientation by birds. In *Current Ornithol. 6* (ed. D.M. Power), Plenum Press, New York, London, pp. 269–321.

Walker, J.M. and Venables, W.A. (1990) Weather and bird migration. *Weather*, **45**, 47–56.

Wallraff, H.G. (1991a) Conceptual approaches to avian navigation systems. In *Orientation in Birds* (ed. P. Berthold), Birkhäuser, Basel, Boston, Berlin, pp. 128–165.

Wallraff, H.G. (1991b) Critical comments on the discussion about olfactory navigation in homing pigeons. *Verh. Dt. Zool. Ges.*, 368.

Walsberg, G.E. (1990) Problems inhibiting energetic analyses of migration. In *Bird Migration: Physiology and Ecophysiology* (ed. E. Gwinner), Springer, Berlin, Heidelberg, New York, pp. 413–421.

Walsberg, G.E. and Thompson, C.W. (1990) Annual changes in gizzard size and function in a frugivorous bird. *Condor*, **92**, 794–795.

Warkentin, I.G. and James, P.C. (1990) Dispersal terminology: changing definitions in midflight? *Condor*, **92**, 802–803.

Wassmann, R. (1989) Territoriales Verhalten des Pirols (*Oriolus oriolus*) auf dem Zuge. *Ökol. Vögel*, **11**, 283–285.

Weindler, P. and Wiltschko, W. (1991) Angeborene und erlernte Verhaltensmechanismen junger Gartengrasmücken, *Sylvia borin*, auf ihrem ersten Zug. *Verh. Dt. Zool. Ges.*, 369.

Weise, C.M. (1956) Nightly unrest in caged migratory sparrows under outdoors conditions. *Ecology*, **37**, 274–287.

Weise, C.M. (1967) Castration and spring migration in the white-throated sparrow. *Condor*, **69**, 49–68.

Wells, D.R. (1990) Migratory birds and tropical forest in the Sunda region. In *Biogeography and Ecology of Forest Bird Communities* (ed. A. Keast), SPB Academic Publ., The Hague, pp. 357–369.

Westman, B. (1990) Environmental effect on dominance in young great tits *Parus major*: a cross-fostering experiment. *Ornis Scand.*, **21**, 46–51.

Whittow, G.C. (1986) Regulation of body temperature. In *Avian Physiology* (ed. P.D. Sturkie), Springer, New York, pp. 221–252.

Wiedner, D.S., Kerlinger, P., Sibley, D.A., Holt, P., Hough, J. and Crossley, R. (1992) Visible morning flight of neotropical landbird migrants at Cape May, New Jersey. *Auk*, **109**, 500–510.

Wiley, R.H. (1990) Prior-residence and coat-tail effects in dominance relationships of male dark-eyed juncos, *Junco hyemalis*. *Anim. Behav.*, **40**, 587–596.

Williams, T.C. (1991) Constant compass orientation for north American autumnal migrants. *J. Field Ornithol.*, **62**, 218–225.

Williams, T.C. and Williams, J.M. (1978) Orientation of Transatlantic Migrants. In *Animal Migration, Navigation, and Homing* (eds K. Schmidt-Koenig and W.T. Keeton), Springer, Berlin, Heidelberg, New York, pp. 239–251.

Williams, T.C. and Williams, J.M. (1990a) Open ocean bird migration. *IEE Proc.*, **137**, 133–137.

Williams, T.C. and Williams, J.M. (1990b) The orientation of transoceanic migrants. In *Bird Migration: Physiology and Ecophysiology* (ed. E. Gwinner), Springer, Berlin, Heidelberg, New York, pp. 7–21.

Williams, T.C., Williams, J.M., Ireland, L.C. and Teal, J.M. (1977) Autumnal bird migration over the western North Atlantic Ocean. *Amer. Birds*, **31**, 251–267.

Williamson, K. (1961) The concept of 'cyclonic approach'. *Bird Migr.*, **1**, 235–240.

Willis, E.O. (1973) The behavior of occelated antbirds. *Smithsonian Contr. Zool.*, **144**, 1–57.

Willis, E.O. (1986) Vireos, wood warblers and warblers as ant followers. *Gerfaut*, **76**, 177–186.

Willson, M.F., Graff, D.A. and Whelan, C.J. (1990) Color preferences of frugivorous birds in relation to the colors of fleshy fruits. *Condor*, **92**, 545–555.

Wilson, D.S. (1990) Neuroanatomical studies of the flight muscles in the European starling (*Sturnus vulgaris*). *Amer. Zool.*, **30**, 74A.

Wilson, R.P., Wilson, M.-P.T., Link, R., Mempel, H. and Adams, N.J. (1991) Determination of movements of African penguins *Spheniscus demersus* using a compass system: dead reckoning may be an alternative to telemetry. *J. Exp. Biol.*, **157**, 557–564.

Wiltschko, R. (1990a) *Das Orientierungssystem der Vögel*. Habilitationsschr., Univ. Frankfurt.

Wiltschko, R. (1990b) How to return home - navigation in migratory birds. *Baltic Birds*, **5**, 242–257.

Wiltschko, R. (1991) The role of experience in avian navigation and homing. In *Orientation in Birds* (ed. P. Berthold), Birkhäuser, Basel, Boston, Berlin, pp. 250–269.

Wiltschko, R. (1992) Das Verhalten verfrachteter Vögel. *Vogelwarte*, **36**, 249–310.

Wiltschko, W. (1972) The influence of magnetic total intensity and inclination on directions preferred by migrating European robins (*Erithacus rubecula*). In *Animal Orientation and Navigation* (eds S.R. Galler, K. Schmidt-Koenig, G.J.

Jacobs and R.E. Belleville), NASA, Washington, pp. 459–570.

Wiltschko, W. and Wiltschko, R. (1991a) Magnetic orientation and celestial cues in migratory orientation. In *Orientation in Birds* (ed. P. Berthold), Birkhäuser, Basel, Boston, Berlin, pp. 16–37.

Wiltschko, W. and Wiltschko, R. (1991b) Der Magnetkompaß als Komponente eines komplexen Richtungsorientierungssystems. *Zool. Jb. Physiol.*, **95**, 437–446.

Wiltschko, W. and Wiltschko, R. (1991c) Orientation by the earth's magnetic field in migrating birds and homing pigeons. In *Effects of Atmospheric and Geophysical Variables in Biology and Medicine* (ed. H. Lieth), Progress in Biometeorology, **8**, SPB Academic Publishers, The Hague, pp. 31–43.

Wiltschko, W. and Wiltschko, R. (1992) Migratory orientation: magnetic compass orientation of garden warblers (*Sylvia borin*) after a simulated crossing of the magnetic equator. *Ethology*, **91**, 70–74.

Wiltschko, W., Munro, U., Ford, H. and Wiltschko, R. (1993) Red light disrupts magnetic orientation of migratory birds. *Nature*, **364**, 525–527.

Wingfield, J.C. and Farner, D.S. (1978) The annual cycle of plasma irLH and steroid hormones in feral populations of the white-crowned sparrow, *Zonotrichia leucophrys* gambelii. *Biol. Reprod.*, **19**, 1046–1056.

Wingfield, J.C. and Farner, D.S. (1993) The endocrinology of feral species. In *Avian Biology 9* (eds D.S. Farner, J.R. King and K.C. Parkes), Academic Press, New York, San Francisco, London, pp. 163–327.

Wingfield, J.C., Schwabl, H. and Mattocks, P.W. Jr (1990) Endocrine mechanisms of migration. In *Bird Migration: Physiology and Ecophysiology* (ed. E. Gwinner), Springer, Berlin, Heidelberg, New York, pp. 232–256.

Winkel, W. (1993) Zum Migrationsverhalten von Kohl- und Blaumeise. *Jber. Inst. Vogelforschung*, **1**, 9.

Winker, K., Rappole, J.H. and Ramos, M.A. (1990) Population dynamics of the wood thrush in southern Veracruz, Mexico. *Condor*, **92**, 440–460.

Winker, K. and Warner, D.W. (1991) Unprecedented stopover site fidelity in a Tennessee warbler. *Wilson Bull.*, **103**, 512–514.

Winkler, H. and Leisler, B. (1992) On the ecomorphology of migrants. *Ibis*, **134**, Suppl. 1, 21–28.

Witter, M.S. and Cuthill, I.C. (1993) The ecological costs of avian fat storage. *Phil. Trans. R. Soc. Lond.*, B, **340**, 73–92.

Wolfson, A. (1945) The role of the pituitary, fat deposition, and body weight in bird migration. *Condor*, **47**, 95–127.

Wood, B. (1992) Yellow wagtail *Motacilla flava* migration from West Africa to Europe: pointers towards a conservation strategy for migrants on passage. *Ibis*, **134**, 66–76.

Wunderle, J.M. and Waide, R.B. (1993) Distribution of overwintering nearctic migrants in the Bahamas and Greater Antilles. *Condor*, **95**, 904–933.

Wuorinen, J.D. (1992) Do arctic skuas *Stercorarius parasiticus* exploit and follow terns during the fall migration? *Ornis Fenn.*, **69**, 198–200.

Yesou, P. (1989) Site fidelity of migrating and overwintering cormorants *Phalacrocorax carbo*. *Oiseau Rev. Française Ornithol.*, **59**, 175–178.

Yokoyama, K. (1976) Hypothalamic and hormonal control of photoperiodically induced vernal functions in the white-crowned sparrow, *Zonotrichia leucophrys gambelii*. I. The effects of hypothalamic lesions and exogenous hormones. *Cell Tissue Res.*, **174**, 391–416.

Yokoyama, K. (1977) Hypothalamic and hormonal control of photoperiodically induced vernal functions in the white-crowned sparrow, *Zonotrichia leucophrys gambelii*. II. The effects of hypothalamic implantation of testosterone propionate. *Cell Tissue Res.*, **176**, 91–108.

Yong, W. and Moore, F.R. (1993) Relation between migratory activity and energetic condition among thrushes (*Turdinae*) following passage across the Gulf of Mexico. *Condor*, **95**, 934–943.

Young, B.E. (1991) Annual molts and interruption of the fall migration for molting in lazuli buntings. *Condor*, **93**, 236–250.

Yrjölä, R., Routasuo, P., Mikala, A., Mikkola, M. and Laurila, A. (1989) Rytikerttunen ei ole läskimaha. *Lintumies*, **24**, 132–133.

Zalakevicius, M. (1990) The theory of controlling seasonal bird migration. *Acta Ornithol. Lituanica*, **2**, 3–35.

Zamora, R. (1992) Seasonal variations in foraging behaviour and substrate use by the black redstart (*Phoenicurus ochruros*). *Rev. Ecol.*, **47**, 67–84.

Zink, G. (1962) Eine Mönchsgrasmücke (*Sylvia atricapilla*) zieht im Herbst von Oberösterreich nach Irland. *Vogelwarte*, **21**, 222–223.

Zink, G. (1973–85) *Der Zug europäischer Singvögel*. Vogelzug, Möggingen.

Zwarts, L. (1990) Increased prey availability drives premigration hyperphagia in whimbrels and allows them to leave the Banc d'Arguin, Mauritania, in time. *Ardea*, **78**, 279–300.

Zwarts, L., Blomert, A.-M. and Hupkes, R. (1990) Increase of feeding time in waders preparing for spring migration from the Banc d'Arguin, Mauritania. *Ardea*, **78**, 237–256.

Zwarts, L., Blomert, A.-M. and Wanink, J.H. (1992) Annual and seasonal variation in the food supply harvestable by knot *Calidris canutus* staging in the Wadden Sea in late summer. *Mar. Ecol. Progr. Ser.*, **83**, 129–139.

Zwarts, L., Ens, B.J., Kersten, M. and Piersma, T. (1990) Moult, mass and flight range of waders ready to take off for long-distance migrations. *Ardea*, **78**, 339–364.

# Species index

Page numbers in **bold** refer to figures.

# COMMON NAMES

# Subject index

Page numbers in *italic* refer to tables and **bold** to figures

Ice-fields   182
Inclination compass   252, 296
Infrasound   254–5, 258
Infundibular nucleus   99, 100, 287
Inhalation   145
Initiation factor model   207
Insemination   243
Instantaneous migration speed   27
Instinct birds   26
Insulin   98
Intermediate/intermittent migratory
    movements   5–6
Interrupted moult   237, 238–9, 295
Iopanoic acid   94, 96
IRI   23
Irruptions   6, 25, 199
  and food availability   207–9
Isepipteses   191
Isochrones   191
Isochronuous lines   191
Isophenes   191

Juvenile moult   233–6, 237
  genetic control   79
Juveniles
  choice of migratory direction
    247–9
  development
    and migration   233–6
    and photoperiod   86–7
  fat deposition   121
  misorientation   269
  mortality   11

Landfalls   197–8
Landmarks   249, 295
  acoustic   258
Leading-line effects   249, 272, 295
Leap-frog migration   8
LH   **94**, 241
Lift:drag ratio   118
Light intensity   88–9
Lighthouses   271
Linoleic acid   112, *113*, 288
Linolenic acid   *113*
Lipid *see* Fat
Lipogenesis   101–4
Lipolysis   106–7
Lipoprotein lipase   **102**, 103
Locomotion, control of mode and
    speed   163–73

Locomotor activity
  changes in diurnal patterns   152–6
  splitting   51, 93
Long-distance migration   7
  and fat deposition   119
Long-stage migrants   114
Loop migration   191
Loxodromes   247, **248**, 295
Luteinizing hormone   **94**, 241

MacCready tangents   172
Magnetic compass   250–3, 265–6,
    268–9, 295–6
  disturbance   271
  and navigation   262–3
Magnetite   256, 257
Magnetocline compass   263, 296
Magnetoreception   256–7, 262
Main migration routes   26
Male-biased dispersal   7
Mass accidents   11, 197–9
Mass migration ways   7
Mechanoreceptors   258
Median eminence   99, 100, 287
Medium-distance migration   7
Melanin   256
Melatonin   98–9, 256–7, 287
Mesotocin   107
Metabolic adaptations   100–14
Metabolic rate   138
Metabolic studies   22
Metabolically limited migrants   105
Meteorological factors *see* Weather
Methylthiouracil   94
Mettnau–Reit–Illmitz program   14,
    221–4, **225**
Microevolutionary changes   273–83,
    297
Middle-distance migration   7
Migration corridors   7
Migration divide   26
Migration speed   172–3, 291–2
  and stopovers   174–6
Migratoriness
  extension of   273
  reduction in   273
Migratory disposition   25–6, 285
Migratory distance   8
  changes in   274, 280–1
Migratory drive   26
  genetic control mechanisms   67–72
  in partial migrants   68–72